Biology of Damselfishes

Biology of Damselfishes

Editors

Dr. Bruno Frédérich
Laboratoire de Morphologie Fonctionnelle et Évolutive
AFFISH – Research Center
Université de Liège
Liège
Belgium

Prof. Eric Parmentier
Laboratoire de Morphologie Fonctionnelle et Evolutive
AFFISH – Research Center
Université de Liège
Liège
Belgium

CRC Press
Taylor & Francis Group
Boca Raton London New York

CRC Press is an imprint of the
Taylor & Francis Group, an **informa** business

A SCIENCE PUBLISHERS BOOK

CRC Press
Taylor & Francis Group
6000 Broken Sound Parkway NW, Suite 300
Boca Raton, FL 33487-2742

First issued in paperback 2021

ISBN 13: 978-0-367-78288-7 (pbk)
ISBN 13: 978-1-4822-1209-9 (hbk)

Library of Congress Cataloging-in-Publication Data

Names: Frédérich, Bruno. | Parmentier, Eric.
Title: Biology of damselfishes / [edited by] Bruno Frédérich and Eric Parmentier.
Description: Boca Raton : Taylor & Francis, 2016. | "A CRC title." | Includes bibliographical references and index.
Identifiers: LCCN 2015048045 | ISBN 9781482212099 (hardcover : alk. paper)
Subjects: LCSH: Pomacentridae. | Coral reef fishes.
Classification: LCC SF458.P66 B56 2016 | DDC 639.3/772--dc23
LC record available at http://lccn.loc.gov/2015048045

Foreword

Damselfishes have long captured the attention of taxonomists, ecologists, and especially students of reef fish behaviour. The very large size of this family in combination with their occurrence over a wide range of habitats and very interesting breeding biology, have served to continually keep this group in the limelight. Moreover, these fishes are highly conspicuous, diurnal inhabitants of mainly shallow reef areas, lending themselves especially well to intense study. Another asset is their ready acceptance of aquarium conditions, making them excellent subjects for captive laboratory studies involving various aspects of physiology, reproduction, growth and general behaviour.

I first became involved with this fascinating family in the mid-1960s during graduate studies at the University of Hawaii under the tutelage of Dr. Jack Randall. The topic of my PhD thesis was the classification and biology of the anemonefishes. They seemed a perfect target for research activities. Firstly, the classification was confused and presented a huge challenge and secondly there was scant documentation of their intriguing natural history. After graduation my interest in the family took a quantum leap when I accepted a fisheries job at Palau on the edge of what we now refer to as the Coral Triangle, the global centre of reef fish diversity. Compared to Hawaii and the Marshall Islands, the site of my graduate field studies, Palau harboured an amazingly rich fauna. It proved an excellent location for honing field identification skills and an opportunity for a close-up and personal approach to the study of pomacentrids. The die was cast and from that point onwards this intriguing family became a focal point of research that continues to this day, nearly 50 years later.

It is most gratifying to see that my passion for pomacentrid fishes is now shared by so many others. When I look back to the early years of graduate school, there was a general scarcity of taxonomic information. Weber and De Beaufort's classical *Indo-Australian Fishes* was the primary source of knowledge for the species-rich East Indian region. The Pomacentridae was covered in Volume VIII (1940), which included brief descriptions of 86 species in 11 genera. Now, 73 years later, the most recent tally (Allen and Erdmann 2012) is 187 species belonging to 21 genera. Worldwide we are fast approaching the 400 species mark, which represents the fourth richest marine-fish group after the Gobiidae, Serranidae, and Labridae. Modern genetic techniques in particular have opened a new chapter for our understanding of natural relationships and the existence of what appears to be many as yet undescribed cryptic species.

The advent of the Catalog of Fishes, the product of William N. Eschmeyer, former Curator of Fishes at the California Academy of Sciences has greatly facilitated the taxonomic process, resulting in a high level of accuracy while saving huge amounts of time. I vividly recall the painstaking process of compiling a complete inventory of taxonomic literature associated with the genus *Amphiprion* while working on my PhD thesis. Hours were spent for days on end tracking down various references, which were then hand-written on 3 x 5 cards. Now, thanks to the computerised Catalog, it takes a single click of the mouse and about five seconds to produce an annotated list of every nominal species (about 800) of Pomacentridae. The extreme value of this taxonomic tool cannot be overstated!

Back in the 1960s there was little accurate knowledge of pomacentrid natural history. Sex reversal, now well documented for *Amphipiron*, was still unknown. Not surprising, prior to the widespread use of scuba gear by diving scientists there was a general lack of information pertaining to damselfish behaviour and ecology. One notable exception was the detailed aquarium study of *Amphiprion* and their special relationship with sea anemones conducted at Java by Verwey (1930) in the 1920s. Thankfully, scuba diving has given us almost unlimited access to the previously unknown, private world of damselfishes, resulting in a large body of knowledge, particularly related to their ecology and various aspects of biology, including the fascinating realm of reproduction. Fast forward to 2015 and there is phenomenal interest in pomacentrids and a rapidly expanding literature base that now includes hundreds of published studies. The present volume is particularly timely and gratifying. It represents an excellent compilation of the current state of knowledge on a wide range of topics by leading researchers in the field. Highly informative, state of the art chapters cover a wide range of topics relating to crucial aspects of damselfish taxonomy, life history, ecology and habitat use, as well as the latest word on a variety of other subjects.

Chapter I introduces the damselfishes, presenting an insightful historical background of the systematics and general biological information, setting the stage for the highly informative chapters that follow. In Chapter II the taxonomic relationships of damselfishes are explored, with emphasis on important changes to traditional classification resulting from recent genetic investigations. Reproduction and larval recruitment are the focus of Chapter III, with particular emphasis on the pronounced habitat selection exhibited by damselfishes and the postsettlement mortality process. Chapter IV provides a review of damselfish mating systems, with special emphasis on genera that exhibit protandrous and protogynous sex change. The key concepts underpinning these sex-change strategies are discussed in the context of social behaviour and habitat utilization. In Chapter V habitat use is examined more closely, particularly the relative levels of ecological specialisation in coral reef pomacentrids with special emphasis on coral-dwelling species and anemone fishes. This chapter also addresses the hypothesis that highly specialised coral reef species have lower abundance and generally narrower geographic ranges compared to damsels with more generalised requirements. The fascinating topic of algal farming by damselfishes has been documented over several decades and Chapter VI presents a concise review. Discussion topics include the implications for the damselfish farmers, in the form of their digestion and behaviour, and the effects on other members of the reef community. Our current knowledge about farmer effects on benthic communities, especially algae, corals, and other invertebrates is also included. The tropic ecology of damselfishes is explored in Chapter VII with the goal of reviewing the diversity of trophic niches present in this diverse family. A brief overview of the methods commonly used to investigate feeding habits is provided, as well as a description of the different trophic groups in the family. Geographic and ontogenetic variation in diet is also discussed, along with the role of detritus in the diet of certain species, and individual variation/specialization within populations of damselfishes. Chapter VIII covers the ontogeny and early life-history stages of damselfishes, including a review of the growth and development from fertilization to post-settlement stages, illustrating some of the better known variations among taxa. The family includes species that exhibit unusually large variations in the duration of early life history stages, as well as behavioural variations in comparison to other marine fishes. Chapter IX describes the wide range of ecomorphological diversity, discussing the adaptive significance of specific aspects of functional morphology, particularly with relation to feeding and swimming behaviour. The reticulated pattern of evolutionary diversification in damselfish ecomorphology is also discussed in light of results from current phylogenetic studies. Sound production is well documented in damselfishes and forms the subject of Chapter X with a focus on the ethological, physiological and morphological aspects of acoustic communication. This informative review is nicely complimented by Chapter XI, which provides a succinct summary of our knowledge of hearing capabilities in pomacentrids. Perhaps no other group of damselfishes has attracted the attention of researchers as much as "clownfishes" of the genera *Amphirion* and *Premnas*, which forms the subject of Chapter XII. The intimate relationship between the fishes and their anemone hosts is reviewed with emphasis on recent observations of this intriguing symbiosis. Vision and colour diversity in damselfishes are detailed in Chapter XIII. Vision plays a major role in the damselfish life cycle, beginning from the moment of hatching and importantly when new recruits are faced with a gauntlet of predatory fishes. The underwater light environment, which sets the limits for vision and determines the spectrum of possible colours, is detailed. The chapter also reviews current knowledge of colour patterns and their functions and includes a comprehensive summary of what is known about pomacentrid visual systems. The book concludes with Chapter XIV, which discusses the pharyngeal jaws and cerato-mandibular ligament in damselfishes, particularly in relation to feeding and sound production, and the possible important role of the cerato-mandibular ligament in the impressive evolutionary radiation of this family.

Although certainly not the last word on this fascinating group of fishes, this book provides an excellent summary of our current knowledge and therefore represents an indispensable tool for both students and experienced researchers.

November, 2015

Gerald R. Allen
Perth, Australia

Preface

Thanks to their bright colors, their abundance on coral reefs and some representatives such as clownfishes (including Nemo®), Pomacentridae is one of the universally known groups of marine vertebrates. They are widely used in aquarium trade and are found in all public aquariums around the world as representative of coral reef fishes. Besides their highly conspicuous aspect, the damselfish family shows also an important set of biological characteristics (algal farmer, sex change, symbiosis, sound production, high diversity, etc.) and has consequently interested biologists from many disciplines. As such, it seems that Pomacentridae form a group of highly interesting fishes and it is opportune to propose a book concerning the knowledge about this taxa. The central questions that have motivated the book could be: how is it possible to find a so high diversity of fishes in the same biotope? How do they share their environment? What kind of communication channel do they use? The gathering of various information in a common place should provide the basis required for going further and/or provoke additional works and questioning. We hope to share our fascination for this fish group with readers and highlight the huge number of their intriguing aspects.

The major aim of this book is to simultaneously provide a solid series of reviews that should help to globally understand the different aspects of the way of life of pomacentrids. The book provides important insights on different evolutionary aspects and adaptations that help to understand their important diversity and their various uses of the habitat. The book will be useful to graduate and undergraduate students who wish to obtain in-depth knowledge. It could also be used by senior scientists working not only on pomacentrids but on fishes in general. In many cases, we have tried to underline the connections between the different chapters and hope this book is more than a compilation of independent contributions.

As is usually the case in science, solidarity, cooperation and common wish to go beyond is required for this kind of initiative. We requested leading researchers from around the globe to contribute to this work. We are particularly grateful for their enthusiastic responses and the work they have done. We wish also recognize all people that have collected data on pomacentrids since many years.

November, 2015

Bruno Frédérich
Eric Parmentier
Liège, Belgium

Acknowledgements

Firstly we would like to thank Pierre Vandewalle (University of Liège) who initiated our interest in damselfishes, and introduced us to functional morphology and field studies. We thank all the people from the lab of Functional and Evolutionary Morphology at the University of Liège (ULg), who contributed directly or indirectly to the production of this book: Mariella Lunetta, Nicole Decloux, Antoine Tack, Philippe Compère, Orphal Colleye, Loïc Kéver, Frédéric Bertucci, Damien Olivier, Laura Gajdzik, Geoffrey Mélotte, Séverine Henry, Mélissan Trévisan et France Collard.

All the chapters of this book were reviewed by different internationally renowned scientists, whose comments were helpful and constructive. We sincerely think that their work helped to improve the different contributions. So, we acknowledge Lia Aguilar-Medrano, Gerry Allen, Giacomo Bernardi, Roberta Bonaldo, Daniella Ceccarelli, W. James Cooper, Christopher Goatley, John Godwin, Hiroki Hata, Anthony Herrel, Nicolai Konow, David Lecchini, Gilles Lepoint, Glenn Litsios, Julian Partridge, Phillip Lobel, Karen Maruska, Claire Paris, Jian Qin, Aaron N. Rice, Milly Sharkey, Timothy Tricas, Tom Trnsky and Mark Westneat for their helpful criticisms, advice and assistance in the review process.

We gratefully acknowledge Mark Erdmann for providing beautiful photos of damselfishes illustrating the cover of this book.

Contents

List of Contributors

Aguilar-Medrano, Rosalía
Instituto de Ecología Aplicada, Universidad Autónoma de Tamaulipas, 356 División del Golfo, Col. Libertad, Ciudad Victoria, Tamaulipas, México, 87029.
Email: liabiol@gmail.com

Allen, Gerald R.
Western Australian Museum, Locked Bag 49, Welshpool DC, Perth, Western Australia 6986, Australia.
Email: gerald.allen@wa.gov.au

Beldade, Ricardo
USR 3278 CRIOBE CNRS-EPHE-UPVD, Laboratoire d'Excellence 'CORAIL', 66860 Perpignan, France. Universidade de Lisboa, Faculdade de Ciências, Centro de Oceanografia, 1749-016 Lisboa, Portugal.
Email: rbeldade@gmail.com

Casadevall, Margarida
Environmental Sciences Department, Sciences Faculty, Girona University, 17071 Girona, Spain.
Email: margarida.casadevall@udg.edu

Ceccarelli, Daniela M.
ARC Centre of Excellence for Coral Reef Studies, James Cook University, Townsville, Australia.
Email: dmcecca@gmail.com

Colleye, Orphal
Laboratoire de Morphologie Fonctionnelle et Evolutive, AFFISH - Research Center, University of Liège, Quartier Agora, Allée du six Août 15, Bât. B6C, 4000 Liège (Sart Tilman), Belgium.
Email: o.colleye@ulg.ac.be

Cooper, W. James
School of Biological Sciences, Washington State University, Pullman, WA, 99164, USA.
Email: jim.cooper@wsu.edu

Frédérich, Bruno
Laboratoire de Morphologie Fonctionnelle et Evolutive, AFFISH - Research Center, University of Liège, Quartier Agora, Allée du six Août 15, Bât. B6C, 4000 Liège (Sart Tilman), Belgium.
Email: bruno.frederich@ulg.ac.be

Gajdzik, Laura
Laboratoire de Morphologie Fonctionnelle et Evolutive, AFFISH - Research Center, University of Liège, Quartier Agora, Allée du six Août 15, Bât. B6C, 4000 Liège (Sart Tilman), Belgium.
Email: laura.gajdzik@ulg.ac.be

Hata, Hiroki
Graduate School of Science and Engineering, Ehime University, Bunkyo 2-5, Matsuyama, Ehime, Japan 790-8577.
Email: hata@sci.ehime-u.ac.jp

Hattori, Akihisa
Faculty of Liberal Arts and Education, Shiga University, 2-5-1 Hiratsu, Otsu, Shiga 520-0862, Japan.
Email: hattori@edu.shiga-u.ac.jp

Hobbs, Jean-Paul A.
Department of Environment and Agriculture, Curtin University, Perth, WA 6845, Australia.
Email: jp.hobbs@curtin.edu.au

Hoey, Andrew S.
ARC Centre of Excellence for Coral Reef Studies, James Cook University, Townsville, Australia.
Email: andrew.hoey1@jcu.edu.au

Iwata, Eri
College of Science and Engineering, Iwaki Meisei University, 5-5-1 Chuoudai, Ihino, Iwaki, Fukushima 970-8032, Japan.
Email: asealion@iwakimu.ac.jp

Kavanagh, Kathryn
Biology Department, University of Massachusetts, Dartmouth, USA.
Email: kkavanagh@umassd.edu

Kéver, Loïc
Laboratoire de Morphologie Fonctionnelle et Evolutive, AFFISH - Research Center, University of Liège, Quartier Agora, Allée du six Août 15, Bât. B6C, 4000 Liège (Sart Tilman), Belgium.
Email: loic.kever@ulg.ac.be

Lecchini, David
USR 3278 CNRS-EPHE-UPVD, CRIOBE, Laboratoire d'Excellence 'CORAIL',
98729 Moorea, French Polynesia.
Email: lecchini@univ-perp.fr

Mann, David A.
Loggerhead Instruments, 6576 Palmer Park Circle, Sarasota, FL 34238, USA.
Email: dmann@loggerhead.com

Mills, Suzanne C.
USR 3278 CRIOBE CNRS-EPHE-UPVD, Laboratoire d'Excellence 'CORAIL',
66860 Perpignan, France.
Email: suzanne.mills@univ-perp.fr

Olivier, Damien
Laboratoire de Morphologie Fonctionnelle et Evolutive, AFFISH - Research Center,
University of Liège, Quartier Agora, Allée du six Août 15, Bât. B6C, 4000 Liège
(Sart Tilman), Belgium.
Email: dolivier@ulg.ac.be

Parmentier, Eric
Laboratoire de Morphologie Fonctionnelle et Evolutive, AFFISH - Research Center,
University of Liège, Quartier Agora, Allée du six Août 15, Bât. B6C, 4000 Liège
(Sart Tilman), Belgium.
Email: e.parmentier@ulg.ac.be

Pratchett, Morgan S.
ARC Centre of Excellence for Coral Reef Studies, James Cook University, Townsville,
Australia.
Email: morgan.pratchett@jcu.edu.au

Santini, Francesco
Dipartimento di Scienze della Terra, Università degli Studi di Torino, 10125 Torino,
Italy.
Email: francesco.santini@alumni.utoronto.ca

Siebeck, Ulrike E.
School of Biomedical Sciences and Global Change Institute, The University of
Queensland, St. Lucia 4072, QLD, Australia.
Email: u.siebeck@uq.edu.au

Wilson, Shaun K.
Marine Science Program, Department of Parks and Wildlife, Kensington, WA 6151,
Australia.
Email: shaun.wilson@dpaw.wa.gov.au

Meet the Damselfishes

Eric Parmentier* and Bruno Frédérich

Introduction

Various well illustrated books such as "Damselfishes of the World" (Allen 1991), and "Damselfishes & Anemonefishes" (Scott 2008) have summarized general information contributing towards a better understanding of pomacentrid life history strategies. These books and others are also useful for identifying the numerous species in the family. Nevertheless, pomacentrids have been the subject of an increasing number of studies over the past few decades, covering multiple disciplines. These studies include different aspects of the biology such as diet (or trophic diversity), sound production, habitat description, sexual selection, phylogeny, evolutionary trends, ecomorphology, sex change, hearing, vision, etc. However, data are scattered throughout the literature without any attempt to consolidate a global view of the life history of this interesting group. Consequently, the goal of this book is to gather and consolidate the latest information relevant to the biology of this large and diverse fish family. The contributing authors have kindly accepted the challenge of summarizing and synthesizing their special research areas, and therefore most of the chapters offer the latest information, including previously unpublished data. In this introductory chapter we introduce the damselfishes, first presenting a historical background of the systematics and then providing more specialized chapters covering different aspects of the way of life (biogeography, behavioral diversity, sociology, sexuality, biological life cycle and ontogeny) and other specialized subjects.

Laboratoire de Morphologie Fonctionnelle et Evolutive, AFFISH - Research Center, University of Liège, Quartier Agora, Allée du six Août 15, Bât. B6C, 4000 Liège (Sart Tilman), Belgium.
* Corresponding author: e.parmentier@ulg.ac.be

Diagnosis

Chiefly marine; rare in brackish water. All tropical seas, mainly Indo-Pacific. Small percoid fishes (to about 25 cm); body elongated to rounded in lateral profile, and laterally compressed. Eye medium-sized, generally exceeding snout length, mouth small, jaws equal and oblique. One nostril on each side of head; double nostrils in some species of *Chromis* and *Dascyllus*. Incomplete and interrupted lateral line. Anal fin with usually 2 spines, very rarely 3. Small conical to incisiform teeth in jaws in a single row, sometimes with an additional row of slender teeth behind primary row; teeth absent on vomer and palatines. Total gill rakers on first gill arch 10 to 85. A single dorsal fin with VII to XVII stout spines and 9 to 21 soft rays; anal fin with II spines and 9 to 16 soft rays; pelvic fins with I spine and 5 soft rays; pectoral fins with 14 to 22 rays; caudal fin rounded to forked. Scales comparatively large to moderate-sized and ctenoid, extending onto head and median fins; cheeks and operculum scaly; lateral-line scales 12 to 60. Damselfishes lay elliptical demersal eggs that are guarded by the males (Allen 1991, Nelson 2006). 393 valid species (Eschmeyer 2015).

Historical Background

In book IV, Chapter VIII, Aristotle noted: "Some people maintain that fishes have the best hearing capacities than any other animal [...]. The best hearing abilities are [...] chromis and fishes of this order". Later in Chapter IX, he noted: "Fishes do not have any voice because they do not have lungs, windpipe and pharynx. Some of them produce some noise and grindings that look like a voice, such as the "lyre" and the "chromis" that emit a kind of grunt [...]".

The first damselfish to attract the attention of scholars was probably *Chromis chromis*, which forms large aggregations in the shallow coastal waters of the Mediterranean Sea. In 1557, Rondelet reported (in old French): "[...] Chromis n'ha point de nom latin (Chromis does not have Latin noun) [...] C'eft poiffon de riuage (this is a coastal fish) [...] ce poiffon eft peit et vil (this fish is small and lively) [...] tout le corps eft noiraftre (all the body is blackish) [...] il ha des pierres au cerueau, il fait des oeufs vne fois l'an, comme efcrit Ariftote, é eft de ceux qui oient fort bien, é gronde (it has otoliths, spawns once a year, has good hearing abilities and is able to make sounds)". Rondelet (1557) also reports that the Italian name (from Geneo more precisely) is *Caftagno* (the current French vernacular name of *Chromis chromis* is "castagnolle") meaning chestnut in reference to the deep brown color of this fish. Lacépède also noted that this fish corresponds to the chromis of Aristotle and furthermore he added in his comments that some authors (he did not mention their names) think that this fish could be the species collected during a universally well-known fish party in Galilee (de Lacépède 1798). However, Cuvier and Valenciennes (1830) estimated that the chromis of Aristotle was a meagre and rejected the view of Rondelet (1557).[1] On the other hand, Theodore Gill stated that the chromis of Aristotle was probably the sciaenid (drum) *Umbrina steindachneri* (Gill 1911). On the other hand, Linneaus first described the pomacentrid *Chromis chromis* as

---

[1] Cuvier and Valencienne however described a small fish they called castagneau.

Sparus chromis (Linnaeus 1758). Following Linnaeus, different authors such as Lacépède or Risso placed this species in the Sparidae (de Lacépède 1798, Risso 1810). Desmarest (1814) reported that, according to the observations of Cuvier, *Sparus chromis* has more affinities for Labridae than for Sparidae and proposed to create a new genus: *Chromis* (that included also the cichlid *Oerochromis niloticus*). The Mediterranean pomacentrid was then called *Chromis castanea* and was placed in the Labridae (Cuvier 1815) before being incorporated in the "pomacentre" (Müller 1843).

As *Chromis chromis*, various species were assigned to different family groups by early authors meaning the recognition of Pomacentridae as a natural group has followed a tortuous path. For example, *Amphiprion* were first placed in holocentrid, *Pomacentrus* and *Dascyllus* in chaetodon, *Premnas* in serranid or in scorpaneid groups respectively (Cuvier and Valenciennes 1830). The noun "pomacentre" appeared for the first time in the book "Histoire naturelle des poissons" (de Lacépède 1798). The latter author gave this diagnosis "…ceux dont l'opercule est denté, qui n'ont qu'une dorsale" (= those having a serrated opercle and only one dorsal fin). This brief description corresponds obviously to the etymology because in Greek the word "*poma*" refers to the fish opercles and "*kentron*" is for stings, referring to the serration of the opercles.

The type genus for the family, collected in the Indian Ocean and first described as *Chaetodon pavo*, is *Pomacentrus pavo* (Bloch 1787) meaning "peacock pomacentre" because Bloch estimated that the fish's color pattern had similarities with the bird and the black circle at the level of the upper part of the opercle reminded the eyes of the bird's tail. However, Cuvier and Valenciennes (1828) noted (rightly) that Bloch took some liberties with the color pattern of this fish. Cuvier and Valenciennes also noted that the "pomacentre" of Lacepède was erroneously placed in the Chaetodontidae. After the examination of the "pomacentre paon" (= *Pomacentrus pavo*), they decided to keep the noun as a new type but gave a new description to the fish because they did not agree with the one of Lacépède. They also noted these fish are close to "amphiprions" and "premnades". Cuvier and Valenciennes (1828) were the first to publish a compilation of the damselfishes. However, both authors proposed to create a new family without naming it; they just noted these fish are "Scienoïde having less than seven branchiostegal rays and an interrupted lateral line". The characteristics were short: oval body, swim bladder without appendix, palate without teeth, presence of pyloric caeca and serrations at the level of the opercles. They distinguished seven genera: "Glyphisodon" (*Abudefduf*), "Amphiprion" (*Amphiprion*), "Premnade" (*Premnas*), "Pomacentre" (*Pomacentrus*), "Dascylles" (*Dascyllus*), Etrople and Héliases (corresponding to some species of *Chromis*). The noun Pomacentrini appeared for the first time in the work of Bonaparte (1831), who followed Cuvier and Valenciennes in placing this taxa in the Sciaenidae. He gave a diagnosis but did not report which species are placed in this group. The genus *Pomacentrus* is the officially recognized as the type genus for the family, of which the family name was first used by Bonaparte in 1831 (Van Der Laar et al. 2014). Ernst Heckel transferred them in the Labroidei but did not distinguish Pomacentridae from Cichlidae (Heckel 1840).

Among the teleosts, Müller distinguished a taxa called Pharyngognathi acanthopterygii with the following characters: the lower pharyngeal bones are coalesced; part of the rays of dorsal, anal and ventral fins are not articulated forming spines, and the swim bladder is deprived of pneumatic duct (Müller 1843). He divided this order into three families: Labroidei ctenoidei (pomacentrid), Labroidei cycloidei (labrid) and Chromide (cichlid). He also discussed that *Chromis castaneus* (= *Chromis chromis*) should be placed among the Labroidei ctenoidei and not in the Chromide (Müller 1844). His Labroidei ctenoidei possess 7 branchiostegal rays and an interrupted lateral line. Müller gave later for the first time the name Pomacentridae to the family Labroidei ctenoidei (Müller 1844). This classification was largely admitted by different authors (Günther 1862, Owen 1866, Playfair and Günther 1866, Klunzinger 1870). For example Günther (1862) listed 143 species from different genera: *Amphiprion* (17 species), *Premnas* (1 species), *Dascyllus* (8 species), *Lepidozygus* (1 species), *Pomacentrus* (44 species), *Glyphidodon* (52 species), *Parma* (4 species) and *Heliastes* (16 species). In his study on pomacentrids from the Indian archipelago, Bleeker (1877) did not completely follow Müller and considered the "Pomacentroid" to be marine species of the Chromide clade (= Pomacentridae + Cichlidae) and justified this group's contents as "acanthopterygian fishes with single nostril". Pomacentroid possess anal fin with two spines, two anal fins and Cichlidae have more than two spines. He divided the Pomacentroid into two taxa: Prochlini (grouping *Amphiprion* and *Premnas*) and Glyphidodontini (all the other species). More recently, Allen (1975, 1991) classified the damselfishes into four subfamilies: Amphiprioninae, Chrominae, Lepidozyginae, and Pomacentrinae. Based on molecular phylogenetic hypotheses, Cooper et al. (2009) suggested a classification consisting of 5 subfamilies (Lepidoziginae, Stegastinae, Chrominae, Abudefdufinae and Pomacentrinae). Further details on the intra-familial relationships can be found in Chapter II.

All these considerations did not take into account the different vernacular names. The German "riffbarsch" (barsch = perch and riff = reef) and the Dutch "rifbaarzen" names remind that these fish are perciforms living on the reef. In handbooks, the demoiselle (= damsel in English, damisela in Spanish) noun has been given to different fish species since more than 250 years ago (Valmont-Bomare 1754, Aubert de La Chesnaye des Bois 1799) but the descriptions do not enable us to be sure that these vernacular names are used only for pomacentrid species. Generally speaking, the vernacular name is mainly due to the fish's small size, brilliant colors and pleasant aspect.

Familial Relationships

The Pomacentroid clade of Bleeker (1877) was a member of Müller's (1843) Pharyngonathi acanthopterygii that include taxa with at least (1) united left and right lower jaw elements (fifth ceratobranchials), (2) a muscular sling that directly connects the underside of the neurocranium with the lower pharyngeal jaw; and (3) a mobile diarthrotic articulation of the upper pharyngeal jaws with the neurocranium (Stiassny 1981, Kaufman and Liem 1982, Wainwright et al. 2012). These morphological and functional characters were proposed as synapomorphies uniting an expanded Labroidei clade (Liem and Greenwood 1981) that included Cichlidae, Embiotocidae, Labridae, Odacidae and Scaridae. One year later, Kaufman and Liem (1982) added Pomacentridae to the Labroidei. However, recent molecular phylogenetic studies revealed that the Labroidei is not a monophyletic clade (Mabuchi et al. 2007). Labridae are separated from the remainder of the traditional labroid lineages (Cichlidae, Embiotocidae, and Pomacentridae) and these three families are now included in a clade of 40 families and more than 4800 species which were named Ovalentaria for their characteristic demersal, adhesive eggs with chorionic filaments (Wainwright et al. 2012).

Morphology

Our goal is to highlight the main morphological characteristics of Pomacentridae, which will create a better understanding of their way of life with emphasis on key subjects such as feeding or sound production, both behaviors being intimately interconnected and detailed in subsequent chapters (Chapters VI, VII, IX, X, XIV). Pomacentrids are perciform fishes distinguished by a single nostril on each side of the head, two anal spines, an incomplete lateral line, a single dorsal fin, small teeth and a smooth palate (Allen 1991, Cooper et al. 2009). Apparently the damselfish skeleton does not show highly specialized or distinctive structures with the exception of the cerato-mandibular ligament (Chapter XIV) (Fig. 1). Despite their similar overall morphology (Allen 1991), the cephalic skeleton reveals considerable differences in the relative proportions of individual bones and also in tooth morphology (Frédérich et al. 2006). Suction feeding is particularly well developed in pomacentrids as in other perciform fishes. However, the suction mechanism is more efficient when the buccal cavity is shaped like a large cone rather than a tube (Liem 1978, Lauder 1980, Lauder and Lanyon 1980). Having

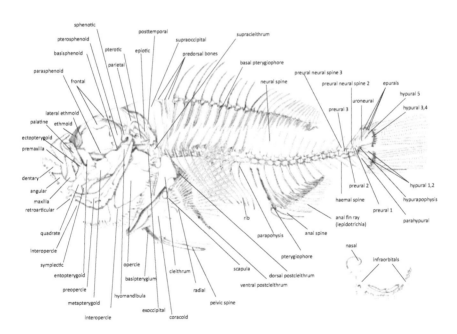

Fig. 1. Left lateral view of the skeleton in *Amphiprion akallopisos*. Circumorbital bones are found on the right bottom. Intermuscular bones are not shown.

a large cone requires for example long ascending process of the premaxilla (Gosline 1987) and/or high suspensoria and opercles (Liem 1993). It appears pomacentrids mainly use differences in the size and shape of their skeletal pieces to accommodate different diets (Chapter IX).

Pharyngognathy concerns the fusion of the two fifth ceratobranchial bones into a single functional unit (Liem 1973, Stiassny and Jensen 1987). Although this anatomical feature is found in different percomorphs, it apparently has evolved multiple times in these taxa (Wainwright et al. 2012). In pomacentrids, the characteristics of the lower pharyngeal jaw (LPJ) are a Y-shape (and width is greater than the length), no trace of a central sutural union, a well-developed median keel on the ventral face of the bone, and tooth rows arranged radially across the LPJ with teeth located over the median region of the jaw (Kaufman and Liem 1982, Stiassny and Jensen 1987). Although it is not found in all the pomacentrids (e.g., *Microspathodon* and *Chromis*), pharyngo-cleithral articulations can join the expanded lateral horns of the LPJ to the cleithrum (Liem 1973, Liem and Greenwood 1981, Kaufman and Liem 1982, Stiassny and Jensen 1987). Liem (1973) proposed that pharyngognathy has major implications for feeding performance because the fusion of the LPJ in a single plate allows the combination of the force exerted by right and left muscle and leads to the enhancement of force and the ability to crush hard prey (Wainwright et al. 2012). Moreover, the support provided by the articulation of the LPJ with the shoulder girdle can increase the total biting force that can be exerted on prey (Galis and Snelderwaard 1997).

The main morphological characteristic in damselfish concerns the cerato-mandibular ligament (c-md) that joins the ceratohyal of the hyoid bar to the lower jaw, at the level of the coronoid process (Chapter XIV). Although secondarily lost in a few species (Frédérich et al. 2014), this ligament appears to be a synapomorphic trait within Pomacentridae (Stiassny 1981). Recent studies have highlighted that it is involved in at least two major functions of the damselfish behavior: sound production (Parmentier et al. 2007, Colleye et al. 2012) and feeding (Olivier et al. 2014, 2015). It means that it impacts two fundamental tasks that can be considered to be the main axes of diversification in vertebrates (Streelman and Danley 2003). Both behaviors are based on the same principle: the c-md allows rapidly closing the lower jaws in a few milliseconds, without the help of the *adductor mandibulae* muscles. The slam of the oral jaws causes teeth collision creating a vibrational wave and the resulting sounds (Colleye et al. 2012).

This movement corresponds also to a second major mouth closing mechanism that can be used in different ways according to the species. In farmer species, it enables the fish to perform accurate strikes on small filamentous algae and the resulting farming activity allows these damselfishes to occupy distinct niches (Olivier et al. 2014). In no-grazing species, it provides additional ways of feeding (Olivier et al. 2015). This fast jaw closing mechanism and the related accurate strike can also be a key element related to reproduction because males most probably use it to clean places on rocks or coral where the eggs will be laid. We here would like to stress the importance of this cerato-mandibular ligament, which we consider to be a fundamental key to the origin of pomacentrid diversification process. The advantages it provides in different behaviors (communication, feeding, and reproduction) is of primordial importance for the taxa success.

Table 1. Geographic distribution of the damselfishes. The species are classified by three main regions and appropriate combinations of the provinces defined by Spalding et al. (2007). N refers to the number of species.

Region	N	Provinces (Spalding et al. 2007)	N
Indo-West Pacific	329	Central & Eastern Indo-Pacific	23
		Central, Eastern Indo-Pacific & Temperate Australasia	1
		Central Indo-Pacific	130
		Central Indo-Pacific & Temperate Australasia	10
		Eastern Indo-Pacific	27
		Temperate Australasia	11
		Temperate Northern Pacific & Central Indo-Pacific	7
		Western & Central Indo-Pacific	34
		Western Indo-Pacific	63
		Western, Central & Eastern Indo-Pacific	22
		Western, Central & Eastern Indo-Pacific, & Temperate Australasia	1
Eastern Pacific	26	Temperate Northern Pacific and Tropical Eastern Pacific	7
		Tropical Eastern Pacific	12
		Temperate South America (Pacific)	3
		Tropical Eastern Pacific and Temperate South America	2
Atlantic	34	Temperate Northern and Tropical Atlantic	2
		Temperate Northern Atlantic (Mediterranean Sea)	1
		Tropical Atlantic	31
TOTAL			387

Biogeography

The damselfishes are distributed worldwide, inhabiting tropical and temperate seas (Allen 1991). Most species (329) are known from the Indo-West Pacific province including 85% of the world's total (Table 1). The other major regions harbor relatively low numbers: 24 endemic species in the Eastern Pacific, 33 in the Atlantic, and one in the Mediterranean Sea (Table 1). At the subfamily level, Lepidoziginae, Stegastinae, Chrominae and Abudefdufinae have representatives in the Indian, Pacific and Atlantic Ocean. Conversely, the Pomacentrinae have not colonized the Atlantic and the Eastern Pacific (Allen 1991, Cooper 2008).

The Central Indo-Pacific region, including the "coral triangle" (the well-known roughly triangular area of the tropical marine waters of Indonesia, Malaysia, Papua New Guinea, Philippines, Solomon Islands and Timor-Leste), has the highest number of damselfish species (181 species, Table 1). The second most species-rich province is the Western Indo-Pacific, including 120 species whereas the Eastern Indo-Pacific harbors 74 species (Table 1).

There are three basic patterns of damselfish distribution. The first includes species that are widely distributed over a huge area, for example *Dascyllus aruanus* or *Chromis viridis* that occur over much of the Indo-West Pacific region. A second group contains species that have more restricted sub-regional distributions, for example *Pomacentrus sulfureus* and *Pomacentrus trilineatus* living in the Western Indian Ocean or *Chromis iomelas* living in the Central and Western Pacific. The third category includes species that are restricted (i.e., endemic) to small areas such as a single island, an archipelago, a sea or a gulf. Many species belong to this final group: for example, *Abudefduf abdominalis* living in the Hawaiian Islands, *Amphiprion fuscocaudatus* inhabiting the Seychelles, *Similiparma hermani* occurring at the Cape Verde Islands, *Amblyglyphidodon flavilatus* and *Dascyllus marginatus* living in the Red Sea, and *Pomacentrus arabicus* from the Gulf of Oman.

Resources Diversity

The partitioning of resources may be viewed as one of the key factors in the diversifying process, which promotes the coexistence of closely related and ecologically equivalent species (Colwell and Fuentes 1975). Both habitat and food are generally the most studied ecological resources.

The habitat diversity and some ecological factors explaining the number of damselfish species at a given locality will be detailed in Chapter V. However, a short overview of the habitats, where damselfishes are encountered, is provided hereafter. The great majority of damselfishes occurs in the vicinity of coral reef environments. Most of the tropical species live amongst living or dead coral formations on the barrier reef (outer reef slope, reef flat) and in the lagoon (micro-atolls, coral heads, fringing reef). The habitat of numerous species can be restricted to one zone. For example, *Pomacentrus sulfureus* lives only behind the reef crest (Ormond et al. 1996). *Stegastes nigricans* and *Dascyllus aruanus* are strictly occurring in the lagoon (Meekan et al. 1995, Holbrook et al. 2000, Lecchini and Galzin 2005). On the other hand, some species can be encountered in various reef zones: *Pomacentrus amboinensis* and *Pomacentrus bankanensis* at Lizard Island (Meekan et al. 1995), and *Stegastes albifasciatus* at Moorea Island (Lecchini and Galzin 2005). Finally, relatively few species live everywhere on the reef environment, i.e., both on the barrier reef and in the lagoon: *Pomacentrus trichourus* in the Red Sea (Ormond et al. 1996), *Pomacentrus wardi* at Heron Island (Robertson and Lassig 1980) and *Abudefduf sexfasciatus* at Moorea Island (Lecchini and Galzin 2005). The distribution of the species at small spatial scales is mainly related to the depth, the presence/absence of conspecific, the presence/absence of predators and/or the kind of substrates (see Chapter V).

Few pomacentrids are encountered around mangroves and seagrass beds at the adult stage. For example, the presence of some species of the genus *Dischistodus* was reported in a seagrass bed of a fringing reef at Iriomote Island, Southern Japan (Nakamura et al. 2003). We also observed that *Chrysiptera annulata* lives in seagrass beds of the Great Barrier Reef of Toliara (Madagascar). Some species are frequently encountered in coastal freshwater streams or brackish estuaries: *Stegastes otophorus*, *Pomacentrus taeniometopon* and *Neopomacentrus taeniurus* (Allen 1991). Others may frequent shallow inshore areas including harbors or protected embankments where sandy or silty conditions prevail or visibility is reduced (e.g., juveniles of *Abudefduf vaigiensis* or *A. sexfasciatus*, pers. observation). The damselfishes of temperate zones mainly occur in rocky areas (e.g., *Chromis chromis* in the Mediterranean Sea) and some live closely associated with kelp forests (e.g., *Hypsypops rubicundus* and *Chromis punctipinnis*).

Chapter VII will be devoted to the trophic diversity of damselfishes. Generally, they are grouped in three major trophic guilds: zooplanktivorous, algivorous and omnivorous species (Allen 1991, Frédérich et al. 2009). Most of the *Dascyllus*, *Chromis*, *Abudefduf* species or *Lepidozygus tapeinosoma* mainly feed on zooplankton. The genera *Stegastes* and *Plectroglyphidon*, *Neoglyphidodon* include mainly benthic feeders grazing filamentous algae and picking small sessile and mobile invertebrates. A large number of *Chrysiptera* and *Pomacentrus* species are considered as omnivorous, feeding on variable proportions of zooplankton, filamentous algae and small benthic invertebrates.

Behavioral Diversity, Sociology and Sexuality

The Pomacentridae include solitary and gregarious species (Fishelson 1998). The great majority of solitary species are highly territorial, defending a small area against conspecific and others organisms such as fishes and mobile invertebrates. Some of these territorial species are considered as algal farmers (Hata and Kato 2002, Ceccarelli 2007). To our knowledge, this farming behavior is only encountered in the damselfishes and Chapter VI deals with this special ecological trait. Non-territorial species may form small groups of less than 10 individuals or larger ones composed of more than 50 fish. The gregarious damselfishes show diverse social structure, including species with either monogamous or polygamous adults. The best examples of monogamous species are most of the clownfishes (*Amphiprion* spp.), which form permanent reproductive pairs with a high level of fidelity (see Chapters IV & XII). Like most of its relatives, the Humbug damselfish (*Dascyllus aruanus*) is a permanent polygamous species, forming haremic groups consisting of one or two males and numerous females. Some species of the genus *Amblyglyphidodon* form discrete reproductive units only during spawning periods (Fishelson 1998). The varied social structures include different sex-change strategies. For example, species in the genus *Amphiprion* are characterized by protandrous sex change in contrast to *Dascyllus*, which exhibits protogynous sex transformation. Although we have provided a short list of examples, Chapter IV presents an overview of the socio-ecology and related sex-changes in damselfishes.

The brightly colored clownfishes (*Amphiprion* and *Premnas*) are well known for their symbiosis with tropical sea anemones. This peculiar way of life is associated with various behavioral and socio-ecological traits, which are described in Chapter XII. Like many other members of the family they are also well known sound producers. Indeed, damselfishes are one of the best-studied families for the use of acoustic communication. They produce pulsed sounds, which accompany a variety of behavioral modes, including nuptial signal jumping, mating, chasing, fighting, and nest cleaning (Mann and Lobel 1998, Parmentier et al. 2010). It seems that sound communication is a key component of the diversity of damselfishes and both Chapters X & XI focus on sound production and hearing abilities.

Damselfishes exhibit a great diversity of color patterns. Color-based signals are extremely important in the behavioral regime of many animals and color-related cues appear to be essential in damselfishes as well. Recently, Siebeck and colleagues showed that coral reef fishes, and damselfishes especially, appear to be sensitive to ultraviolet light (Marshall 2000, Siebeck and Marshall 2001) and at least one damselfish, *Pomacentrus amboinensis*, utilizes ultraviolet signals for communication (Siebeck 2004). Chapter XIII summarizes our current knowledge of color diversity and color vision in damselfishes.

Biological Life Cycle and Ontogeny

As the great majority of coral reef fishes, the damselfishes possess a complex life cycle divided into two distinct phases: a pelagic larval phase, and a demersal juvenile and adult phase associated with the coral reef environment (Leis and McCormick 2002). The biological life cycle of pomacentrids consists of five discrete stages: egg, larva, settling larva, juvenile and adult. Damselfishes have demersal eggs and show paternal care of the eggs (Allen 1991). Nest guarding males are extremely aggressive and frequently attack much larger fishes. During nuptial periods, including courtship and nesting, many damselfishes exhibit dramatic, instantaneous color pattern changes, frequently associated with sound production (Chapter X). Chapter III reviews reproductive traits of damselfishes, including the biology of the larval stage and habitat selection of juveniles at the time of settlement. The early development of skeleton and sensory abilities in damselfishes is discussed in Chapters III and VIII. Kavanagh and Frédérich (Chapter VIII) primarily focus on the morphogenesis and the ontogenetic shape changes of damselfishes during the larval stage and their post-settlement ontogeny. The early development of skeletal features and morphological transformations are discussed in an ecomorphological context. For example, some studies show that most damselfish larvae feed on planktonic copepods in contrast to adults, which exhibit varied diets. Consequently, diet shift is expected in some species and is generally associated with morphological transformations and behavioral adaptations (Chapter VIII).

References

Allen, G. 1975. Damselfishes of the South Seas. T.F.H. Publications, Neptune City, N.J.

Allen, G. 1991. Damselfishes of the World. Mergus, Melle, Germany.

Aubert de La Chesnaye des Bois, F.A. 1799. Dictionnaire raisonné et universel des animaux, ou le règne animal (Vol. 2). Paris, Bauche.

Bleeker, P. 1877. Mémoire sur les Chromides marins ou Pomacentroïdes de l'Inde archipélagique, par P. Bleeker: Loosjes.

Bloch, M.E. 1787. Ichtyologie, ou histoire naturelle, générale et particulière des poissons. Berlin.

Bonaparte, C.L. 1831. Saggio di una distribuzione metodica degli animali vertebrati. Presso. Roma: Antonio Boulzaler.

Ceccarelli, D.M. 2007. Modification of benthic communities by territorial damselfish: a multi-species comparison. Coral Reefs 26(4): 853–866.

Colleye, O., M. Nakamura, B. Frédérich and E. Parmentier. 2012. Further insight into the sound-producing mechanism of clownfishes: what structure is involved in sound radiation? J. Exp. Biol. 215(13): 2192–2202.

Colwell, R.K. and E.R. Fuentes. 1975. Experimental studies of the niche. Annu. Rev. Ecol. Evol. Syst. 6: 281–310.

Cooper, J.W. 2008. The Biogeography of Damselfish Skull Evolution: A Major Radiation throughout the Indo-West Pacific Produces no Unique Skull Shapes. 11th International Coral Reef Symposium, Ft. Lauderdale, Florida.

Cooper, W., L. Smith and M. Westneat. 2009. Exploring the radiation of a diverse reef fish family: Phylogenetics of the damselfishes (Pomacentridae), with new classifications based on molecular analyses of all genera. Mol. Phylogenet. Evol. 52(1): 1–16.

Cuvier, G. 1815. Suite des observations et recherches techniques sur différents poissons de la Méditerranée, et à leur occasion sur des poissons d'autres mers, plus ou moins liés avec eux. Mémoires du Museum d'Histoire Naturelle 1: 353–363.

Cuvier, G. and A. Valenciennes. 1828. Histoire naturelle des poissons, Tome I. Strasbourg, F.G. Levrault.

Cuvier, G. and A. Valenciennes. 1830. Histoire naturelle des poissons, Tome V. Strasbourg, F.G. Levrault.

de Lacépède, B.G. 1798. Histoire naturelle des poissons, Tome IV. Paris, Plassan.

Desmarest, A.G. 1814. Sur le petit castagneau, appelé *Sparus chromis* par tous les auteurs, qui doit devenir le type d'un nouveau genre nommé *Chromis*, et appartenant à la famille des Labres. Bull. Sci. Soc. Philom. Paris 1(3): 88–89.

Eschmeyer, W.N. 2015. Catalog of Fishes, http://research.calacademy.org/research/ichthyology/catalog/fishcatmain.asp. Electronic version accessed March, 2015.

Fishelson, L. 1998. Behaviour, socio-ecology and sexuality in damselfishes (Pomacentridae). Ital. J. Zool. 65: 387–398.

Frédérich, B., E. Parmentier and P. Vandewalle. 2006. A preliminary study of development of the buccal apparatus in Pomacentridae (Teleostei, Perciformes). Anim. Biol. 56(3): 351–372.

Frédérich, B., G. Fabri, G. Lepoint, P. Vandewalle and E. Parmentier. 2009. Trophic niches of thirteen damselfishes (Pomacentridae) at the Grand Récif of Toliara, Madagascar. Ichthyol. Res. 56(1): 10–17.

Frédérich, B., D. Olivier, G. Litsios, M.E. Alfaro and E. Parmentier. 2014. Trait decoupling promotes evolutionary diversification of the trophic and acoustic system of damselfishes. Proc. R. Soc. B-Biol. Sci. 281(1789).

Galis, F. and P. Snelderwaard. 1997. A novel biting mechanism in damselfishes (Pomacentridae): the pushing up of the lower pharyngeal jaw by the pectoral girdle. Neth. J. Zool. 47: 405–410.

Gill, T. 1911. A New Translation of Aristotle's "History of Animals". Science 33: 730–738.

Günther, A. 1862. Catalogue of the Fishes in the Bristish Museum (Vol. 4). The Trustees, London.

Hata, H. and M. Kato. 2002. Weeding by the herbivorous damselfish *Stegastes nigricans* in nearly monocultural algae farms. Mar. Ecol. Prog. Ser. 237: 227–231.

Heckel, J.J. 1840. Johann Natterer's neue Flussfische Brasilien's nach den Beobachtungen und Mittheilungen des Entdeckers. Annalen des Wiener Museums der Naturgeschichte 2: 325–470.

Holbrook, S., G. Forrester and R. Schmitt. 2000. Spatial patterns in abundance of a damselfish reflect availability of suitable habitat. Oecologia 122: 109–120.

Kaufman, L.S. and K.F. Liem. 1982. Fishes of the Suborder Labroidei (Pisces, Perciformes): phylogeny, ecology and evolutionary significance. Breviora 472: 1–19.

Klunzinger, C.B. 1870. Synopsis der Fische des Rothen Meeres. Vienna, C. Ueberreuter'sche Buchdruckerei.

Lecchini, D. and R. Galzin. 2005. Spatial repartition and ontogenetic shifts in habitat use by coral reef fishes (Moorea, French Polynesia). Mar. Biol. 147(1): 47–58.

Leis, J.M. and M.I. McCormick. 2002. The biology, behaviour and ecology of the pelagic, larval stage of coral reef fishes. pp. 171–199. *In*: P.F. Sale (ed.). Coral Reef Fishes: Dynamics and Diversity in a Complex Ecosystem. Academic Press, San Diego.

Liem, K.F. 1973. Evolutionary strategies and morphological innovations: cichlid pharyngeal jaws. Syst. Zool. 22: 425–441.

Liem, K.F. 1978. Modulatory multiplicity in the functional repertoire of the feeding mechanism in cichlid fishes. I. Piscivores. J. Morphol. 158(3): 323–360.

Liem, K.F. 1993. Ecomorphology of the teleostean skull. pp. 423–452. *In*: J. Hanken and B.K. Hall (eds.). The Skull, Vol. 3. Functional and Evolutionary Mechanisms. The University of Chicago Press, Chicago.

Liem, K.F. and P.H. Greenwood. 1981. A functional approach to the phylogeny of the pharyngognath teleosts. Am. Zool. 21: 83–101.

Linnaeus, C. 1758. Systema naturæ per regna tria naturæ, secundum classes, ordines, genera, species, cum characteribus, differentiis, synonymis, locis 1 (10th ed.). Stockholm: Impensis Laurentii Salvii.

Mabuchi, K., M. Miya, Y. Azuma and M. Nishida. 2007. Independent evolution of the specialized pharyngeal jaw apparatus in cichlid and labrid fishes. BMC Evol. Biol. 7(1): 10.

Mann, D. and P.S. Lobel. 1998. Acoustic behaviour of the damselfish *Dascyllus albisella*: behavioural and geographic variation. Environ. Biol. Fishes 51: 421–428.

Marshall, N. 2000. Communication and camouflage with the same 'bright' colours in reef fishes. Philos. T. Roy. Soc. B. 355: 1243–1248.

Meekan, M., A. Steven and M. Fortin. 1995. Spatial patterns in the distribution of damselfishes on a fringing coral-reef. Coral Reefs 14: 151–161.

Müller, J. 1843. Beiträge zur Kenntnis der natürlichen Familien der Fische. Archiv für Naturgeschichte: 292–381.

Müller, J. 1844. Uber den bau und die grenzen der Ganoiden und über das natürliche system der fische. Abhandlungen der Königlichen Akademie der Wissenschaften: 117–216.

Nakamura, Y., M. Horinouchi, T. Nakai and S. Sano. 2003. Food habits of fishes in a seagrass bed on a fringing coral reef at Iriomote Island, southern Japan. Ichthyol. Res. 50(1): 15–22.

Nelson, J.S. 2006. Fishes of the World (4th ed.). John Wiley & Sons, Inc., Hoboken.

Olivier, D., B. Frédérich, M. Spanopoulos-Zarco, E. Balart and E. Parmentier. 2014. The cerato-mandibular ligament: a key functional trait for grazing in damselfishes (Pomacentridae). Front. Zool. 11: 63.

Olivier, D., B. Frédérich, A. Herrel and E. Parmentier. 2015. A morphological novelty for feeding and sound production in the yellowtail clownfish. J. Exp. Zool. Part A 323: 227–238.

Ormond, R., J. Roberts and R. Jan. 1996. Behavioural differences in microhabitat use by damselfishes (Pomacentridae): Implications for reef fish biodiveristy. J. Exp. Mar. Biol. Ecol. 202: 85–95.

Owen, R. 1866. On the Anatomy of Vertebrates, Fishes and Reptiles, Vol. 1. Longmans, Green and Co., London.

Parmentier, E., O. Colleye, M. Fine, B. Frédérich, P. Vandewalle and A. Herrel. 2007. Sound production in the clownfish *Amphiprion clarkii*. Science 316: 1006.

Parmentier, E., L. Kéver, M. Casadevall and D. Lecchini. 2010. Diversity and complexity in the acoustic behaviour of *Dacyllus flavicaudus* (Pomacentridae). Mar. Biol. 157(10): 2317–2327.

Playfair, R.L. and A.C. Günther. 1866. Fishes of Zanzibar: Acanthopterygii/Pharyngognathi. London: John van Voorst, Paternoster Row.

Risso, A. 1810. Ichtyologie de Nice, ou Histoire naturelle des poissons du département des Alpes-Maritimes. Paris: F. Schoell.

Robertson, D. and B. Lassig. 1980. Spatial distribution patterns and coexistence of a group of territorial damselfishes from the Great Barrier Reef. Bull. Mar. Sci. 30: 187–203.

Scott, W.M. 2008. Damselfishes & Anemonefishes. T.F.H. Publications, Neptune City, N.J.

Siebeck, U.E. 2004. Communication in coral reef fish: the role of ultraviolet colour patterns in damselfish territorial behaviour. Anim. Behav. 68: 273–282.

Siebeck, U.E. and N.J. Marshall. 2001. Ocular media transmission of coral reef fish-can coral reef fish see ultraviolet light? Vision Res. 41(2): 133–149.

Stiassny, M.L.J. 1981. The phyletic status of the family Cichlidae (pisces, perciformes): a comparative anatomical investigation. Neth. J. Zool. 31: 275–314.

Stiassny, M.L.J. and J. Jensen. 1987. Labroid interrelationships revisited: morphological complexity, key innovations, and the study of comparative diversity. Bull. Mus. Comp. Zool. 151: 269–319.

Streelman, J.T. and P.D. Danley. 2003. The stages of vertebrate evolutionary radiation. Trends Ecol. Evol. 18(3): 126–131.

Valmont-Bomare, J.C. 1754. Dictionnaire raisonné universel d'histoire naturelle (Vol. 2). Paris: Didot.

Wainwright, P.C., W.L. Smith, S.A. Price, K.L. Tang, J.S. Sparks, L.A. Ferry, K.L. Kuhn, R.I. Eytan and T.J. Near. 2012. The evolution of pharyngognathy: a phylogenetic and functional appraisal of the pharyngeal jaw key innovation in labroid fishes and beyond. Syst. Biol. 61(6): 1001–1027.

A Revised Damselfish Taxonomy with a Description of the Tribe Microspathodontini (Giant Damselfishes)

W. James Cooper[1],* and *Francesco Santini*[2]

Damselfish taxonomy is as old as taxonomy itself. It formed a part of the establishment of our modern system of zoological nomenclature, as descriptions of the two damselfish species of found in the Mediterranean (*Abudefduf saxatilis* and *Chromis chromis*) were included in Linnæus's *Systema Naturae* (1758), though at the time they were referred to as *Chaetodon saxatilis* and *Sparus chromis*. *Chromis chromis* had been previously described in Peter Artedi's *Ichthyologia* (1738), which is often considered the foundational text for the formal discipline of ichthyology. "*Chromis*" (χρομισ) is also mentioned in the first great western biology text: Aristotle's History of Animals (4th century BCE). Although Aristotle's "*Chromis*" was probably not a damselfish, his *korakinos* (κορακινοσ) was likely *Chromis chromis* itself (Gill 1911). One could therefore consider the study of damselfishes to be as old as the science of biology (see Chapter I for further details).

There are currently 396 recognized damselfish species (Eschmeyer and Fricke 2015). John Randall's work at the Bishop Museum in Honolulu and elsewhere has contributed descriptions of 47 of these (Allen and Randall 1974) as part of an

---

[1] School of Biological Sciences, Washington State University, Pullman, Washington, 99164, USA.
[2] Dipartimento di Scienze della Terra, Università degli Studi di Torino, Via Valperga Caluso 35, 10125 Torino, Italy.
* Corresponding author: jim.cooper@wsu.edu

extraordinary body of work in fish taxonomy. His frequent co-author Gerald Allen has not only described more damselfish species (>100) and more currently accepted genera than any other author, but in 1975 he also organized the Pomacentridae into four subfamilies in his book *Damselfishes of the South Seas*. This work included keys for identifying both the subfamily (Amphiprionae, Chrominae, Lepidozyginae or Pomacentrinae) and genus of all the damselfishes known to inhabit the Indo-West Pacific at that time. This was followed in 1991 by his *Damselfishes of the World*, which included photographs, or in a very limited number of cases, illustrations, of every currently known species. Both of these texts are critical must-haves for any damselfish biologist, all of whom are greatly in his debt. Allen's taxonomic classifications became critically important hypotheses of damselfish evolutionary relationships that would eventually be tested using molecular phylogenetic methods.

Molecular Phylogenetics and Damselfish Taxonomy

The placement of organisms within taxa has always been based on the characters or traits that they have in common with other organisms. Taxonomy predates our modern understanding of evolution, but from a system based strictly on similarity we have moved to (or at least toward) a classification system in which taxonomy is at least largely synonymous with evolutionary relationships. Similar anatomical characters present in multiple species can, of course, arise through descent from a common ancestor (these are known as *synapomorphies*, e.g., the presence of 2 spines on the anal fin of all damselfishes) or via convergent evolution (this represents *homoplasy*, or *homoplastic character* states, e.g., the presence of procurrent spines on the caudal peduncle of species in the genera *Chromis* and *Acanthochromis*). Synapomorphies are useful for reconstructing evolutionary history, while homoplasy is misleading.

Phylogenetic systematics is a modern field that uses mathematical analyses to construct phylogenies (i.e., evolutionary trees) based on synapomorphies (Hennig 1950, 1966). Advances in molecular analyses and computational speed now allow us to sequence large amounts of genetic data and use each nucleotide position as a character in phylogenetic analyses. In the late 1900's numerous authors were publishing molecular phylogenies and using them to reorganize the taxonomy of a large number of species. Some of the results were surprising and some were spectacularly wrong and many experienced, traditional taxonomists voiced strong concerns. It was certainly true that some young scientists (and some that weren't so young) were going from tissue samples to molecules to phylogeny without taking the time to learn enough about their taxa. As a result they didn't necessarily have a good sense for when their phylogenetic trees looked suspicious. It was also true that advances in molecular systematics were showing us that some previous taxonomic hypotheses, and it should be remembered that taxonomic groupings are *never more than hypotheses*, were clearly wrong.

The use of molecular phylogenetic analyses to examine the evolutionary relationships of damselfishes and to depict these relationships as phylogenetic trees was first accomplished by Kevin Tang in 2001. He identified the sequences of 1510 base pairs of DNA from 3 mitochondrial genes in 23 damselfishes representing 14 genera. In many cases all of the species that one might examine using molecular methods will exhibit the same character state for most nucleotide positions. In this particular case either 435 or 444 characters, depending on the type of mathematical analysis used, were informative in building hypotheses of evolutionary relationships among species. Tang's study showed that his best-supported molecular phylogenetic hypotheses conflicted with important parts of Gerald Allen's taxonomic classifications (Allen 1975, 1991).

As molecular phylogenetic techniques improved, the number of damselfish species and genera sequenced climbed rapidly (e.g., McCafferty et al. 2002, Quenouille et al. 2004, Tang et al. 2004, Santini and Polacco 2006, Cooper et al. 2009, Bernardi 2011, Frédérich et al. 2013). Simultaneously, the number of sequenced genes, the length of these sequences, the kinds of genes examined (nuclear and mitochondrial), and the number of analytical techniques used to examine the genetic data all increased very quickly as well. A significant advance was made by Quenouille et al. in 2004 when they examined more than 100 species from 18 damselfish genera. They nearly doubled the number of species examined (though Jang-Liaw et al. also examined 18 genera in 2002), nearly tripled the amount of analyzed sequence data that was useful for reconstructing trees and were the first to examine genes from nuclear as well as mitochondrial DNA. In 2009 Cooper et al. also examined over 100 species using both mitochondrial and nuclear DNA from representatives of all 29 pomacentrid genera (though sequences obtained from formalin-fixed specimens of *Altrichthys* and *Nexilosus* were somewhat fragmentary). Frédérich et al. recently contributed significantly to this body of work by analyzing sequence data from 208 damselfishes (over half of the described species) from all genera but *Nexilosus* (Frédérich et al. 2013).

The three largest molecular studies of damselfish relationships (Quenouille et al. 2004, Cooper et al. 2009, Frédérich et al. 2013) are largely in agreement. We now have strong reasons to believe that the broad patterns of damselfish evolution are well resolved (Fig. 1), although filling in the remaining gaps will represent useful contributions. The most uncertain aspect is the placement of the very long branch leading only to *Lepidozygus tapeinosoma*, which is the sole representative of the subfamily Lepidozyginae. Long branches are notoriously hard to place, but there is strong evidence that the *Lepidozygus* lineage diverged at least 20 Ma ago if not much earlier.

The comparison of molecular and anatomical data has shown that the damselfishes have undergone a tremendous amount of convergent morphological evolution (Cooper and Westneat 2009, Frédérich et al. 2013). They are a robust example of a highly successful adaptive radiation (Schluter 2000) that has undergone repeated ecological divergence. This radiation is unusual in that instead of invading an ever-increasing number of new niches, it has progressed by repeatedly evolving similar morphological states, such that newly evolved damselfish lineages have converged on a small number of niches multiple times. This pattern was described as a "reticulate adaptive radiation" by Cooper and Westneat in 2009 and the iterative nature of this radiation was examined in greater detail by Frédérich et al. in 2013 and also in Chapter IX here. This pattern of repeated convergence was originally described for skull morphology, but it is also seen in the morphological characters traditionally used to describe fish taxa.

When the characters or combinations of anatomical characters used to define taxa arise repeatedly and independently in closely related species, then those characters

Evolutionary relationships of damselfish subfamilies, tribes and genera - I

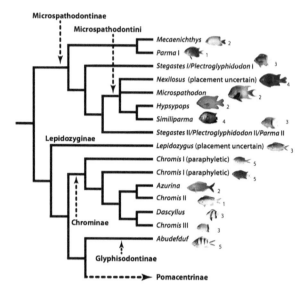

Evolutionary relationships of damselfish subfamilies, tribes and genera - II

The Pomacentrinae

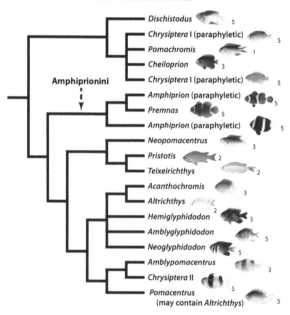

Fig. 1. Phylogenetic relationships of damselfish subfamilies, tribes and genera. Photography credit key: (1) John Randall; (2) Gerald Allen; (3) Richard Winterbottom; (4) W. James Cooper and (5) Jeffrey Williams.

do not represent synapomorphies that are reliable for identifying branches of the tree of life. The reverse is also true, in that a newly evolved lineage need not necessarily exhibit strong, nor indeed any, differences in morphology relative to its sister group/s or immediate ancestor. All of the following represent character states that have evolved multiple times in the damselfishes: small, spine-like rays (spiniform procurrent rays) at the upper and lower base of the caudal fin; various tooth morphologies and arrangements; the degree to which the forward part of the head is covered with scales; the number of spines and rays in the various fins; the number of lateral line scales; the degree to which the body is elongate or orbiculate and whether or not the opercular or suborbital bones have smooth, crenelated or serrated margins. Unfortunately damselfish taxonomy has relied heavily on just these characters.

Taxonomic Revisions of the Pomacentridae

Of Allen's four pomacentrid subfamilies, only two are actually monophyletic and one of these is the monotypic Lepidozyginae. Both the Chrominae and the Pomacentrinae (*sensu* Allen 1975) included species from multiple lineages that had convergently evolved similar combinations of characters. The Chrominae only required the removal of the genera *Acanthochromis* and *Altrichthys* to become monophyletic and since these genera contain only three species, this is perhaps not much of a rearrangement. The Pomacentrinae exhibited a very different pattern and that subfamily required much more extensive reorganization. The anemonefish subfamily (Amphiprioninae), which is one of the most distinct groups of damselfishes, survived intact, but merited demotion in "status" from subfamily to tribe.

Cooper et al. made an attempt to reorganize the subfamilial taxonomy of the damselfishes in 2009, but there were two important errors in their nomenclature that need to be corrected here, as well as an additional issue (lack of formal anatomical taxonomic descriptions) that has interfered with the acceptance of these monophyletic taxonomic groupings. The International Commission on Zoological Nomenclature has strict rules about the precedence of taxonomic names. It is not acceptable to give a taxon a new name if a previously existing one is appropriate.

The members of the genera *Hypsypops, Mecaenichthys, Microspathodon, Nexilosus, Parma, Plectroglyphidodon, Similiparma* and *Stegastes* form a clear monophyletic group of species (Cooper et al. 2009, Frédérich et al. 2013) and this lineage was given the subfamily name "Stegastinae" by Cooper et al. (2009). However, the subfamilial name Microspathodontinae had been established by Jordan and Evermann in 1898 and although at the time it only included *Microspathodon*, the name still has precedence and must supercede "Stegastinae". A similar issue is found with "Abudefdufinae" (Cooper et al. 2009), which should instead be Glyphisodontinae (Richardson 1844). Even though the genera within these two subfamilies that have the oldest descriptions are indeed *Stegastes* (Jenyns 1842) and *Abudefduf* (Forsskål 1775), neither of these names were used to derive family or subfamily names before Microspathodontinae and Glyphisodontinae were established. Cooper et al. (2009) also suggested synonymizing the genus *Azurina* (within Chrominae) with *Chromis*, since quantitative morphological analyses showed that the morphology of *Azurina* was extremely similar to that of its closest relatives in *Chromis* (e.g., *Chromis multilineata*), with the exception of possessing a longer lateral line (Cooper 2006). This suggestion is retracted here.

Abudefduf luridus was found by both Cooper et al. (2009) and Frédérich et al. (2013) to be the sister species to *Similiparma hermani* and it has recently been re-assigned to that genus as *Similiparma lurida* (Cooper et al. 2014). This allows both the genus *Abudefduf* and the subfamily Glyphisodontinae to be monophyletic. The lineage that includes *Similiparma* and the genera *Hypsypops*, *Microspathodon*, and *Nexilosus* merits recognition as a taxon because these fishes' large size (they include the only damselfishes to reach 30 cm; Allen 1991) and characteristic morphology (Figs. 2–4) distinguish them from other pomacentrids. Jordan and Evermann's (1898) subfamily Microspathodontinae was based on *Microspathodon dorsalis*. We apply the tribe name Microspathodontini to this lineage of "Giant Damselfishes" that includes all species in the genera *Hypsypops*, *Microspathodon*, *Nexilosus* and *Similiparma*. Although the genetic evidence for concluding that this group includes *Nexilosus latifrons* (the genus is monotypic) is limited (see Cooper et al. 2009), the morphological, ecological and geographical evidence overwhelmingly supports its inclusion in the Microspathodontini (Cooper 2009, Cooper and Westneat 2009, Aguilar-Medrano et al. 2011, Frédérich et al. 2013; Fig. 3). A detailed description of this tribe is given below.

Descriptions of damselfish subfamilies and tribes

Subfamily Chrominae (revised from Allen 1975, 1991, Cooper et al. 2009)

Type genus: *Chromis* Cuvier, 1814

All species within the genera *Azurina*, *Chromis*, and *Dascyllus* or within genera that may result from the subdivision of these groups. Dorsal fin spines XII–XIV; teeth conical, either biserial or multiserial; scales in longitudinal series 30–34; opercular bone margins smooth to weakly serrate or with limited crenulations near posterior-ventral angle. Presence of 2–3 procurrent spines on upper or lower edge of caudal fin base is not a reliable character. Body may be elongate to orbiculate.

Subfamily Glyphisodontinae (derived from Allen 1991, Cooper et al. 2009 and Forsskål 1775)

Type genus: *Abudefduf* Forsskål, 1775

All species within the genus *Abudefduf* or within genera that may result from its subdivision. This following sentence is a direct quote of Allen's (1991) description of the genus *Abudefduf*: Margin of suborbital and preopercle smooth; notch between pre- and suborbital absent; teeth uniserial, usually with notched or flattened tips at front of jaw; relatively deep-bodied, depth of adults usually 1.6–1.9 in SL; dorsal fin spines XIII; soft dorsal rays 11–16; color pattern often composed of a series of dark cross bars on sides.

Subfamily Lepidozyginae (derived from Allen 1975, 1991 and Günther 1862)

Type genus: *Lepidozygus* Günther, 1862

All species within the genus *Lepidozygus* (currently monotypic) or within genera that may result from the subdivision of this genus. This following sentence is a direct quote of Allen's (1991) description of the genus *Lepidozygus*: Row of papilla-like structures on rear edge of eye socket (posterior circumorbitals); extremely elongate body shape, depth about 2.9–3.0 in SL; rear edge of preopercle finely serrated; teeth of jaws uniserial; edge of suborbital hidden by scales; 33–36 scales in longitudinal series from upper edge of operculum to base of caudal fin; dorsal fin spines XII.

Subfamily Microspathodontinae

Type genus: *Microspathodon* Günther, 1862

All species within the genera *Hypsypops, Mecaenichthys, Microspathodon, Nexilosus, Parma, Plectroglyphidodon, Similiparma* and *Stegastes* or within genera that may result from the subdivision of these groups. Margins of suborbital and preopercle usually smooth or poorly developed, but may be crenulated in *Parma* and are serrated in *Stegastes*. No notch between preorbital and suborbital except in *Microspathodon*. Preorbital may be broad. Teeth uniserial. Body orbiculate. Body depth (BD) 1.6–2.3 in SL. Dorsal-fin spines XII or XIII; dorsal fin rays 14–21; anal fin spines 2; anal fin rays 12–18; pectoral fin rays 17–32; lateral line scales 17–35; gill rakers on first branchial arch ≥10.

Tribe Microspathodontini (the giant damselfishes)

Type genus: *Microspathodon* Günther, 1862

Diagnosis—Dorsal fin spines XII or XIII; dorsal fin rays 14–19; anal fin spines 2; anal fin rays 12–14; pectoral fin rays 20–25; lateral line scales 19–24; gill rakers on first branchial arch ≥13; BD 1.7–2.2 in SL.

Description—A tribe within the damselfish subfamily Microspathodontinae. All species within the genera *Hypsypops, Microspathodon, Nexilosus* and *Similiparma* or within genera that may result from the subdivision of these groups. This tribe is confined to the tropical and temperate near-shore waters of the Atlantic and Eastern Pacific. All microspathodontine species are strongly associated with rocky reefs, except for *M. chrysurus*, which inhabits coral reefs (Allen 1991). These are large damselfishes, with the smallest species (*Similiparma lurida*) reaching 125 mm SL. The four largest microspathodontine species (*Hypsypops rubicundus, Microspathodon bairdii, M. dorsalis* and *Nexilosus latifrons*), which are also the largest damselfish species, reach standard lengths of 250–300 mm (Allen 1991). Their skull anatomy (Figs. 2–4) is characterized by large jaw adductor muscles, high supraoccipital crests, eyes that are dorsal to the mouth and moderately sized (eye diameter of no more than 1/4th of head length). Their head morphology is strongly consistent with a primarily benthic feeding lifestyle that includes foraging on algae and all species are strongly associated with the benthos (Cooper and Westneat 2009). Adult body coloration includes: uniformly grey/dark brown (*M. bairdii, M. dorsalis, M. frontatus*); uniform bright orange (*H. rubicundus*); overall dark brown with a bright blue axial spot and bright blue markings on the head, pectoral fins and anal fin (*S. lurida*); overall dark grey/black with a bright white caudal fin (*S. hermani*) and overall dark brown with bright blue spots on the dorsal portion of the body and a bright yellow caudal fin (*M. chrysurus*).

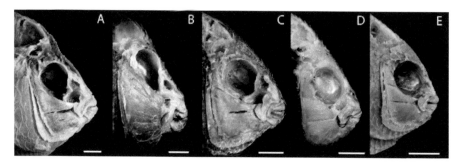

Fig. 2. Comparative of head anatomy of representative Giant Damselfishes: tribe Microspathodontini. (A) *Hypsypops rubicundus.* (B) *Microspathodon dorsalis.* (C) *Nexilosus latifrons.* (D) *Similiparma hermani.* (E) *Similiparma lurida.* All scale bars are 1 cm.

The results of a morphological study that compares the Microspathodontini with other damselfishes is presented in a principal components score plot (Fig. 3) in which the proximity of taxa to one another represents morphological similarity. This is an extension of the work presented in Cooper and Westneat (2009) and Cooper et al. (2014), which can be referenced for specific details of the analyses. In brief, we performed a geometric morphometric analysis of the 16 anatomical landmarks on the head of damselfish specimens (Fig. 3) that represent 44 species (>11% of the described damselfish species) and all extant genera. Calculations were based on the shape means of multiple specimens per species.

Bayesian phylogenetic analyses of molecular data have consistently returned 100% posterior support values for the inclusion of *Hypsypops, Microspathodon* and *Similiparma* in a single clade (Cooper et al. 2009, Frédérich et al. 2013). Although there is not yet any high-quality DNA sequence data for *Nexilosus latifrons*, partial gene sequence data (Cooper et al. 2009) is consistent with placing this species within the Microspathodontini. Further molecular genetic evidence is clearly needed to test this hypothesis. Similarities in body size, head anatomy and biogeography (Cooper and Westneat 2009, Aguilar-Medrano et al. 2011, Cooper et al. 2014) also indicate that *N. latifrons* is a member of this tribe. We hypothesize that *Nexilosus* and *Hypsypops* are sister lineages.

The Microspathodontini have radiated within one of the extreme regions of pomacentrid "shape space" (Fig. 3) while also evolving the largest damselfish body sizes. The trophic morphology of all species facilitates the production of strong, but relatively slow bites. Their extreme morphology is reflected in their restricted feeding ecology (i.e., very little planktivory; Cooper and Westneat 2009, Aguilar-Medrano et al. 2011). While damselfish evolution is strongly characterized by rapid transitions

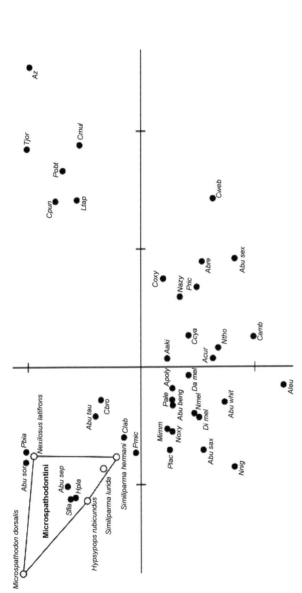

Fig. 3. Principal component score plot generated from a geometric morphometric analysis of damselfish cranial anatomy (all damselfish genera and 44 select species were sampled). PC 1 (X axis) and PC 2 (Y axis) account for 42.14% and 16.98% of the total head shape data in the dataset, respectively. **Microspathodontini:** *Hypsypops rubicundus, Microspathodon dorsalis, Nexilosus latifrons, Similiparma hermani, Similiparma lurida.* **Key to other pomacentrids examined ● :** *Abudefduf bengalensis* (Abu beng), *Abudefduf saxatilis* (Abu sax), *Abudefduf septemfasciatus* (Abu sep), *Abudefduf sexfasciatus* (Abu sex), *Abudefduf sordidus* (Abu sor), *Abudefduf taurus* (Abu tau), *Abudefduf whileyi* (Abu whit), *Acanthochromis polyacanthus* (Apoly), *Amblyglyphidodon curacao* (Acur), *Amblyglyphidodon leucogaster* (Aleu), *Amblypomacentrus breviceps* (Abre), *Amphiprion akindynos* (Aaki), *Azurina hirundo* (Az), *Cheiloprion labiatus* (Clab), *Chromis amboinensis* (Camb), *Chromis multilineata* (Cmul), *Chromis punctipinnis* (Cpun), *Chromis weberi* (Cweb), *Chrysiptera brownriggii* (Cbro), *Chrysiptera cyanea* (Ccya), *Chrysiptera oxycephala* (Coxy), *Dascyllus melanurus* (Damel), *Dischistodus melanotus* (Dimel), *Hemiglyphidodon plagiometopon* (Hpla), *Lepidozygus tapeinosoma* (Ltap), *Mecaenichthys immaculatus* (Mimm), *Neoglyphidodon melas* (Nmel), *Neoglyphidodon nigroris* (Nnig), *Neoglyphidodon oxyodon* (Noxy), *Neoglyphidodon thoracotaeniatus* (Ntho), *Neopomacentrus azysron* (Nazy), *Parma microlepis* (Pmic), *Plectroglyphidodon lacrymatus* (Plac), *Pomacentrus alexanderae* (Pale), *Pomachromis richardsoni* (Pric), *Premnas biaculeatus* (Pbia), *Pristotis obtusirostris* (Pobt), *Stegastes flavilatus* (Sfla), *Teixeirichthys jordani* (Tjor).

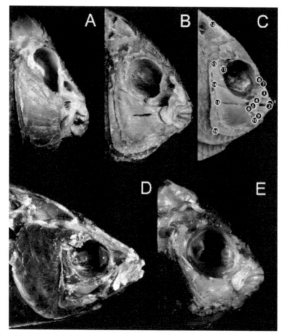

Fig. 4. Pictorial descriptions of the morphometric axes depicted in Fig. 3. (A) *Microspathodon dorsalis.* (B) *Nexilosus latifrons.* (C) *Similiparma lurida* showing the landmarks analyzed in the shape analysis. (D) *Azurina hirundo.* (E) *Neoglyphidodon nigroris.* The head shape differences between the specimens in plates A and D describe the morphological variation associated with the X axis (PC 1) in Fig. 3. Specimens with low PC 1 scores (those towards the left of the X axis) have heads that are compact in terms of anterior-posterior extension. They also have larger *pars rictalis* divisions of the *adductor mandibulae*, shorter jaw bones, smaller eyes in a more dorsal position, taller supraoccipital crests, more subterminal mouths and shorter ascending processes on the premaxillae that prohibit extensive jaw protrusion. The head shape differences between the specimens in plates B and E describe the morphological variation associated with the Y axis (PC 2) in Fig. 3. Specimens with low PC 2 scores (those towards the bottom of the Y axis) have smaller *pars malaris* divisions of the *adductor mandibulae*. Anatomical landmarks depicted in plate C: 1 = Tip of the anterior-most tooth on the premaxilla; 2 = Tip of the anterior-most tooth on the dentary; 3 = Maxillary-palatine joint (upper rotation point of the maxilla); 4 = Insertion of the *pars malaris* division of the *adductor mandibulae* on the maxilla; 5 = Maxillary-articular joint (lower point of rotation of the maxilla); 6 = Insertion of the *pars rictalis* division of the *adductor mandibulae* on the primordial process of the articular; 7 = Posterior tip of the ascending arm of the premaxilla; 8 = Joint between the nasal bone and the neurocranium; 9 = Articular-quadrate joint (lower jaw joint); 10 = Insertion of the interopercular ligament on the articular; 11 = Most posterio-ventral point of the eye socket; 12 = Dorsal-most tip of the supraoccipital crest on the neurocranium; 13 = Most dorsal point on the origin of the *pars stegalis* division of the *adductor mandibulae* on the hyomandibula and preopercle; 14 = Most dorsal point on the origin of the *pars malaris* division of the *adductor mandibulae* on the hyomandibula and preopercle; 15 = Most dorsal point on the origin of the *pars rictalis* division of the *adductor mandibulae* on the hyomandibula and preopercle; 16 = Posterio-ventral corner of the preopercle. Muscle nomenclature after Datovo and Vari (2013). The *pars malaris* constitutes the dorso-lateral portion of the *adductor mandibulae* in damselfishes and inserts on the maxilla. The *pars rictalis* constitutes the ventro-lateral portion of the *adductor mandibulae* in damselfishes and inserts on the primordial process of the articular in the lower jaw. The *pars stegalis* constitutes the medial portion of the *adductor mandibulae* in damselfishes and inserts on the medial surface of the articular within the dento-articular groove (the medial surface of the articular excurvation). The *pars malaris, pars rictalis* and *pars stegalis* are synonymous with the A1, A2 and A3 subdivisions of the *adductor mandibulae* (respectively), *sensu* Vetter (1878) and Winterbottom (1973). Osteological nomenclature after Barel et al. (1976).

between benthic and pelagic feeding niches (Cooper and Westneat 2009, Frédérich et al. 2013), this clade exhibits an alternative pattern in which extensive pelagic feeding has not arisen. Whether this pattern is due to the low species diversity of the group, limited ecological opportunities, developmental constraints, or simply chance alone is not known, but it is clear that the Microspathodontini represent a unique part of the damselfish adaptive radiation.

At least some, if not all members of the genus *Microspathodon* have unusual anchoring of the oral dentition. The teeth in the upper jaw are not strongly attached to the premaxilla and the upper-jaw teeth as a group are therefore able to conform to irregular surfaces as these fishes scrape algae and other matter from hard substrates. It is very likely that this significantly improves their feeding efficiency. Ciardelli (1967) reported this unusual anatomy in *M. chrysurus* and it was apparent in dissections of *M. dorsalis* (pers. obs.). Ciardelli (1967) observed that the upper-jaw teeth of *M. chrysurus* are anchored by relatively loose connective tissue within an unusual anterio-ventral fossa on the premaxilla.

Subfamily Pomacentrinae

Type genus: *Pomacentrus* Lacepède, 1802

All species within the genera *Acanthochromis*, *Amphiprion*, *Altrichthys*, *Amblyglyphidodon*, *Amblypomacentrus*, *Cheiloprion*, *Chrysiptera*, *Dischistodus*, *Hemiglyphidodon*, *Neoglyphidodon*, *Neopomacentrus*, *Pomacentrus*, *Pomachromis*, *Premnas*, *Pristotis*, and *Teixeirichthys* or within genera that may result from their subdivision. The necessarily broad ranges of characters in the following description serve to outline the problem of using traditional morphological characters for describing important branches of the damselfish phylogenetic tree as formal taxa. Dorsal fin spines VIII–XVII; preopercle and subopercle edges smooth, weakly serrate, finely serrate, or strongly serrate; edges of suborbital exposed or hidden by scales; notch between orbital and preorbital either present or absent; preorbital and suborbital scaled or scaleless; teeth uniserial, weakly biserial, irregularly biserial or biserial; inner teeth may or may not be present; scales large to small; scales in longitudinal series < 30–76; body highly orbiculate to elongate; BD 1.5–3.0 in SL. Distribution restricted to the Indo-West Pacific. Given the paucity of pomacentrid fossils (see below), assigning dates to the origin of particular branches of the pomacentrid phylogenetic tree is problematic. However, the wide distribution of the Pomacentrinae throughout the tropical Indo-West Pacific and their absence from the Atlantic allows for reasonably accurate dating of their divergence at just before the closing of the Tethys Seaway 12–18 Ma (Cooper 2009).

Tribe Amphiprionini (derived from Allen 1975, 1991 and Cooper et al. 2009)

Type genus: *Amphiprion* Bloch and Schneider, 1801

All species within the genera *Amphiprion* and *Premnas* or within genera that may result from their subdivision. Opercle and interopercle serrated (serrations may be strong or weak); subopercle serrated in *Amphiprion*, but not serrated in *Premnas*; dorsal fin spines VII–XI; scales small, about 47–76 in a longitudinal row from the upper edge of operculum to the base of caudal fin; suborbital usually possess a distinctive elongate spine in *Premnas*. All species are protandrous (Chapter IV) and symbiotic with sea anemones in the wild.

Fossil Damselfish Species

In spite of a long evolutionary history that dates back to the early Neogene (Cowman and Bellwood 2012, Frédérich et al. 2013, Near et al. 2013), damselfishes have a relatively scarce fossil record, which is currently limited to 6 described taxa. The known pomacentrid fossil record was reviewed by Bellwood and Sorbini (1996), Bellwood (1999), Carnevale and Landini (2000) and Bannikov and Bellwood (2014). The brief review included here will rely heavily upon their findings and conclusions.

The oldest fossils that can unquestionably be assigned to the family Pomacentridae are *Palaeopomacentrus orphae* Bellwood and Sorbini (1996), *Lorenzichthys olihan* Bellwood (1999), and *Sorbinichromis francescoi* Bannikov and Bellwood (2014). All three taxa are from the Ypresian deposits of Monte Bolca (Middle Eocene, 50 Ma; Bellwood and Sorbini 1996, Bellwood 1999, Bannikov and Bellwood 2014, Carnevale et al. 2014).

Palaeopomacentrus orphae is currently known from two specimens: the holotype, located in the Museo Civico di Storia Naturale in Verona, Italy, and a paratype kept in the Natural History Museum in London, UK. *Lorenzichthys olihan* is known from a single specimen, also housed in the Verona Museum. No additional counterparts are known for any of these three fossils. *Sorbinichromis francescoi* is also known from a single specimen (available in part and counterpart) preserved in the collection of the Museo Civico di Storia Naturale in Verona. All known specimens for the three species are of very small size (less than 30 mm SL).

The Eocene fossil pomacentrids are diagnosed by an ovate body with a continuous and prominent dorsal fin; 10 dorsal spines (11 in *Sorbinichromis*); 14 or 15 caudal vertebrae; 2 supernumerary anal fin spines; and ctenoid scales. *Palaeopomacentrus orphae* also possess pharyngeal jaws with the lower pharyngeal bone in close association with the cleithrum; serrated margins of the preopercular bone; small, caniniform and unicuspid teeth; 11 abdominal vertebrae; 3 supraneurals; 10 soft dorsal rays; 7–8 soft anal rays; three epurals; 1+8–7+1 caudal fin rays. All *Palaeopomacentrus* specimens show pigmented regions. The pigmentation of the choroid layer of the eye is preserved as a dark pigmented central region of the orbit, while the lining of the peritoneal cavity is preserved as dark coloring of the anteroventral region of the abdominal cavity. Furthermore, the holotype preserves two dark bands of pigmentation on the body, while small, brown irregular dots (which have been interpreted as chromatophores) are visible on the body of the paratype (Bellwood and Sorbini 1996). *Lorenzichthys olihan* is also diagnosed by 8 abdominal vertebrae; 12 soft dorsal rays; 12 soft anal rays; 1+7–6+1 caudal fin rays (Bellwood 1999). The head of *Lorenzichthys olihan* is very poorly preserved. The spiny portion of the dorsal fin and the soft portion of the anal fin, as well as the pelvic fin rays, are bordered at their distal margins by a narrow band of pigment (Bellwood 1999) and patches of pigment can also be seen on all these fins. *Sorbinichromis francescoi* is diagnosed by, among other features, a deep and pointed supraoccipital crest; the articulation of the lower jaw that is found under the middle of the orbit, a mouth that is relatively small; the presence of 10 abdominal and 15 caudal vertebrae; 2 supraneurals; the presence of a continuous dorsal fin with 11 spines and 12 soft rays and of an anal fin with 2 supernumerary spines and 10 soft rays. *Sorbinichromis* also has relatively large scales that are thick around the dorsal and ventral edges of the body (Bannikov and Bellwood 2014).

According to Bellwood and Sorbini (1996) the body shapes of both *Palaeopomacentrus orphae* and *Lorenzichthys olihan* resemble those of many extant species of pomacentrids, while the teeth of *P. orphae* appear similar to those of living *Chromis* species. Bellwood and Sorbini (1996) assigned *Palaeopomacentrus orphae* to the pomacentrids on the basis of the presence of two supernumerary anal spines, which are considered a synapomorphy of the family, as well as the body shape, the overall osteology of the head and the pectoral girdle, the number of vertebrae, epurals and supraneurals, the form of the caudal skeleton, a number of meristic values, the form of pharyngeal teeth and bones, the form of the oral jaw teeth and the presence of ctenoid scales. According to Bellwood (1999) the support for including *Lorenzichthys olihan* within the pomacentridae is not as strong as that for *Palaeopomacentrus orphae*, due to the poor preservation of the head and the presence of only 22 vertebrae (3 less than the lowest value known from recent pomacentrids). In spite of this, the presence of two supernumerary anal spines, as well as the overall body shape, in combination with the presence of pigmentation on the fins which resembles that found in juveniles of some living damselfishes (e.g., *Stegastes*; Bellwood 1999) supports the inclusion of *Lorenzichthys* in the Pomacentridae. According to Bannikov and Bellwood (2014) the presence of the two supernumerary anal-fin spines, of a putative pharyngeal jaw and the general body shape support the inclusion of *Sorbinichromis* in the pomacentrids, even though its meristic characters do not allow assignment to any extant genus, in spite of some overlap with *Chrysiptera* (e.g., XI to XIV dorsal spines, 9 to 15 dorsal rays, II anal spines, 10 to 16 anal rays). Bannikov and Bellwood (2014) hypothesize that the shape of the body and fins of *Sorbinichromis*, as well as the presence of small teeth, resemble the morphology of recent schooling species that live around reefs, such as some members of the genera *Amblyglyphidodon*, *Chromis* and *Chrysiptera*, and that *Sorbinichromis* may have been an omnivore or planktivore living in proximity to a hard substratum.

After the Monte Bolca fossils, the fossil record of pomacentrids features a gap of over 30 Ma, with the next oldest fossil being *Izuus nakamurai* Tokunaga and Saito (1938). *Izuus nakamurai* is the only species currently assigned to its genus and this species is also known from only a single specimen, in this case from the Lower Miocene of Central Japan. *Izuus nakamurai* also possessed two supernumerary anal spines, but although it is not clear if the specimen possessed 23 or 24 vertebrae (Bellwood and Sorbini 1996), all other meristic counts are consistent with those of extant damselfishes, as is the overall body shape.

Originally described by Arambourg (1927) and known from two specimens from the Late Miocene of Oran, Algeria (~6.5 Ma), *Chromis savornini* is the only fossil species currently assigned to an extant pomacentrid genus. In their review of the pomacentrid fossil record Bellwood and Sorbini (1996) determined that *Chromis savornini's* possession of two supernumerary anal spines represents the key synapomorphy that supports its identification as a damselfish, while its placement within the genus *Chromis* was supported by both the body shape and several meristic features.

An additional specimen represents an unnamed and highly incomplete pomacentrid fossil from the Late Miocene (~7 Ma) of Abruzzo, Italy (Carnevale and Landini 2000). The remains of this specimen are composed of a partial skull, 5 abdominal vertebrae, part of the pectoral girdle and the anterior region of the dorsal fin. The fossil was tentatively assigned to the Pomacentridae on the basis of the shape of the head, the osteology of the skull (especially the shape and structure of the pharyngeal jaws) and the presence of ctenoid scales.

A number of other fossils have in the past been described as damselfishes (*Odonteus* Agassiz, 1838; *Priscacara* Cope, 1877; *Cockerellites* Jordan, 1923; *Palaeochromis* sauvage, 1907), but it has been demonstrated that none of these taxa belong within the Pomacentridae (Bellwood and Sorbini 1996). The lack of fossil species currently assigned to the pomacentrids is likely due to a number of factors. The lifestyle of pomacentrids, which tend to live in habitats that are not prone to fossilization, such as coral or rocky reefs, is likely an important factor. The lack of a comprehensive morphological pomacentrid phylogeny, coupled with the non-monophyly of the Labroidei (the suborder in which damselfishes were formally placed; Betancur-R et al. 2013, Near et al. 2013) also makes it difficult to confidently assign many fossil taxa that are currently placed *incertae sedis* within the labroids (e.g., Bannikov and Carnevale 2012). As our understanding of the higher level relationships of percomorph fishes improves, it is likely that many fossils tentatively assigned to the "labroids" will be re-evaluated in the light of the new fish trees and that newer pomacentrid fossils will be identified.

Future Goals for Damselfish Taxonomy

As the morphological description of the Pomacentrinae attests, rampant convergence on similar morphological states makes formal descriptions of damselfish taxa a serious problem. Although accurate, the description of the Pomacentrinae provided here is completely useless in regard to being able to place a specific specimen within that subfamily.

Molecular techniques have repeatedly shown with a very high level of confidence that the subfamilies and tribes Chrominae, Glyphisodontinae, Lepidozyginae, Microspathodontinae, Pomacentrinae, Amphiprionini and Microspathodontini represent real and important branches of the damselfish tree (Quenouille et al. 2004, Cooper et al. 2009, Frédérich et al. 2013). This does not, however, guarantee that there are morphological synapomorphies that can be used to describe them taxonomically. Such anatomical characters almost certainly existed at one point in the evolution of the Pomacentrinae (for example), but they have either been lost or have yet to be discovered. The examination of literally thousands of specimens from all pomacentrid genera suggests that such characters are not easily identified (if they exist at all), though it should be noted that most of these morphological analyses have focused largely on the skull and that more expansive examinations could provide additional characters of taxonomic importance.

Though anatomical synapomorphies may be difficult or impossible to identify, it does not follow that the search for them should cease. The determination of taxonomically important morphological characters will always remain an extremely important part of systematic studies, but it is now clear that we cannot always assume that this is a reliable way to proceed. Quantitative morphometric analyses (i.e., shape analyses), such as the one used here to describe the Microspathodontini, can provide a useful complement to traditional, qualitative anatomical descriptions of taxa. As with many molecular phylogenetic techniques, this is another area that is seeing extremely rapid technological advancement that offers several sets of tools to systematists.

Taxonomy is not explicitly directed at naming the branches of the tree of life. Gerald Allen noted in 1975, for example, that the Pomacentrinae were likely to be a group made of multiple lineages. However, now that phylogenetic systematics has advanced to its current level of sophistication the existence of polyphyletic taxa has become increasingly unacceptable. We are now compelled to ask, "What is the utility of taxa that are not synonymous with evolutionary lineages?" If taxa do not represent monophyletic, or at least paraphyletic, groups of organisms, then they have no biological utility whatsoever other than to point out cases of convergent evolution. Because their phylogenetic relationships have been examined intensely, the damselfishes help to identify a problem that is very common for many groups of organisms. The evolutionary divergence of lineages does not mean that new morphologies must have evolved in each branch, nor does it require that such synapomorphies will be maintained if they have evolved. This difficulty is by no means necessarily restricted to species-poor or recently evolved lineages. Although this problem is easy to identify, it can be difficult to fix.

In the absence of clear morphological characters that can be used to describe a taxon, it is possible to take a "species up" approach and define taxa, at least partially, by the species that they are known to contain based on molecular phylogenetic analyses. We have done so here because this makes a great deal of sense where the damselfishes are concerned. The obvious flaw with using this method alone is that we won't know the taxonomy of a species until it is sequenced. As noted earlier, molecular data are not useful in regard to determining the taxonomic placement of a specimen in the field or a museum collection and with extraordinarily few exceptions this is an insurmountable problem for paleontological specimens. Morphological descriptions of taxa are extraordinarily useful, but overwhelming evidence now indicates that they will not always be effective for describing every branch of the tree of life. In those cases where important lineages are not morphologically distinct from their close relatives we still need names for the branches.

Although the major divisions of the damselfish lineage have now been named, there remains considerable work to be done, particularly among such polyphyletic taxa as the genera *Chrysiptera*, *Plectroglyphidodon*, *Stegastes*, and possibly *Parma* (Quenouille et al. 2004, Cooper et al. 2009, Frédérich et al. 2013). If taxa are to be monophyletic, then several species in *Plectroglyphidodon* will need to be reassigned to *Stegastes* (e.g., *P. lacrymatus*) and *vice versa* (e.g., *S. altus*, *S. apicalis*, *S. fasciolatus*, and *S. obreptus* reassigned to *Plectroglyphidodon*), but since there are several species within these two genera that have yet to be sequenced, we have suggested no changes here. Untangling *Chrysiptera* may call for even more drastic measures (see Quenouille et al. 2004, Cooper et al. 2009, Frédérich et al. 2013). The very large and paraphyletic genus *Chromis* is a likely candidate for division into new genera, with the species in the sister lineage to the genus *Dascyllus* offering a tempting target for a new genus (e.g., *C. agilis, C. amboinensis, C. atripes, C. delta, C. dimidiata, C. iomelas, C. margaritifer, C. nigrura, C. ovatiformis, C. retrofasciata, C. vanderbilti* and other species that have not yet been sequenced; Quenouille et al. 2004, Cooper et al. 2009, Frédérich et al. 2013).

In lineages such as the damselfishes where the traits that have traditionally been used to describe taxa have repeatedly converged on similar character states, we suggest that the determination of genetic synapomorphies and the use of quantitative shape analyses (such those provided here for the Microspathodontini) can provide a useful complement to traditional descriptive approaches. We particularly recommend the use of geometric morphometric methods for quantifying the morphology of a lineage and comparing its anatomy to that of closely related species. We suggest that the wider adoption of modern analytical techniques would strongly complement systematic studies of a wide range of organisms.

Acknowledgements

We wish to thank Bruno Frédérich and Eric Parmentier for the invitation to contribute to this book and for their critical insights, as well as two anonymous reviewers for comments that helped improve the manuscript. We also wish to thank Giorgio Carnevale (University of Torino) for help with the stratigraphic age of some of the fossils, as well as for pointing out to us the recently described *Sorbinichromis*, and Alex Bannikov (Borisyak Paleontological Institute of the Russian Academy of Sciences) for making available before final publication the preprint of the description of *Sorbinichromis*. We would like to thank Leo Smith (The University of Kansas) for assistance with damselfish taxonomy.

References

Agassiz, L. 1833–1843. Recherches sur les poissons fossiles. Tome IV. Petitpierre, Neuchâtel, Switzerland.
Aguilar-Medrano, R., B. Frédérich, E. De Luna and E.F. Balart. 2011. Patterns of morphological evolution of the cephalic region in damselfishes (Perciformes: Pomacentridae) of the Eastern Pacific. Biol. J. Linnean Soc. 102(3): 593–613.
Allen, G.R. 1975. Damselfishes of the South Seas. T.F.H. Publications, Neptune City, New Jersey.
Allen, G.R. 1991. Damselfishes of the World. Mergus, Melle, Germany.
Allen, G.R. and J.E. Randall. 1974. Five new species and a new genus of damselfishes (family Pomacentridae) from the South Pacific Ocean. Tropical Fish Hobbyist 22: 36–46, 48–49.
Arambourg, C. 1927. Les poissons fossiles d'Oran. Materiaux pour la carte geologique de l'Algerie, serie 1. Paleontologie 6: 1–218.
Aristotle. 1910. History of Animals. D.W. Thompson (translator). Clarendon Press, Oxford, United Kingdom.
Artedi, P. 1738. Ichthyologia. C. Linnæus (ed.). Conradum Wishoff, Leyden, The Netherlands.
Bannikov, A.F. and G. Carnevale. 2012. *Frippia labroiformis* n. gen. n. sp., a new perciform fish from the Eocene of Pesciara di Bolca, Italy. Boll. Soc. Paleontol. Ital. 51: 155–165.
Bannikov, A. and D.R. Bellwood. 2014. A new genus and species of pomacentrid fish (Perciformes) from the Eocene of Bolca in northern Italy. Studi e Ricerche sui Giacimenti Terziari di Bolca 12: 7–14.

Barel, C., F. Witte and M. van Oijen. 1976. The shape of the skeletal elements in the head of a generalized *Haplochromis* species: *H. elegans* Trewavas 1933 (Pisces: Cichlidae). Neth. J. Zool. 26(2): 163–265.

Bellwood, D.R. 1999. Fossil pharyngognath fishes from Monte Bolca, Italy, with a description of a new pomacentrid genus and species. Studi e Ricerche sui Giacimenti Terziari di Bolca 8: 207–217.

Bellwood, D.R. and L. Sorbini. 1996. A review of the fossil record of the Pomacentridae (Teleostei: Labroidei) with a description of a new genus and species from the Eocene of Monte Bolca, Italy. Zool. J. Linnean Soc. 117: 159–174.

Bernardi, G. 2011. Monophyletic origin of brood care in damselfishes. Mol. Phylogenet. Evol. 59: 245–248.

Betancur-R, R., R.E. Broughton, E.O. Wiley, K. Carpenter, J.A. López, C. Li, N.I. Holcroft, D. Arcila, M. Sanciangco, J.C. Cureton II, F. Zhang, T. Buser, M.A. Campbell, J.A. Ballesteros, A. Roa-Varon, S. Willis, W.C. Borden, T. Rowley, P.C. Reneau, D.J. Hough, G. Lu, T. Grande, G. Arratia and G. Ortí. 2013. The tree of life and a new classification of bony fishes. PLOS Currents Tree of Life. doi: 10.1371/currents.tol.53ba26640df0ccaee75bb165c8c26288.

Bloch, M.E. and J.G. Schneider. 1801. Systema Ichthyologiae. Self-published, Berlin, Germany.

Carnevale, G. and W. Landini. 2000. A fossil damselfish (Pisces, Pomacentridae) from the Late Miocene of Central Italy. Biological and biogeographical consideration. Paleontographia Italica 87: 65–70.

Carnevale, G., A.F. Bannikov, G. Marrama, J.C. Tyler and R. Zorzin. 2014. The Pesciara-Monte Postale Fossil-Lagerstaette: 2. Fishes and other vertebrates. pp. 37–63. *In*: C.A. Papazzoni, L. Giusberti. G. Carnevale, G. Roghi, D. Bassi, R. Zorzin. The Bolca Fossil-Lagerstaette: A window into the Eocene world. Rendiconti della Societa' Paleontologica Italiana 4, 2014.

Ciardelli, A. 1967. The anatomy of the feeding mechanism and the food habits of *Microspathodon chrysurus* (Pisces: Pomacentridae). Bull. Mar. Sci. 17(4): 845–883.

Cooper, W.J. 2006. The Evolution of the Damselfishes: Phylogenetics, Biomechanics, and Development of a Diverse Coral Reef Fish Family. Ph.D. Dissertation, The University of Chicago.

Cooper, W.J. 2009. The biogeography of damselfish skull evolution: a major radiation throughout the Indo-West Pacific produces no unique skull shapes. Proc. 11th Int. Coral Reef Symp. 2: 1370–1374.

Cooper, W.J. and M.W. Westneat. 2009. Form and function of damselfish skulls: rapid and repeated evolution into a limited number of trophic niches. BMC Evol. Biol. 9: 24.

Cooper, W.J., L.L. Smith and M.W. Westneat. 2009. Exploring the radiation of a diverse reef fish family: Phylogenetics of the damselfishes (Pomacentridae), with new classifications based on molecular analyses of all genera. Mol. Phylogenet. Evol. 52: 1–16.

Cooper, W.J., R.C. Albertson, R.E. Jacob and M.W. Westneat. 2014. Re-description and reassignment of the damselfish *Abudefduf luridus* (Cuvier, 1830) using both traditional and geometric morphometric approaches. Copeia 2014(3): 473–480.

Cope, E. 1877. A contribution to the knowledge of the ichthyological fauna of the Green River shales. Bulletin of the U.S. Geological and Geographical Survey of the Territories 3(4): 807–819.

Cowman, P.F. and D.R. Bellwood. 2011. Coral reefs as drivers of cladogenesis: expanding coral reefs, cryptic extinction events, and the development of biodiversity hotspots. J. Evol. Biol. 24: 2543–2562.

Cuvier, G. 1814. Observations et recherches critiques sur différens poissons de la Méditerranée et, à leur occasion, sur des poissons des autres mers plus ou moins liés avec eux. Memoirs du Museum National d'Histoire Naturelle 1: 226–241.

Datovo, A. and R.P. Vari. 2013. The jaw adductor muscle complex in teleostean fishes: evolution, homologies and revised nomenclature (Osteichthyes: Actinopterygii). PLoS ONE 8(4): e60846.

Eschmeyer, W.N. and R. Fricke (eds.). Catalog of Fishes. http://research.calacademy.org/research/ichthyology/catalog/fishcatmain.asp. Electronic version accessed September 29, 2015.

Forsskål, P. 1775. Descriptiones animalium, avium, amphibiorum, piscium, insectorum, vermium; quae in itinere orientali observavit Petrus Forskål. C. Niebuhr (ed.). Mölleri, Copenhagen, Denmark.

Frédérich, B., L. Sorenson, F. Santini, G.J. Slater and M.E. Alfaro. 2013. Iterative ecological radiation and convergence during the evolutionary history of damselfishes (Pomacentridae). Am. Nat. 181: 94–113.

Gill, T. 1911. A new translation of Aristotle's "History of Animals". Science 33: 730–738.

Günther, A. 1862. Catalogue of the Fishes in the British Museum, Volume 4: Catalogue of the Acanthopterygii, Pharyngognathi and Anacanthini in the Collection of the British Mueseum. British Museum Press, London, United Kingdom.

Hennig, W. 1950. Grundzüge einer Theorie der phylogenetischen Systematik. Deutscher Zentralverlag, Berlin, Germany.

Hennig, W. 1966. Phylogenetic Systematics. D. Davis and R. Zangerl (translators). University of Illinois Press, Urbana, United States.

Jang-Liaw, N.H., K.L. Tang, C.F. Hui and K.T. Shao. 2002. Molecular phylogeny of 48 species of damselfishes (Perciformes: Pomacentridae) using 12S mtDNA sequences. Mol. Phylogenet. Evol. 25: 445–454.

Jenyns, L. 1842. Fish Part 4 of The Zoology of the Voyage of H.M.S. Beagle. C. Darwin (ed.). Smith, Elder & Co., London, United Kingdom.

Jordan, D.S. 1923. A classification of fishes including families and genera as far as known. Stanford University Publications, University Series, Biological Sciences 3: 77–243.

Jordan, D.S. and B.W. Evermann. 1898. The fishes of North and Middle America: a descriptive catalogue of the species of fish-like vertebrates found in the waters of North America, north of the Isthmus of Panama. Part II. Bull. U.S. Natl. Mus. 47: 1241–2183.

Linnæus, C. 1758. Systema naturae per regna tria naturae, secundum classes, ordines, genera, species, cum characteribus, differentiis, synonymis, locis. 10th edition. Laurentius salvius, Stockholm, Sweden.

McCafferty, S., E. Bermingham, B. Quenouille, S. Planes, G. Hoelzer and K. Asoh. 2002. Historical biogeography and molecular systematics of the Indo-Pacific genus *Dascyllus* (Teleostei: Pomacentridae). Mol. Ecol. 11: 1377–1392.

Near, T.J., A. Dornburg, R.I. Eytan, B. Keck, W.L. Smith, K.L. Kuhn, J.A. Moore, S.A. Price, F.T. Burbrink, M. Friedman and P.C. Wainwright. 2013. Phylogeny and tempo of diversification in the superradiation of spiny-rayed fishes. Proc. Natl. Acad. Sci. USA 110: 12738–12743.

Quenouille, B., E. Bermingham and S. Planes. 2004. Molecular systematics of the damselfishes (Teleostei: Pomacentridae): Bayesian phylogenetic analyses of mitochondrial and nuclear DNA sequences. Mol. Phylogenet. Evol. 31: 66–88.

Santini, S. and G. Polacco. 2006. Finding Nemo: molecular phylogeny and evolution of the unusual life style of anemonefish. Gene 385: 19–27.

Sauvage, H. 1907. Sur les poissons de la famille des Cichlidés trouvés dans le terrain tertiaire de Guelma. Compte Rendu De L'académie Des Sciences Paris 145: 360–361.

Schluter, D. 2000. The Ecology of Adaptive Radiation. Oxford University Press, New York.

Tang, K.L. 2001. Phylogenetic relationships among damselfishes (Teleostei: Pomacentridae) as determined by mitochondrial DNA data. Copeia: 591–601.

Tang, K.L., K.M. McNyset and N.I. Holcroft. 2004. The phylogenetic position of five genera (*Acanthochromis*, *Azurina*, *Chrysiptera*, *Dischistodus*, and *Neopomacentrus*) of damselfishes (Perciformes : Pomacentridae). Mol. Phylogenet. Evol. 30: 823–828.

Tokunaga, S. and K. Saito. 1938. A fossil pomacentrid fish from the Miocene of Izu, Japan. Jap. J. Geol. Geogr. 15: 83–86.

Vetter, B. 1878. Untersuchungen zur vergleichenden Anatomie der Kiemen und Kiefermusculatur der Fische. Theil II. Jen. Zeitschr. Naturwis. 12: 431–550.

Winterbottom, R. 1973. A descriptive synonymy of the striated muscles of the teleostei. Proc. Acad. Nat. Sci. Phila. 125: 225–317.

Reproduction and Larval Recruitment in Damselfishes

David Lecchini,[1,2,] Suzanne C. Mills[2,3,a] and Ricardo Beldade[2,3,4,b]*

Introduction

As with most marine organisms (i.e., corals, crustaceans, fishes), damselfishes have structured life histories with two distinct stages, a relatively sedentary benthic stage (usually juveniles and adults) and a pelagic larval stage capable of long-distance dispersal (for review see Leis and McCormick 2002). The only current known exception to this pomacentrid pattern involves *Acanthochromis polyacanthus* and the two species of *Altrichthys* that do not have a pelagic phase. In spite of their diversity, most pomacentrids have common reproductive traits including oviparity, substrate-attached eggs (demersal), paternal care of the eggs and a free pelagic phase (Allen 1991). Thus, damselfish reproduction varies, but all species lay demersal eggs that are guarded by the male. Adult, juvenile and larval traits render this species-rich family one of the most interesting and varied in terms of reproduction and the first ontogenetic developmental stages. In this chapter, we describe the reproductive traits including the mating system, reproductive behavior, fecundity, spawning location and timing and

[1] USR 3278 CNRS-EPHE-UPVD, CRIOBE, BP1013 Papetoai, 98729 Moorea, French Polynesia.
[2] Laboratoire d'Excellence 'CORAIL'.
[a] Email: suzanne.mills@univ-perp.fr
[b] Email: rbeldade@gmail.com
[3] USR 3278 CRIOBE CNRS-EPHE-UPVD, CBETM de l'Université de Perpignan, 66860 Perpignan, France.
[4] Universidade de Lisboa, Faculdade de Ciências, Centro de Oceanografia, Campo Grande, 1749-016 Lisboa, Portugal.
* Corresponding author: lecchini@univ-perp.fr

discuss the evolutionary implications of these traits. We also review the duration of the pelagic stage, development of sensory and swimming abilities, as well as dispersal traits including distribution of larval stages in the ocean and dispersal kernels.

Having dispersed, a larva has the choice of settlement place, and the behavior involved in the process of settlement is usually described as habitat selection (Huntingford 1984). Moreover, a species-specific change in morphology and physiology, called metamorphosis, occurs at settlement during which the fish lose many of the characteristics that enhance survival in the plankton and develop other features suited to their new benthic environment (McCormick et al. 2002, Parmentier et al. 2004, Frédérich et al. 2012). During the settlement phase, fish larvae are thus subjected to strong selective pressures to choose a suitable reef habitat that will promote post-settlement survival and growth (for review see Doherty 2002). Up to 90% of fish larvae may be removed by predation during the first post-settlement days if they do not select a suitable habitat (e.g., Doherty et al. 2004, Lecchini et al. 2007). Thus, many reef fish species show marked selectivity in habitat choice at settlement based on the presence of specific substrates and/or conspecifics, and on the absence of predators or competitors for food and space (Doherty 2002). After several weeks or months at the settlement habitat, juveniles are integrated into the adult population and become sexually mature (for review see Lecchini and Galzin 2003). In this chapter, we also review habitat selection of pomacentrid juveniles at settlement and the post-settlement mortality process.

1. Reproductive Patterns in Pomacentridae

Mating system: polygamy vs. monogamy

In spite of the common bond formation observed in many species during the breeding season (Allen 1975), as well as the presence of territoriality and parental care, monogamy is not the prevalent reproductive strategy of pomacentrids (*sensu* Whiteman and Côté 2004). Most pomacentrids are polygamous. Examples of polygamous species include *Amblyglyphidodon leucogaster* (Goulet 1995), *Microspathodon chrysurus* (Pressley 1980) and *Chrysiptera cyanea* (Gronell 2010). In most polygamous species, a territorial male allows only one female to deposit eggs in his nest at any one time, actively chasing away uninvited females while its nest is occupied. Nonetheless multiple clutches from the same pair may be present within the nest during the spawning episode (e.g., *Dascyllus aruanus*, Mizushima et al. 2000). Such behavior may be justified as a strategy to avoid sneaking or to avoid disturbing the female that is spawning. Robertson et al. (1990) studied 15 damselfish species in Caribbean Panama and Pacific Mexico. They found that the number of clutches that a nest could hold during each single spawning episode varied among species (average 1.7 clutches/nest in *M. chrysurus* to 6.8 clutches/nest in *Stegastes diencaeus*), but also within species (e.g., varying from 4.5 to 7.1 clutches/nest in *Stegastes partitus*).

Mating strategies may also be labile, for example polygamy best describes the mating system for *Dascyllus marginatus*, but monogamy is also known to occur rarely or in particular locations (Fricke 1980). There is evidence of facultative monogamy in other pomacentrids such as the spiny chromis *Acanthochromis polyacanthus* where there is bi-parental care (Robertson 1973) and *Stegastes fasciolatus* where there is paternal care (Rasa 1969). The best-studied examples of monogamy come from the anemonefishes (*Amphiprion* spp.). They have small territories that both females and males defend and these species show paternal care in clutch defense and aeration (Fautin and Allen 1992, Chapter XII). Examples of monogamous anemonefishes include: *Amphiprion clarkii, A. frenatus, A. perideraion* (Hirose 1995). Only two known exceptions to monogamy among anemonefishes have been reported for *A. clarkii* and *A. akallopisos*, which are two of the most mobile anemonefishes (Moyer and Bell 1976).

Spawning habitat

Nest site. One common feature of pomacentrid reproductive biology is the choice and preparation of a specific nesting site by the male. The variety of habitats that species occupy involves a wide variety of substrates that are used for harboring the benthic-attached eggs. Examples of nest substrates include dead corals and gorgonians for *Amblyglyphidodon* spp.; the underside of rocks and inside the valves of giant clams for *Chrysiptera* spp.; bare rock, wharf pilings, etc. (for *Abudefduf* spp.) (Allen 1975), and algae for *Stegastes nigricans* (Karino and Nakazono 1993). Anemonefishes generally lay eggs on cleared rock surfaces that are just under the tentacles of the host anemones (Fautin and Allen 1992), but *Amphiprion polymnus* that is sometimes associated with to sand-dwelling anemones (Fautin and Allen 1992) may bring sand dollars, empty shells and even dead fronds to the vicinity of the anemone to use as nesting sites. The presence of nesting sites in a territory would influence how fast that territory is colonized (e.g., *S. diencaeus*; Cheney and Côté 2003).

Territoriality. Territoriality is prevalent among pomacentrids (Allen 1991), however it may be permanent or restricted to periods of reproductive activity. Examples of permanently territorial pomacentrids include the anemonefishes (Fautin and Allen 1992) and *Abudefduf saxatilis* (Fishelson et al. 1974). At the opposite end of the spectrum, most other species such as: *Plectroglyphidodon lacrymatus, Plectroglyphidodon leucozoma, Pomacentrus albicaudatus* and *P. trichourus* (Fishelson et al. 1974) only hold territories during reproductive episodes.

Territory size is in many cases related to reproductive success. For example, the size of the coral head is linked to the size of the group in *D. aruanus* where one or more males hold harems (Sale 1972). On the other hand, there may not be great variation in territory size for some species such as *S. diencaeus* (Cheney and Côté 2003). Territory size can vary between the sexes and among individuals of the same sex and may also account for differences in reproductive success. In *Stegastes leucostictus*, larger female territories and the distance travelled are correlated to female body length, allowing for greater mate assessment and probably better mate choice (Horne and Itzkowitz 1995).

Territory location can have a clear influence on reproductive success (e.g., Leese et al. 2009). Herbivorous pomacentrids that "grow" or "farm" their own gardens (Chapter VI) provide one of the best-studied examples of how the position of a territory (quality) within a cluster of joint territories can influence reproductive success. Meadows (2001) studied the variation in behavior and fitness with spatial position in a group (centre vs. edge) in the permanently territorial damselfish *Stegastes planifrons*. Fitness correlates varied with group position; survival and age did not differ between fish resident in the center or edge, but central fish were larger and grew faster than edge fish and central males also received more clutches, eggs/clutch, and cumulative number of eggs to defend than edge males (Meadows 2001). Thus, the positional differences in fitness measures and territory size may be due to the lower energetic costs of territory defense for central fish, permitting a greater investment in growth and reproduction.

Spawning may match tidal cycles in some species (see section on spawning timing), thus species that synchronize their spawning with the lunar cycle should prefer shallower habitats, whilst species that show a less pronounced synchronization with the lunar cycle should prefer deeper sites (e.g., *Chrysiptera biocellata*, Thresher and Moyer 1983).

Reproductive behavior

Reproductive behavior includes many types of displays, including courtship and mate defense, which may vary from species to species. Courtship displays may include for example "dips" and "caudal display" in the female's territory as in the case of *S. nigricans* (Karino and Nakazono 1993). In *C. biocellata* courtship consists of rapid chasing of males and stationary shaking movements with erected fins, followed by the male repeatedly escorting the female to the nest site for egg laying (Allen 1975). In *Pomacentrus nagasakiensis*, a first phase of "signal jumping" is used to attract females to the nest site. Once a female approaches, the enticing phase starts with directional swimming towards the female, "quivering and swimming with exaggerated motions of the caudal and pectoral fins" and a fast return to the nest when close to the female. Finally, the female is led to the nest and begins to spawn (Moyer 1975). The complexity of reproductive behaviors varies among species, and this variation may be related to the utilization of other cues that allow mates (females) to choose their partners. Tresher and Moyer (1983) found a behavioral complexity gradient in three pomacentrids (*Chrysiptera* spp.) that vary in the degree of dichromatism: lowest in the permanently dichromatic species, highest in the permanently monochromatic one, and intermediate in the temporarily dichromatic species. Sexual selection in other damselfish species, the majority being monochromatic or temporarily presenting color alterations, is principally based on the vigor of male courtship and is usually accompanied by sound production (Chapter X). Recently the role of UV patterns as well as the ability to perceive them, found in at least 40 pomacentrid species, have been shown to have implications in reproductive and territorial defense behaviors (Siebeck and Marshall 2001, Siebeck 2004; Chapter XIII). The trade-off between signaling to potential breeding partners and attracting potential predators by doing so appears to be resolved by UV communication. Many coral reef fish possess UV-absorbing ocular media (Siebeck and Marshall 2001) so that, even when they are close to their predator, UV signaling is unlikely to attract their attention. Thus, Siebeck (2004) showed that the use of this 'secret communication channel' may therefore enable reef fish such as *P. amboinensis* to communicate effectively with bright UV colors without the associated danger of attracting the attention of predators. Reproductive behaviors such as courting, mate guarding and parental care often entail a cost of increased predation risk, either to the individual itself or to the offspring under its care. Figueira and Lyman (2007) demonstrated that male *S. partitus* show context-dependent courting behavior and seem to balance risk and reward in courting decisions. Reproductive behaviors are thus curtailed depending upon the magnitude of potential risks and benefits.

Fecundity

Damselfish fecundity (number of eggs per nest) can vary substantially depending mostly on species, female size, reproductive system (monogamous or polygamous), nest quality and season. The lowest fecundities among damselfishes have been reported for anemonefishes (e.g., *Amphiprion melanopus* and *A. polymnus* which spawn as few as 200 to 400 eggs/nest; Ross 1978). Thus, all species from this family that have a pelagic larval phase are demersal spawners. It should not be surprising that the female would have more difficulty in laying eggs with increasing wave action. Seasonal fecundity has been reported by Cowen (2002) in the Caribbean where winds are stronger during the dry season. However recent work on *Amphiprion chrysopterus* shows that there can be a large range of fecundities for the same species at the same geographical location (from 400 to over 3,500 eggs/nest; Beldade et al. 2012, R. Beldade unpublished data). The variability in fecundity appears to be linked to the size of the female as well as the site quality and environmental fluctuations. Intraspecific difference in fecundity tied to environmental fluctuations was also found in *Pomacentrus vaiuli*, at 2 sites that differed in terms of the distance to a passing cyclone (McIlwain 2002). Significant differences in the initial brood size and the number of broods per pair per year among sites have also been found in *Acanthochromis polyacanthus* (Thresher 1985). Such differences may be due to differential food availability and predation pressure (Thresher 1983). Other damselfishes can have much higher fecundities. *M. chrysurus* can have as many as 19,000 eggs in a nest (Pressley 1980) and *Dascyllus albisella* may lay as many as 25,000 eggs per day (in several nests) (Danilowicz 1995). Even between species in the same genus, there may be striking differences in fecundity, e.g., *Pomacentrus flavicauda* and *P. wardi* which may spawn 2,300,000 and 133,000 eggs over 9 reproductive bouts respectively (Thresher 1985).

Spawning synchronization and spawning timing

Egg laying among female Pomacentridae in colonies tends to happen at the same time (i.e., synchronous) and is widespread. Social facilitation of breeding cues in larger colonies, for example visual and auditory cues of courtship displays in nearby reef fishes, may directly increase breeding synchronization. Thus population or colony size would be expected to influence the degree of spawning synchronization. Support for this hypothesis in coral reef fishes was reported in two colonies of *D. aruanus*; a large colony spawned synchronously while a smaller colony spawned asynchronously (Fricke 1980 *in* Danilowicz 1995). On the other hand, Danilowicz (1995) reported mixed results for *Dascyllus albisella*, with spawning synchronization being independent of sub-population size in some cases, and in others, smaller sub-populations were clearly less synchronous. Asynchronous spawning has been reported for a number of species including *Abudefduf troschelii* in Pacific Mexico (Robertson et al. 1990), *Abudefduf vaigiensis* in Taiwan (Jan and Ormond 1992) and also *D. albisella* (Danilowicz 1995).

Spawning synchronization is beneficial for eggs and helps in hatching propagules including aiding egg survival via the oversaturation of egg predators (Foster 1987), and the linkage of spawning to optimal settlement times (Tzioumis and Kingsford 1995). Further benefits of spawning synchronization include decreased filial egg cannibalism (Manica 2002), increased larval survival via predator swamping (Johannes 1978), reduced intraspecific competition for food (Allen 1975), and facilitation of off-reef advection from synchronization with spring tides (Mizushima et al. 2000). Moreover, spawning synchronization is also beneficial for adults by facilitating migration to the spawning sites. Brooding males of synchronized species, in particular, may show cooperative defense against predators and also have the opportunity to recover from the energetic costs associated with territory defense and brood care (Robertson 1990). In addition, females that lay eggs in multiple nests in order to reduce the probability of egg cannibalization by parental males (Mizushima et al. 2000) will increase a brooding male's opportunity to pass on his genetic inheritance.

It has been hypothesized that spawning timing at multiple scales (seasonal, monthly and diel) maximizes offspring survival at hatching in order to reduce reef predation on the eggs (Robertson 1990). In general, the best time of the year to spawn should coincide with the lowest predation pressure, and highest available food resources. Tropical damselfish may display a variety of spawning patterns ranging from protracted, year-round, daily events (Robertson et al. 1990) to limited, discrete events restricted to specific months (Mizushima et al. 2000). For example, a marked seasonality in spawning activity has been found in three damselfish species throughout the year: *Stegastes dorsopunicans*, *Stegastes leucostictus* and *Abudefduf saxatilis* (Robertson et al. 1990). However, differences or fluctuations in spawning seasonality may also occur among populations from different locations. Richardson et al. (1997) found that the tropical anemonefish *Amphiprion akindynos* living in a subtropical area had a spawning seasonality that was much closer to its temperate relatives (*Amphiprion clarkii* and *A. lantezonatus*) than to other tropical anemonefishes. Among the extrinsic causes of changing spawning intervals are food availability, waves and temperature which were shown to affect the reproductive success of *C. cyanea* (Bapary et al. 2012). Finally, certain stochastic phenomena may interrupt the breeding season of pomacentrids, lending support to the match between favorable environmental conditions (for the survival of the offspring) and breeding seasonality.

At a smaller temporal scale, most damselfishes spawn on a lunar or semi-lunar cycle (Asoh 2003). Spawning in pomacentrids varies from acyclic, as was reported for *A. leucogaster* from the Red Sea (Goulet 1995) to an increasing temporal lag in spawning episodes, within a period of 2 days for *Stegastes variabilis*, *S. partitus* and *S. flavilatus*; or 8 days for *Chromis multilineata* in the Caribbean Panama (Robertson et al. 1990). Semi-lunar spawning periodicity is likely the prevalent pattern and has been reported for a number of species including *A. troschelii* in Pacific Panama (Foster 1987); *Pomacentrus flavicauda* and *P. wardi* (Doherty 1983); *S. leucostictus* and *A. saxatilis* (Robertson et al. 1990). Lastly, Pressley (1980) showed a lunar periodicity in the spawning of *M. chrysurus*, a tropical-Atlantic damselfish that spawned mainly between the full and new moon, but which also spawned several times within this period.

Regardless of the seasonal or monthly patterns, many pomacentrids spawn at sunrise or early morning and the process usually lasts around one hour (Doherty 1983). However, *D. albisella* in Hawaii and the orange-fin anemonefish *A. chrysopterus* in the French Polynesia, serve as examples of the variability in the diel spawning timing as both species spawn mostly in the morning but also throughout the day (Asoh and Yoshikawa 2002, R. Beldade unpublished data). Interestingly while in *Dascyllus*, females may spawn in several nests within the same episode, which precludes a narrow spawning window, *Amphiprion* are monogamous and thus the length and timing of spawning are likely differently constrained. There may be trade-offs associated with dawn spawning. A study on the yellowtail damselfish *M. chrysurus* in Barbados, showed that especially at dawn, when there is a peak in ectoparasite infestations, females seek cleaning stations more often during spawning than outside spawning periods or immediately after (Sikkel et al. 2005).

Incubation period

The incubation period, between the laying of eggs and hatching of larvae, in tropical pomacentrids does not vary by more than a few days. For example, Robertson et al. (1990) provided incubation times for 16 Caribbean pomacentrids which varied between 3 days for *Chromis multilineata* and 4 to 5 days for *Abudefduf saxatilis*, *A. troschelii*, *Microspathodon chrysurus*, *M. dorsalis*, *M. bairdii*, *Stegastes diencaeus*, *S. dorsopunicans*, *S. leucostictus*, *S. partitus*, *S. planifrons*, *S. variabilis*, *S. acapulcoensis*, *S. flavilatus*, and *S. rectifraenum*. In Hawaii, Doherty (1983) observed very similar incubation times in two Pomacentrus: *P. chrysurus* (5 days) and *P. wardi* (4 days).

The length of the incubation period is nonetheless dependent on temperature. For example, while *A. melanopus* showed only a small variation in incubation timing (7.5 to 8.5 days), *A. clarkii*, in the temperate waters of Japan, varied between 6.5 and 13.5 days where temperature ranged between 19 and 27.5°C (Ross 1978). In tropical regimes the duration and range in incubation is likely more reduced. For example, in Moorea (French Polynesia) the seasonal variation of incubation time and water temperature for *A. chrysopterus* extends from 6 days (at 29°C) to 8 days (at 27°C) (Beldade pers. comm.).

During incubation, most reef benthic spawning pomacentrids aerate their eggs using their fins over the nests. Fanning increases the oxygenation of the water around the eggs and has been shown in other fish families to be costly for parents. There are many examples of pomacentrids that aerate their eggs such as *Amblyglyphidodon aureus* or *Amphiprion melanopus* who spend large amounts of time and energy in fanning eggs (Breder and Rosen 1966, Ross 1978). Fewer cases have been described in which the adult fishes do not fan the eggs, such as *A. polyacanthus* (Robertson 1973).

In pomacentrids the care for the eggs is costly and mostly assured by males. It has been hypothesized that such a sharp division of labor probably allows for greater egg production in females than would otherwise be the case.

2. The Larval Phase in Open Water

During the period that extends from hatching to reef colonization, larvae of pomacentrids are pelagic. The only exceptions found among pomacentrids are *A. polyacanthus* and two species of *Altrichthys* in which the live young remain close to the substrate and are cared for by the parents (Robertson 1973, Allen 1999). As a result of the restricted ability to disperse, genetic differentiation as well as specific color patterns are found among relatively nearby populations (Miller-Sims et al. 2008). However, the vast majority of pomacentrids lay benthic eggs (Allen 1975) that hatch and live freely in the plankton before settling.

Pelagic larval duration

Specific markings in the otoliths' of fish larvae allow us to back calculate the duration of the pelagic larval stage (e.g., Wilson and McCormick 1997). Most pomacentrids have a wide range of pelagic larval durations, from a virtually absent dispersive period (e.g., *A. polyacanthus*; Robertson 1973), to over a month (e.g., 37 days for *D. albisella*; Danilowicz 1997). The most comprehensive list of pelagic larval durations in pomacentrids published to date by Wellington and Victor (1989) covers 100 species across 8 genera. Pelagic larval durations (PLD) within these genera were similar within the same geographical area, a pattern that was not observed in other fish families such as Labridae (Victor 1986). In spite of the small intraspecific differences in PLD, species in certain regions appear to have consistently longer PLDs than in others; such as Hawaii and the eastern Pacific (Wellington and Victor 1989). Other studies have nonetheless reported substantial variability in the PLD for the same species. Lobel (1997) reported that the larval duration for *Plectroglyphidodon imparipennis* from Hawaii ranged from 33 to 48 days. Local phenomena capable of influencing the speed of development, such as temperature or food availability can modify the PLD duration (e.g., Green and Fisher 2004, Leis and Clark 2005).

Intuitively the PLD could be a good proxy for how far the larvae of a species can travel. However, correlations between species' geographic ranges and their pelagic larval durations have failed to return a clear link (e.g., Wellington and Victor 1989). While the duration of the pelagic stage in a species may limit an individuals' ability to reach distant locations or even a population's ability to export larvae, such limitations may not necessarily restrict a species' distribution (Luiz et al. 2013).

Distance travelled during dispersal

Recent studies have focused on the distance travelled by individual larvae during their dispersive stage. The first evidence that pomacentrid larvae could be returning to the same natal population, thus covering very small net distances came from a mark recapture study by Jones et al. (1999), in which the larvae of *Pomacentrus amboinensis* were marked prior to dispersing, while still in the eggs, and later were captured in light traps next to the natal reef. Posterior studies on damselfishes also provided minimum and maximum dispersal net distances: *A. chrysopterus* in Moorea, French Polynesia, from just a few dozens of meters to 22 km (Beldade et al. unpublished data); *S. partitus* where on average 25% larvae returned to the isolated island of Barbados (Paris and Cowen 2004); *A. polymnus* in Kimbe Bay (Papua New Guinea) usually under 1 km (Jones et al. 2005); *A. polymnus* at Shulman Island in Papua New Guinea, up to 6 km (Saenz-Agudelo et al. 2009); roughly 60% larvae returned to the same small island and some travelled up to 20 km (Almany et al. 2007, Berumen et al. 2012); and in *A. percula* in Kimbe Bay (PNG) up to 35 km (Planes et al. 2009). The exact travelled distance covered by the larva may nonetheless be superior to the net distance, and considering that anemonefish have one of the narrowest dispersal windows these are impressive distances to have travelled.

The net distance travelled by larvae from the source to a settlement site has also been used to estimate settlement probability relative to distance to the source population, i.e., dispersal kernels (e.g., Pinsky et al. 2010, Buston et al. 2012). Generally, the probability of successful dispersal between populations declines as the distance between populations increases. Depending on different estimates of effective fish densities, larval dispersal kernels in *Amphiprion clarkii* had a spread near 11 km (4–27 km) (Pinsky et al. 2010). These kernels predicted low fractions of self-recruitment in continuous habitats, but the same kernels were consistent with previously reported, high self-recruitment fractions (40–60%) when realistic levels of habitat patchiness were considered. In *Amphiprion percula* from Papua New Guinea, Buston et al. (2012) reported that the larvae were five times as likely to successfully disperse 1 m, as they were to disperse 1 km.

Larval distribution during dispersal

It is generally accepted that for most fish larvae, swimming is not well developed at hatching. Young, poorly developed larval stages are likely able to regulate their transport by depth regulation or entrainment in hydrographic features until they become 'effective' swimmers and control their trajectories and recruit onshore via directed swimming. This transition from ineffective to effective swimmers can occur from within a few days to weeks after hatching in some species (Fisher et al. 2000), but may occur later in others.

Horizontal distribution. The presence of pomacentrid larvae, in a range of sizes, off One Tree Island as well as in the lagoon (during the night) and in the channels connecting the lagoon to the outer-reef, demonstrates that all ontogenetic stages can be present relatively close to the shore (Kingsford 2001). Near reef environments are thought to comprise a higher risk of predation, lower feeding rate and a higher risk of aggressive interactions with other species, such as for the later stage larvae of pomacentrids (Leis and Carson-Ewart 1998). The ability of young larvae to move away from the reef and yet remain close to the coast can be exemplified by the young larvae of *Amblyglyphidodon curacao*, which were shown to possess directional swimming trajectories that enabled them to stay close to Lizard Island (Leis et al. 2007).

Vertical distribution. Vertical distributions of fish larvae can affect essential ecological processes such as transport, feeding, growth, and survival. Vertical gradients in physical factors (e.g., light or pressure, Huebert 2008) as well as biological gradients (e.g., density of predators and prey, Fortier and Harris 1989 and light sensitivity and predation performance, Job and Bellwood 2000) are also important. A common pattern among pomacentrids is the presence of younger larvae in shallower water (e.g., Huebert et al. 2011). Larval transport or semi-passive dispersal can be greatly enhanced by vertical migration. Paris and Cowen (2004) found vertical migration to be a mechanistic explanation for retention of *S. partitus* larvae in the Caribbean. The highest concentrations of pre-flexion *S. partitus* larvae were in the upper 20 m, while those of older larvae were always deeper. The downward migration throughout ontogeny represents a retention mechanism for locally spawned larvae and most of the variability in estimated retention rates occurred during the earliest stages as a result of the dynamic nature of surface currents experienced by larvae prior to the onset of vertical migration (Paris and Cowen 2004). Larvae of tropical pomacentrids appear to have lower light sensitivity as compared to larvae from other families such as Apogonidae, and this can, for example, result in clear predation performance differences at different depths (Job and Bellwood 2000). Based on aquarium experiments, larvae of at least a few pomacentrids appear to be photopositive upon hatching, moving to the surface towards a dim light, but not in darkness, e.g., *A. chrysopterus* and *Amphiprion ocellaris* (R. Beldade unpublished data). Diel vertical larval migrations have been commonly reported for pomacentrids (e.g., Huebert et al. 2010); where the daily vertical stratification is broken down during the night hours (Kingsford 2001). Kingsford (2001) found larvae to be present in higher abundances during the day than at night. Similar behaviors in other settlement-stage pomacentrid larvae, that are older and larger, lend support to the early influence of larval behaviors in their dispersal. Furthermore vertical movement permits more or less passive horizontal movements, thus the two types of movements need to be considered simultaneously throughout development. One such example comes from young larvae of *A. curacao* in Lizard Island which showed a different vertical distribution at leeward and windward sites: modal depth was 2.5–5.0 m and 10.0–12.5 m, respectively (Leis et al. 2007).

Larval Recruitment

In this chapter, larval recruitment is defined in terms of two processes: reef colonization (i.e., transition from a pelagic oceanic environment to a benthic reef environment) and settlement phase (i.e., choice of a suitable habitat based mainly on the characteristics of coral habitat and the presence or absence of conspecifics as well as other species).

Reef colonization

Which larval processes influence how many individuals will join a juvenile population? This is a central question in marine ecology and conservation for fish species with complex life cycles, such as Pomacentridae (Lemberget and McCormick 2009). Reef fish larval colonization occurs through the crossing of the reef crest and this is mostly a nocturnal process (Dufour and Galzin 1993, Lecchini et al. 2004). Reef colonization is often considered to be a bottleneck in the life cycle of coral reef fish (Lecchini and Galzin 2003) because up to 60% of fish larvae may be removed by predation during the night of their reef colonization (Doherty et al. 2004).

Vertical and horizontal distributions of Pomacentridae larvae in open water just before their reef colonization. In contrast to the large number of studies conducted on spatio-temporal variation in larval supply, only a few studies have attempted to describe the vertical and horizontal distributions of fish larvae in open water just before their reef colonization (e.g., Doherty and Carleton 1997, Hendriks et al. 2001, Fisher and Bellwood 2002, Leis 2004, Huebert et al. 2010). To explore the vertical and horizontal distributions of fish larvae, Leis and collaborators have used an *in situ* method for observing larval fish behavior during the day (e.g., Leis and Carson-Ewart 1999, Leis 2004), while the other studies have used light-traps deployed at night.

Pomacentrid larvae are predominantly captured in light-traps set up in sub-surface traps, while other fish families such as Labridae and Apogonidae larvae are usually collected in traps near the seabed (Hendriks et al. 2001). Lecchini et al. (2013) conducted two experiments with six light-traps set at different depths (A) sub-surface, (B) mid-water, and (C) bottom and at different habitat types (reef slope: 50 m horizontal distance from the reef crest; frontier zone: 110 m horizontal distance; and sandy zone: 200 m horizontal distance), on the outer reef slope of Moorea Island, French Polynesia. One species highlighted a common trend for Pomacentridae: *Chromis viridis* larvae were mainly captured with bottom traps in the sandy zone, but with sub-surface traps on the reef slope. Lecchini et al. (2013) concluded that Pomacentridae larvae were more abundant near the seabed when offshore (i.e., sandy zone on which few reef predators live) but were more abundant near the surface when close to the reef (i.e., on the reef slope where many reef predators live). Similarly, Leis (2004) found that *Chromis atripectoralis* larvae swam deeper when they were far from a reef (depth between 5–10 m at 500 and 1000 m from a reef), and swam in sub-surface waters when they were close to a reef (depth 0–5 m at 100 m from a reef). Leis (2004) tried to determine if larvae responded to the distance from the reef or to the depth of the water column. Although depth was correlated to the distance from shore, Leis concluded that depth was probably more important than distance from shore, in part because of differences in the vertical distribution of larvae on the different sides of Lizard Island.

Overall, the different studies conducted on the vertical and horizontal distributions of fish larvae in open water just before their reef colonization allow us to hypothesize that Pomacentridae larvae probably move towards the surface when they approach the reef (i.e., distance of offshore more important that depth; Lecchini et al. 2013). Alternatively, Paris and Cowen (2004) hypothesized that *S. partitus* approach the reef via the bottom topography since the late stages migrate to deep waters as they approach the island of Barbados. Other pomancentrid larvae with heavy pigmentation such as *Abudefduf* spp. have the opposite strategy, i.e., larvae move towards the surface during ontogeny (Cowen 2002). Whatever the strategy towards approaching a reef, the presence of well-developed swim bladders and advanced swimming abilities in Pomacentridae species (Leis and McCormick 2002) would allow larval fish to do rapid vertical movements and perform vertical maneuvering in order to increase the chance of recruitment by reducing reef predation.

Temporal and spatial patterns of reef colonization. Once fish larvae have reached the reef, several studies have shown that temporal variation in larval supply is highly cyclical, with nocturnal (i.e., highest larval supply during the night), lunar (i.e., highest larval supply around the new moon), seasonal (i.e., highest larval supply during the summer) and interannual (i.e., higher larval supply during La Niña than El Niño periods) patterns (references in Leis and McCormick 2002, Lo Yat et al. 2011). For example, Dufour and Galzin (1993) showed that Pomacentridae larvae mainly colonized from the last quarter to the new moon period at Moorea Island (French Polynesia). Lo-Yat (2002) confirmed this general temporal pattern at Rangiroa Atoll (French Polynesia), even if some Pomacentridae species did preferentially colonize around the full moon, such as *Pomacentrus pavo*. The common temporal pattern, i.e., reef colonization around the new moon, might take advantage of these incoming darker nighttime tides for predator avoidance. Predation is so intense that all marine post-larvae arriving on reefs have been said to make "a suicide drop onto the reef" and to face a "wall of mouths" (Doherty 2002).

Other studies have shown little spatial variation in larval distribution at sites separated by 200 m and by 10 km (see references in Lecchini and Galzin 2003). For example, Doherty and McIlwain (1996) and Lecchini et al. (2004) highlighted no variability in larval supply between two sites spaced 200 m apart on the West coast of Australia (Ningaloo Reef) and six sites spaced 200 m apart on the West coast of Moorea Island, further verifying the low spatial variability within sites that are on the same coast. In the study of Lecchini et al. (2004), the larval distribution of *Chromis viridis, Stegastes albifasciatus* and *S. nigricans* was similar between 6 crest nets spaced 200 m apart during a sampling period of 14 days in 1998. Furthermore, Dufour et al. (1996) demonstrated that fish larval supply was homogenous on a much larger scale over the entire reef crest of Moorea Island (five sites spaced between 8–12 km apart). These studies demonstrating simultaneous larval distribution at sites separated by 200 m or by 10 km led to the major question of whether the lack of variability in larval supply reflects the size of the larval patch in the ocean or a dilution of a smaller larval patch in front of the reef crest just before colonization. To validate one of these two hypotheses, future studies should focus on fish larval distribution at different spatial scales whilst also studying passively drifting invertebrate species in order to rule out the active pre-settlement dilution hypothesis and more accurately determining the patch size of planktonic reef dwelling organisms.

Habitat selection at settlement

Selective choice of a suitable habitat. At settlement, marine fish larvae that have led a pelagic life for several weeks must quickly adapt without any prior experience to benthic life. Booth (1991), for example, showed that *D. albisella* larvae adapted to juvenile-style benthic life within 24 hours (i.e., larvae are adapted to the reef life due to a very quick metamorphosis process). Leis and Carson-Ewart (2002) showed that the time taken by the larvae of 13 Pomacentridae species to swim over the reef before settling as well as the distance travelled varied from 0 to a mean of 5.5 min and 43 m at Lizard Island (Great Barrier Reef) respectively. During the settlement phase, fish larvae are subjected to strong selective pressures to choose a suitable reef habitat that will promote post-settlement survival and growth of individuals (for review see Doherty 2002). Up to 90% of fish larvae may be removed by predation during the first post-settlement days if they do not select a suitable habitat (e.g., Doherty et al. 2004, Lecchini et al. 2007). Thus, many Pomacentridae species show marked selectivity in habitat choice at settlement based on the presence of specific substrates and/or conspecifics, and on the absence of predators or competitors for food and space. For example, Leis and Carson-Ewart (2002) studied the settlement patterns of 13 Pomacentridae at Lizard Island. They showed that *C. atripectoralis* only settled into schools of similar-sized individuals (not only conspecifics, but also other pomacentrid species and apogonids) associated with live coral. In contrast, *Chrysiptera rollandi*, *Pomacentrus nagasakiensis* and *P. coelestis* settled exclusively among dead coral. *Pomacentrus wardi* settlement was more-or-less equally divided between dead coral and algae, *P. amboinensis* was divided between live and dead coral, and *D. aruanus* was divided between live coral and holes at the base of yellow gorgonians. Overall, models of habitat selection predict that initial choices at settlement have a major influence on the subsequent patterns of adult abundance among habitats (for review see Gillanders et al. 2003).

Mortality of Pomacentridae at settlement. A fundamental goal in population ecology is to understand the processes that influence individuals to survive in order to reproduce (Kingsolver et al. 2001). Predation is often considered to be an important causal factor in structuring terrestrial and aquatic populations, as it can be a major selective force in the evolution of organism structure and function (e.g., life history traits, morphology, sexual dimorphism, activity times and habitat use; Fenberg and Roy 2008). Several studies have shown that the mortality rate of coral reef fish at settlement can vary markedly according to the characteristics of shelter and/or the presence or absence of conspecifics in the settlement habitat (for review see Doherty 2002). Booth and Beretta (1994) estimated the mortality of newly settled *Chromis cyanea* to be between 54% to 90% during the first post-settlement month. Planes and Lecaillon (2001) estimated the mortality of newly settled *C. viridis, Chrysiptera leucopoma* and *S. nigricans* at 78%, 76%, and 82% respectively. Ault and Johnson (1998) showed that the differences in the mortality of recently settled fish (*Pomacentrus moluccensis*, *P. wardi* and *P. amboinensis*) could be explained by the shape of coral colony (encrusting, digitate, branching, tabulate and massive). Almany (2004) tested whether increased habitat complexity reduced the negative effects of resident predators and competitors on recruitment and survival of *S. leucostictus* in Bahamas Islands. Almany (2004) showed that the effects of habitat complexity on post-settlement predation and competition depend on several key factors, such as the availability of appropriate shelter, behavioral attributes of interacting organisms, and the developmental stage of prey/inferior competitors. Thus, pomacentrid larvae show marked selectivity in the habitats they choose based on the presence of specific benthic substrata and/ or the presence of conspecifics or other species (e.g., Ohman et al. 1998, Leis and Carson-Ewart 2002). The presence of conspecifics could be driven by habitat quality or enhanced survival (study of Booth 1995 conducted on 13 Pomacentridae species on the Great Barrier Reef). For example, McCormick and Meekan (2007) showed that adult males of *P. amboinensis*, indirectly facilitated the increased survival of conspecific juveniles through the territorial defense of their nesting site from potential egg predators. Moreover, they showed that male territoriality resulted in a shift in the selectivity of predation on newly settled juveniles. However, Ben-Tzvi et al. (2009)

highlighted differential strategies of settlers, who do not necessarily join conspecific adults. *Dascyllus aruanus* prefer to settle near (not with) their aggressive adults, and to join them only after gaining in size; whereas *D. marginatus* in densely populated reefs settle independently of their adult distribution. Overall, all these studies confirm that a species-specific suitable habitat is a primordial factor for the settlement success of Pomacentridae larvae in order to reduce post-settlement mortality (Lecchini et al. 2007).

However, it is of particular interest to determine whether post-settlement mortality acts indiscriminately during life stage transitions, or in other words, whether all recently settled juveniles have a similar chance of being removed from the population, or whether a particular phenotype is more susceptible to reef predation. The majority of the literature on selective mortality of Pomacentridae species during the first post-settlement months has emphasized body size, condition or larval growth history (e.g., Vigliola and Meekan 2002, Hoey and McCormick 2004, Holmes and McCormick 2004, Gagliano et al. 2007). For example, in order to identify which individual attributes are the most influential in determining patterns of survival in a cohort of reef fish, Gagliano and McCormick (2007) compared the characteristics of *P. amboinensis* surviving early juvenile stages on the reef with those of the cohort from which they originated. Among the traits examined, planktonic growth history was, by far, the most influential and long-lasting trait associated with juvenile persistence in reef habitats. However, otolith increments suggested that larval growth rate may not be maintained during early juvenile life, when selective mortality swiftly reverses its direction. These changes in selective pressure may mediate growth-mortality trade-offs between predation and starvation risks during early juvenile life. Gagliano and McCormick (2007) concluded that ontogenetic changes in the shape of selectivity may be a mechanism maintaining phenotypic variation in growth rate and size within a population.

Overall, these extreme rates of mortality during and immediately after settlement can be viewed as population size bottlenecks, which may result in a loss of some particular phenotypes or genotypes within populations and ultimately limit the evolution of the species with a potential increase in the risk of extinction (England et al. 2003, Pini et al. 2011).

Sensory abilities of fish larvae at recruitment

Recognition of suitable settlement habitats has been hypothesized to be based on a combination of some or all of acoustic, chemical, visual, sun compass, rheotactic, magnetic, wave motion and thermal cues. However, only the visual, olfactory and auditory senses are known to be functional in coral reef fishes when they settle into their first reef habitat (for review see Myrberg and Fuiman 2002). Thus, one of the greatest challenges facing the majority of marine reef organisms with larval stages that potentially disperse and develop in offshore waters is how to locate the relatively rare patches of coral reef habitat on which they settle and ultimately reside as adults (Myrberg and Fuiman 2002). The answer must lie partly in the sensory modalities of fishes. Two research topics must be distinguished: how do pelagic larvae recognize the reef to colonize, and how do reef larvae recognize their settlement habitat? Therefore, a successful recruitment necessitates, at a range of spatial scales, identification of (1) a coral reef within the boundless oceanic matrix, and then (2) of a suitable microhabitat within this reef (Lecchini and Nakamura 2013).

Detection of a reef to colonize. In the ocean, the larvae of many marine species use swimming behavior, stimulated by chemical or acoustic cues, to control their position within the water column, increasing the probability that they will be transported to a reef (Myrberg and Fuiman 2002). Leis et al. (2003) have shown that larval fishes are able to orientate themselves towards or away from reefs at least 1000 m away, and subsequent studies have shown that macro-scale cues driving this behavior include chemical and acoustic cues (for review, see Leis and McCormick 2002, Leis et al. 2011). For example, Gerlarch et al. (2007) showed that the settling larvae of *P. coelestis* could be capable of olfactory discrimination and preferred the odor of their home reef, thereby demonstrating that nearby reefs smell different. In a field experiment, Leis et al. (2002) showed that settlement-stage Pomacentridae larvae changed their behavior in the presence of broadcast reef sounds. These larvae also behaved differently when exposed to reef sounds or to a mix of pure tones (white noise). Thus, these studies confirm that Pomacentridae larvae are able to hear and smell a reef when they are still in open water (i.e., before reef colonization).

The few studies estimating the distances that fish larvae could potentially detect coastal reef noise provide estimates ranging from 500 m to ~5 km (e.g., Mann et al. 2007, Wright et al. 2010, Radford et al. 2011b). For example, Egner and Mann (2005) showed that *A. saxatilis* have poor hearing sensitivity. Because of the high hearing thresholds found in their study in comparison to recorded ambient reef noise, they concluded that sound could not play a significant role in the navigation of *A. saxatilis* larvae returning to the reef from long distances (>1 km). Similarly, based on fish hearing abilities and ambient sound levels, Mann et al. (2007) have estimated the maximum detection distance of reef by fish larvae to be less than 1 km. Wright et al. (2005, 2008, 2010) took a similar approach as Mann, but attempted to correct for the auditory brainstem response (ABR) data, and concluded that Pomacentridae, Serranidae and Lutjanidae larvae would be capable of detecting sounds from a reef 5–6 km away. Radford et al. (2011b), basing the estimation on sound pressure level recorded in the field, estimated this distance to be several kilometers due to a "reef effect" which would increase the propagation distance of reef noise. These sound propagation distances are also within the range of chemical propagation from ebb tide plumes of lagoon water estimated by Atema et al. (2002), based on temperature measurements and on visual observations of turbidity. Atema et al. (2002) predicted that the transmission distance in the ocean of the odor plume from a coral lagoon could be between 1.6 and 3 km. Lecchini et al. (2014) explored the distance of transmission of chemical cues emitted by live vs. dead coral reefs (High Pressure Liquid Chromatography analyses of water sampling stations at 0, 1 and 2 km away from the reef) at Ishigaki Island, Japan. The results highlighted that a live coral reef produced different and distinct molecules from a dead coral reef, and some of these molecules could be transported to a distance of at least 2 km from the reef with a 14–17-fold reduction in concentration. However, *C. viridis* larvae detected only the chemical cues 1 km away from the live coral reef.

This study shows that chemical cues emitted by a live coral reef are transported further away (at least 2 km) than those from a dead coral reef, and that fish larvae could detect these cues up to 1 km away from the reef.

Overall, pelagic larvae may be induced to change their vertical and/or horizontal movement in response to either chemical cues produced by dominant coral species or other reef organisms, or sound cues produced by the snapping noise of shrimps, scraping noises of herbivores including sea urchins, parrotfishes and surgeonfishes, and vocalizations of fishes (e.g., Myrberg and Fuiman 2002, Simpson et al. 2010). Despite the importance of sensory cues in habitat selection, the distance of detection of these chemical and acoustic cues would be relatively weak (less than 5 km away from a reef).

Detection of a micro-habitat on which to settle. Once a larva of any taxon has located an island, a particular reef patch (e.g., fringing reef, barrier reef, mangrove, seagrass beds) and/or a suitable micro-habitat on which to recruit must be found. Cues allowing larvae to detect the settlement habitat of their species can be emitted by the coral habitat itself (specific shape of coral colony or specific odour of anemone) and/or by the conspecifics already settled.

Visual, chemical and/or sound cues used for reef patch and microhabitat selection or rejection must involve factors directly affecting the fitness of Pomacentridae species (Myrberg and Fuiman 2002), which may use many such cues in order to locate the best available micro-habitat within a given reef for their requirement (for review see Leis and McCormick 2002, Leis et al. 2011).

The influence of reef sound on fish settlement has been studied over the last few years, ever since Stobutzki and Bellwood in 1997 first hypothesized that fish larvae could use sound as a cue for nocturnal orientation. Other studies have since highlighted the attraction of Pomacentridae larvae to reef sound (e.g., Tolimieri et al. 2004, Simpson et al. 2010). For example, the use of sound to locate habitat has also been recently shown in Lizard Island where Pomacentridae were preferentially found on patch reefs broadcasting fringing reef sound than on other treatments (Radford et al. 2011a). Similarly, Berten et al. (in press) showed that Pomacentridae species at Moorea were significantly deterred by mangrove sound but were attracted by barrier reef sound: *D. aruanus* was repulsed by mangrove sound, *Abudefduf sexfasciatus* was repulsed by both fringing reef and mangrove sound and *C. viridis* was neither attracted nor repulsed by any of the tested sounds.

Concerning chemical cues, Lecchini et al. (2005a) studied the visual and chemical abilities of 5 Pomacentridae species (*A. sexfasciatus, A. sordidus, C. leucopoma, P. pavo* and *S. nigricans*) to recognize either conspecifics or their settlement habitat. Conspecifics were recognized visually by 3 out of 5 species (*A. sordidus, C. leucopoma,* and *S. nigricans*), and chemically by 2 out of 5 species (*A. sordidus, P. pavo*). None of the 5 Pomacentridae species recognized their settlement habitat with sensory cues. In an additional study conducted on *C. viridis* at Moorea island (French Polynesia), Lecchini et al. (2005b) showed that *C. viridis* larvae detected reefs containing conspecifics using visual cues at distances <75 cm; detection distances increased to >375 cm when olfactory capacity was present (particularly for reefs located up-current). Dixson et al. (2012) also showed that Pomacentridae larvae could distinguish predatory fishes due to chemical cues of their diet. They manipulated the diet of three nominally nonpiscivorous species and examined the behavioral responses of juvenile anemonefish, *Amphiprion percula*, to the chemical cues of non-predators fed a diet rich in fish products. In pairwise choice trials, they showed that naive *A. percula* were indifferent to chemosensory cues from nonpiscivorous fishes who were fed their usual diet, but significantly avoided chemical cues from piscivorous and nonpiscivorous fishes who were fed a diet containing fish products. The authors concluded that *A. percula* larvae innately distinguish between piscivorous and nonpiscivorous fishes based on chemosensory cues in the diet. Lastly, Lecchini and Nakamura (2013) explored the importance of chemical cues for habitat selection by *C. viridis* and *Dascyllus reticulatus* larvae in a 4-channel choice flume. The spatial scales examined included reef patches (seagrass patch, live coral patch, dead coral patch and control water) within coral reefs (Exp. 1), and micro-habitats (live coral colonies, dead coral colonies, seagrass) or conspecifics within reef patches (Exp. 2) at Ishigaki Island, Japan. In experiment 1, *C. viridis* used chemical cues to move significantly towards the live coral patch water, while *D. reticulatus* were not attracted towards the chemical cues of the reef patches. In experiment 2, *C. viridis* and *D. reticulatus* used chemical cues to move significantly towards the conspecific water. These results suggest that Pomacentridae species can actively select settlement habitats according to olfactory cues, and they prefer the chemical cues of conspecifics over habitat cues.

Such small-scale behavioral responses resulting in patchy distributions can have strong effects on subsequent growth and survival of individuals, and can also reinforce or ameliorate spatial heterogeneity in environmental features (Doherty 2002). The majority of studies conducted on Pomacentridae have shown that larvae recognize their settlement habitat mainly by conspecific cues (for review see Leis and McCormick 2002, Leis et al. 2011). This social aggregation of larvae with older conspecifics may be the result of individuals using conspecific "guides" to find potentially beneficial resources (Childress and Herrnkind 2001, Leis et al. 2002). Thus, Pomacentridae larvae may potentially "search" conspecifics in order to settle on the same habitat, as larvae and juveniles have relatively close morphological and ecological characteristics.

Habitat degradation and perception of information at recruitment. The World Conservation Institute estimates that 20% of coral reefs are already definitively destroyed, another 25% are in great immediate threat, and another 25% will be under threat by 2050 (Wilkinson 2004). In the context of global degradation of coral reefs, variation in the composition of marine communities has largely been attributed to factors affecting the recruitment of marine larvae (e.g., Hughes 1994, Munday et al. 2009a, McCormick et al. 2010). Coral reefs experiencing perturbations often exhibit declines in adult populations, leading to a higher rate of extirpation than in pristine habitats, and the persistence of species in the area becomes reliant on the "rescue" effect of recruitment (Hanski and Gilpin 1997). The potential of the population to be supplemented by recruits depends then on whether pelagic larvae are able to detect an appropriate habitat in that area and then settle and persist (Montgomery et al. 2006). For example, Jones et al. (2004) showed that the decline in adult populations of coral reef fish in degraded habitats had more to do with recruitment failure than with adult mortality. Jones et al. (2004) suggested thus that the rescue effect of recruitment might be completely ineffective in a degraded habitat. However, the mechanisms that determine how marine larvae respond to different stages of coral stress and the extent of coral loss on fish recruitment are poorly understood (for review see Arvedlund and Kavanagh 2009, Leis et al. 2011).

Devine et al. (2012) explored the effects of elevated CO_2 on larval behavior and habitat selection at settlement in *P. amboinensis, P. chrysurus* and *P. moluccensis*. Pair-wise choice tests were performed using a two-channel flume chamber to test olfactory discrimination between hard coral, soft coral and coral rubble habitats. The results showed that exposure to elevated CO_2 disrupted the ability of larvae to discriminate between habitat odors in olfactory trials. Similarly, Lecchini et al. (2013) explored the effects of alternate reef states (coral vs. algal reefs) on the recruitment potential of coral islands in attracting fish larvae (Rangiroa atoll). Experiments with 2-channel choice chambers showed that during the recruitment stage, *C. viridis, C. leucopoma* and *Chrysiptera glauca* preferred water from reefs dominated by coral as compared to reefs dominated by algae. Pomacentridae larvae could thus respond to many different types of chemical cues associated with either (i) coral (and not algae) directly, or (ii) organisms whose abundance changes in response to coral vs. algal cover (Myrberg and Fuiman 2002). Visual surveys reveal that *C. glauca* and *C. leucopoma* were present on both coral and algal reefs. For these two species, the chemical cues from conspecific adults were present in waters from both the coral and algal dominated reefs, yet larvae were only attracted by coral reef water. The efficiency of chemical cues from conspecific adults could vary according to the environment in which they were emitted. Several recent studies concerning the mechanisms that determine how larvae respond to different stages of stress support this hypothesis (e.g., Fisher et al. 2006, Hale et al. 2009, Dixson et al. 2010). Munday et al. (2009b) showed that acidification disrupted the olfactory mechanism by which anemonefish larvae discriminate between cues that may be used in locating suitable adult habitat. Dixson et al. (2010) showed that newly hatched larvae of *A. percula* innately detected predators using olfactory cues and this ability was retained through to settlement. However, when eggs and larvae were exposed to seawater simulating ocean acidification, settlement-stage larvae became strongly attracted to the smell of predators and the ability to discriminate between predators and non-predators was lost.

Overall, these recent studies (e.g., Fisher et al. 2006, Munday et al. 2009b, Dixson et al. 2010, Lecchini et al. 2013) support the assumption of more efficient larval settlement in non-degraded reefs (live coral dominance) as compared to degraded coral reefs (dead coral or algal dominance). Thus, if the settlement potential of coral reefs has decreased, populations of Pomacentridae species will continue their rapid decline, as larval recruitment will not be able to replace and sustain the adult populations on the degraded reefs. Thus, understanding the relationship between reef state and settlement potential will aid management planning in order to maintain coral cover and thus biodiversity on reefs that are increasingly being degraded. Future studies conducted on the sensory abilities of Pomacentridae species at the larval stage should focus more on the effects of local and global changes. Indeed, any disruption to a larva's ability to discriminate a reef's chemical or acoustic stimuli could have far-reaching implications for the sustainability of adult populations, as recruitment would not sustain them.

References

Allen, G.R. 1975. Damselfishes of the South Seas. TFH Publications, Neptune City, New Jersey.

Allen, G.R. 1991. Damselfishes of the World. Mergus Publishers, Melle, Germany.

Allen, G.R. 1999. *Altrichthys*, a new genus of damselfish (Pomacentridae) from Philippine seas with description of a new species. Rev. Fr. Aquariol. 26: 23–28.

Almany, G.R. 2004. Does increased habitat complexity reduce predation and competition in coral reef fish assemblages? Oikos 106: 275–284.

Almany, G.R., M.L. Berumen, S.R. Thorrold, S. Planes and G.P. Jones. 2007. Local replenishment of coral reef fish populations in a marine reserve. Science 316: 742–744.

Arvedlund, M. and K. Kavanagh. 2009. The senses and environmental cues used by marine larvae of fish and decapod crustaceans to find tropical coastal ecosystems. pp. 135–185. *In*: I. Nagelkerken (ed.). Ecological Connectivity among Tropical Coastal Ecosystems. Springer, Berlin.

Asoh, K. 2003. Reproductive parameters of female Hawaiian damselfish *Dascyllus albisella* with comparison to other tropical and subtropical damselfishes. Mar. Biol. 143: 803–810.

Asoh, K. and T. Yoshikawa. 2002. The role of temperature and embryo development time in the diel timing of spawning in a coral-reef damselfish with high-frequency spawning synchrony. Env. Biol. Fishes 64: 379–392.

Atema, J., M.J. Kingsford and G. Gerlach. 2002. Larval reef fish could use odour for detection, retention and orientation to reefs. Mar. Ecol. Progr. Ser. 241: 151–160.

Ault, T.R. and C.R. Johnson. 1998. Relationships between habitat and recruitment of three species of damselfish (Pomacentridae) at Heron Reef, Great Barrier Reef. J. Exp. Mar. Biol. Ecol. 223: 145–166.

Bapary, M.A., M. Nurul Amin and A. Takemura. 2012. Food availability as a possible determinant for initiation and termination of reproductive activity in the tropical damselfish *Chrysiptera cyanea*. Mar. Biol. Res. 8: 154–162.

Beldade, R., S.J. Holbrook, R.J. Schmitt, S. Planes, D. Malone and G. Bernardi. 2012. Larger female fish contribute disproportionately more to self-replenishment. Proc. R. Soc. B-Biol. Sci. 279: 2116–2121.

Ben-Tzvi, O., M. Kiflawi, O. Polak and A. Abelson. 2009. The effect of adult aggression on habitat selection by settlers of two coral-dwelling damselfishes. PLoS One 4: e5511.

Berten, L., D. Lecchini, S.D. Simpson, C.A. Radford and E. Parmentier. Propagation of coral reef noise: a case study from the north coast of Moorea, French Polynesia. Mar. Ecol. Prog. Ser. (in press).

Berumen, M.L., G.R. Almany, S. Planes, G.P. Jones, P. Saenz-Agudelo and S.R. Thorrold. 2012. Persistence of self-recruitment and patterns of larval connectivity in a marine protected area network. Ecol. Evol. 2: 444–452.

Booth, D.J. 1991. The effects of sampling frequency on estimates of recruitment of the domino damselfish *Dascyllus albisella* (Gill). J. Exp. Mar. Biol. Ecol. 145: 149–159.

Booth, D.J. 1995. Juvenile groups in a coral-reef damselfish: density dependent effects on individual fitness and population demography. Ecology 76: 91–106.

Booth, D.J. and G.A. Beretta. 1994. Seasonal recruitment, habitat associations, and survival of pomacentrid reef fish in the US Virgin Islands. Coral Reefs 13: 81–89.

Breder, C.M. and D.E. Rosen. 1966. Modes of Reproduction in Fishes. T.F.H. Publications, Neptune City, New Jersey.

Buston, P.M., G.P. Jones, S. Planes and S.R. Thorrold. 2012. Probability of successful larval dispersal declines fivefold over 1 km in a coral reef fish. Proc. R. Soc. B-Biol. Sci. 279: 1883–1888.

Cheney, K.L. and I.M. Côté. 2003. Habitat choice in adult longfin damselfish: territory characteristics and relocation times. J. Exp. Mar. Biol. Ecol. 287: 1–12.

Childress, M.J. and W.F. Herrnkind. 2001. The guide effect influence on the gregariousness of juvenile Caribbean spiny lobsters. Anim. Behav. 62: 465–472.

Cowen, R.W. 2002. Larval dispersal and retention and consequences for population connectivity. pp. 341–375. *In*: P.F. Sale (ed.). Coral Reef Fishes: Dynamics and Diversity in a Complex Ecosystem. Academic Press, San Diego.

Danilowicz, B.S. 1995. Spatial patterns of spawning in the coral-reef damselfish *Dascyllus albisella*. Mar. Biol. 122: 145–155.

Danilowicz, B.S. 1997. The effects of age and size on habitat selection during settlement of a damselfish. Env. Biol. Fishes 50: 257–265.

Devine, B.M., P.L. Munday and G. Jones. 2012. Rising CO_2 concentrations affect settlement behaviour of larval damselfishes. Coral Reefs 31: 229–238.

Dixson, D.L., P.L. Munday and G.P. Jones. 2010. Ocean acidification disrupts the innate ability of fish to detect predator olfactory cues. Ecol. Lett. 13: 68–75.

Dixson, D.L., M.S. Pratchett and P.L. Munday. 2012. Reef fishes innately distinguish predators based on olfactory cues associated with recent prey items rather than individual species. Anim. Behav. 84: 45–51.

Doherty, P.J. 1983. Diel, lunar and seasonal rhythms in the reproduction of two tropical damselfishes: *Pomacentrus flavicauda* and *P. wardi*. Mar. Biol. 75: 215–224.

Doherty, P.J. 2002. Variable replenishment and the dynamics of reef fish populations. pp. 327–358. *In*: P.F. Sale (ed.). Coral Reef Fishes: Dynamics and Diversity in a Complex Ecosystem. Academic Press, San Diego.

Doherty, P.J. and J. McIlwain. 1996. Monitoring larval fluxes through the surf zones of Australian coral reefs. Mar. Freshw. Res. 47: 383–390.

Doherty, P.J. and J.H. Carleton. 1997. The distribution and abundance of pelagic juvenile fish near Grub Reef, Central Great Barrier Reef. Proceeding of the 8th International Coral Reef Symposium 2: 1155–1160.

Doherty, P.J., V. Dufour, R. Galzin, M.A. Hixon and S. Planes. 2004. High mortality at settlement is a population bottleneck for a tropical surgeonfish. Ecology 85: 2422–2428.

Dufour, V. and R. Galzin. 1993. Colonization patterns of reef fish larvae to the lagoon at Moorea Island, French Polynesia. Mar. Ecol. Progr. Ser. 102: 143–152.

Dufour, V., E. Riclet and A. Lo-Yat. 1996. Colonization of reef fishes at Moorea Island, French Polynesia: temporal and spatial variation of the larval flux. Mar. Freshw. Res. 47: 413–422.

Egner, S.A. and D.A. Mann. 2005. Auditory sensitivity of sergeant major damselfish *Abudefduf saxatilis* from post-settlement juvenile to adult. Mar. Ecol. Prog. Ser. 285: 213–222.

England, P.R., G.H.R. Osler, L.M. Woodworth, M.E. Montgomery, D. Briscoe and R. Frankham. 2003. Effects of intense versus diffuse population bottleneck. Conserv. Genet. 4: 595–604.

Fautin, D.G. and G.R. Allen. 1992. Field Guide to Anemonefishes and their Host Sea Anemones. Western Australian Museum.

Fenberg, P.B. and K. Roy. 2008. Ecological and evolutionary consequences of size-selective harvesting: how much do we know? Mol. Ecol. 17: 209–220.

Figueira, W.F. and S.J. Lyman. 2007. Context-dependent risk tolerance of the bicolour damselfish: courtship in the presence of fish and egg predators. Ani. Behav. 74: 329–336.

Fishelson, L., D. Popper and A. Avidor. 1974. Biosociology and ecology of pomacentrid fishes around the Sinai Peninsula (northern Red Sea). J. Fish Biol. 6: 119–133.

Fisher, H.S., B.B.M. Wong and G.G. Rosenthal. 2006. Alteration of the chemical environment disrupts communication in a freshwater fish. Proc. R. Soc. B-Biol. Sci. 273: 526–537.

Fisher, R. and D.R. Bellwood. 2002. A light trap design for stratum-specific sampling of reef fish larvae. J. Exp. Mar. Biol. Ecol. 269: 27–37.

Fisher, R., D.R. Bellwood and S.D. Job. 2000. Development of swimming abilities in reef fish larvae. Mar. Ecol. Prog. Ser. 202: 163–173.

Fortier, L. and R.P. Harris. 1989. Optimal foraging and density dependent competition in marine fish larvae. Mar. Ecol. Prog. Ser. 51: 19–33.

Foster, S.A. 1987. Diel and lunar patterns of reproduction in the Caribbean and Pacific sergeant major damselfishes *Abudefduf saxatilis* and *A. troschelii*. Mar. Biol. 95: 333–343.

Frédérich, B., O. Colleye, G. Lepoint and D. Lecchini. 2012. Mismatch between shape changes and ecological shifts during the post-settlement growth of the surgeonfish, *Acanthurus triostegus*. Front. Zool. 9: 8.

Fricke, H.W. 1980. Control of differing mating systems in a coral reef fish by one environmental factor. Anim. Behav. 28: 561–569.

Gagliano, M. and M.I. McCormick. 2007. Compensating in the wild: is flexible growth the key to early juvenile survival? Oikos 116: 111–120.

Gagliano, M., M.I. McCormick and M.G. Meekan. 2007. Survival against the odds: ontogenetic changes in selective pressure mediate growth-mortality trade-offs in a marine fish. Proc. Royal Soc. Lond. 274: 1575–1582.

Gerlach, G., J. Atema, M.J. Kingsford, K.P. Black and V. Miller-Sims. 2007. Smelling home can prevent dispersal of reef fish larvae. Proc. Nat. Acad. Sci. USA 104: 858–863.

Gillanders, B.M., K.W. Able, J.A. Brown, D.B. Eggleston and P.F. Sheridan. 2003. Evidence of connectivity between juvenile and adult habitats for mobile marine fauna: an important component of nurseries. Mar. Ecol. Prog. Ser. 247: 281–295.

Goulet, D. 1995. Temporal patterns of reproduction in the Red Sea damselfish *Amblyglyphidodon leucogaster*. Bull. Mar. Sci. 57: 582–595.

Green, B.S. and R. Fisher. 2004. Temperature influences swimming speed, growth and larval duration in coral reef fish larvae. J. Exp. Mar. Biol. Ecol. 299: 115–132.

Gronell, A.M. 2010. Visiting behaviour by females of the sexually dichromatic damselfish, *Chrysiptera cyanea* (Teleostei: Pomacentridae): a probable method of assessing male quality. Ethology 81: 89–122.

Hale, R., S.E. Swearer and B.J. Downes. 2009. Separating natural responses from experimental artefacts: habitat selection by a diadromous fish species using odours from conspecifics and natural stream water. Oecologia 159: 679–687.

Hanski, I. and M.E. Gilpin. 1997. Metapopulation Biology: Ecology, Genetics and Evolution. Academic Press, London.

Hendriks, I.E., D.T. Wilson and M.G. Meekan. 2001. Vertical distributions of late stage larval fishes in the nearshore waters of the San Blas Archipelago, Caribbean Panama. Coral Reefs 20: 77–84.

Hirose, Y. 1995. Patterns of pair formation in protandrous anemonefishes, *Amphiprion clarkii, A. frenatus* and *A. perideraion*, on coral reefs of Okinawa, Japan. Env. Biol. Fishes 43: 153–161.

Hoey, A.S. and M.I. McCormick. 2004. Selective predation for low body condition at the larval-juvenile transition of a coral reef fish. Oecologia 139: 23–29.

Holmes, T.H. and M.I. McCormick. 2004. Location influences size-selective predation on newly-settled reef fish. Mar. Ecol. Prog. Ser. 317: 203–209.

Horne, E.A. and M. Itzkowitz. 1995. Behaviour of the female beaugregory damselfish (*Stegastes leucostictus*). J. Fish Biol. 46: 457–461.

Huebert, K.B. 2008. Barokinesis and depth regulation by pelagic coral reef fish larvae. Mar. Ecol. Prog. Ser. 367: 261–269.

Huebert, K.B., S. Sponaugle and R.K. Cowen. 2010. Predicting the vertical distributions of reef fish larvae in the Straits of Florida from environmental factors. Can. J. Fish. Aqua. Sci. 67: 1755–1767.

Huebert, K.B., R.K. Cowen and S. Sponaugle. 2011. Vertical migrations of reef fish larvae in the Straits of Florida and effects on larval transport. Limn. Ocean. 56: 1653–1666.

Hughes, T.P. 1994. Catastrophes, phase-shifts and large-scale degradation of a Caribbean coral reef. Science 265: 1547–1551.

Huntingford, F.H. 1984. The Study of Animal Behavior. Chapman and Hall, New York.

Jan, R. and R. Ormond. 1992. Spawning of the damselfishes on the northern coast of Taiwan, with emphasis on spawning site distribution. Bull. Inst. Zool. Acad. Sinica 31: 231–245.

Job, S.D. and D.R. Bellwood. 2000. Light sensitivity in larval fishes: Implications for vertical zonation in the pelagic zone. Limn. Ocean. 45: 362–371.

Johannes, R.E. 1978. Reproductive strategies of coastal marine fishes in the tropics. Env. Biol. Fish 3: 65–84.

Jones, G.P., M.J. Milicich, M.J. Emslie and C. Lunow. 1999. Self-recruitment in a coral reef fish population. Nature 402: 802–804.

Jones, G.P., M.I. McCormick, M. Srinivasan and J.V. Eagle. 2004. Coral decline threatens fish biodiversity in marine reserves. Proc. Nat. Acad. Sci. USA 101: 8251–8253.

Jones, G.P., S. Planes and S.R. Thorrold. 2005. Coral reef fish larvae settle close to home. Curr. Biol. 15: 1314–1318.

Karino, K. and A. Nakazono. 1993. Reproductive behavior of the territorial herbivore *Stegastes nigricans* (Pisces: Pomacentridae) in relation to colony formation. J. Ethology 11: 99–110.

Kingsford, M.J. 2001. Diel patterns of abundance of pre-settlement reef fishes and pelagic larvae on a coral reef. Mar. Biol. 138: 853–867.

Kingsolver, J.G., H.E. Hoekstra, J.M. Hoekstra, D. Berrigan, S.N. Vignieri, C.E. Hill and P. Beerli. 2001. The strength of phenotypic selection in natural populations. Am. Nat. 157: 245–261.

Lecchini, D. and R. Galzin. 2003. Influence of pelagic and benthic, biotic and abiotic, stochastic and deterministic processes on the dynamics of auto-recruitment of coral reef fish. Cybium 27: 167–184.

Lecchini, D. and Y. Nakamura. 2013. Use of chemical cues by coral reef animal larvae for habitat selection. Aqua. Biol. 19: 231–238.

Lecchini, D., V. Dufour, J. Carleton, S. Strand and R. Galzin. 2004. Study of the fish larval flux at Moorea Island: is the spatial scale significant? J. Fish Biol. 65: 1142–1146.

Lecchini, D., S. Planes and R. Galzin. 2005a. Experimental assessment of sensory modalities of coral reef fish larvae in the recognition of settlement habitat. Behav. Ecol. Sociobiol. 56: 18–26.

Lecchini, D., J.S. Shima, B. Banaigs and R. Galzin. 2005b. Larval sensory abilities and mechanisms of habitat selection of a coral reef fish during settlement. Oecologia 143: 326–334.

Lecchini, D., S. Planes and R. Galzin. 2007. The influence of habitat characteristics and conspecifics on attraction and survival of coral reef fish juveniles. J. Exp. Mar. Biol. Ecol. 341: 85–90.

Lecchini, D., V.P. Waqalevu, E. Parmentier, C.A. Radford and B. Banaigs. 2013. Fish larvae prefer coral above algal water cues: implications of coral reef degradation. Mar. Ecol. Prog. Ser. 475: 303–307.

Lecchini, D., T. Miura, G. Lecellier, B. Banaigs and Y. Nakamura. 2014. Transmission distance of chemical cues from coral habitats: implications for marine larval settlement in context of reef degradation. Mar. Biol. 161: 1677–1686.

Leese, J.M., J.L. Snekser, A. Ganim and M. Itzkowitz. 2009. Assessment and decision-making in a Caribbean damselfish: nest-site quality influences prioritization of courtship and brood defence. Biol. Lett. 5: 188–90.

Leis, J.M. 2004. Vertical distribution behaviour and its spatial variation in late-stage larvae of coral reef fishes during the day. Mar. Freshw. Behav. Physiol. 37: 65–88.

Leis, J.M. and B.M. Carson-Ewart. 1998. Complex behaviour by coral-reef fish larvae in open-water and near-reef pelagic environments. Env. Biol. Fishes 53: 259–266.

Leis, J.M. and B.M. Carson-Ewart. 1999. *In situ* swimming and settlement behaviour of larvae of an Indo-Pacific coral reef fish, the coral trout Plectropomus leopardus (Pisces: Serranidae). Mar. Biol. 134: 51–64.

Leis, J.M. and B.M. Carson-Ewart. 2002. *In situ* settlement behaviour of damselfish larvae (Pisces: Pomacentridae). J. Fish Biol. 61: 325–346.

Leis, J.M. and M.I. McCormick. 2002. Behavior, dispersal, growth and metamorphosis of the pelagic larvae of coral reef fishes. pp. 171–200. *In*: P.F. Sale (ed.). Coral Reef Fishes: Dynamics and Diversity in a Complex Ecosystem. Academic Press, San Diego.

Leis, J.M. and D.L. Clark. 2005. Feeding greatly enhances swimming endurance of settlement-stage reef-fish larvae of damselfishes (Pomacentridae). Ichth. Res. 52: 185–188.

Leis, J.M., B.M. Carson-Ewart and D.H. Cato. 2002. Sound detection *in situ* by the larvae of a coral-reef damselfish (Pomacentridae). Mar. Ecol. Prog. Ser. 232: 259–268.

Leis, J.M., B.M. Carson-Ewart, A.C. Hay and D.H. Cato. 2003. Coral-reef sounds enable nocturnal navigation by some reef-fish larvae in some places and at some times. J. Fish Biol. 63: 724–737.

Leis, J.M., K.J. Wright and R.N. Johnson. 2007. Behaviour that influences dispersal and connectivity in the small, young larvae of a reef fish. Mar. Biol. 153: 103–117.

Leis, J.M., U. Siebeck and D.L. Dixson. 2011. How Nemo finds home: the neuroecology of dispersal and of population connectivity in larvae of marine fishes. Integ. Comp. Biol. 51: 826–843.

Lemberget, T. and M.I. McCormick. 2009. Replenishment success linked to fluctuating asymmetry in larval fish. Oecologia 159: 83–93.

Lobel, P.S. 1997. Comparative settlement age of damselfish larvae (*Plectroglyphidodon imparipennis*, Pomacentridae) from Hawaii and Johnston atoll. Biol. Bull. 193: 281–283.

Lo-Yat, A. 2002. Variabilité temporelle de la colonisation par les larves de poissons de l'atoll de Rangiroa (Tuamotu, Polynésie française) et utilisation de l'outil "otolithe" de ces larves. PhD Thesis, Université Française du Pacifique, Tahiti, French Polynésia.

Lo-Yat, A., S.D. Simpson, M.G. Meekan, D. Lecchini, E. Martinez and R. Galzin. 2011. Extreme climatic events reduce ocean productivity and larval supply in a tropical reef ecosystem. Glob. Chang. Biol. 17: 1695–1702.

Luiz, O.J., A.P. Allen, D.R. Robertson, S.R. Floeter, M. Kulbicki, L. Vigliola, R. Becheler and J.S. Madin. 2013. Adult and larval traits as determinants of geographic range size among tropical reef fishes. Proc. Nat. Acad. Sci. USA 110: 16498–16502.

Manica, A. 2002. Filial cannibalism in teleost fish. Biol. Rev. 77: 261–277.

Mann, D.A., B.M. Casper, K.S. Boyle and T.C. Tricas. 2007. On the attraction of larval fishes to reef sounds. Mar. Ecol. Prog. Ser. 338: 307–310.

McCormick, M.I. and M.G. Meekan. 2007. Social facilitation of selective mortality. Ecology 88: 1562–1570.

McCormick, M.I., L.J. Makey and V. Dufour. 2002. Comparative study of metamorphosis in tropical reef fishes. Mar. Biol. 141: 841–853.

McCormick, M.I., J.A.Y. Moore and P.L. Munday. 2010. Influence of habitat degradation on fish replenishment. Coral Reefs 29: 537–546.

McIlwain, J. 2002. Link between reproductive output and larval supply of a common damselfish species, with evidence of replenishment from outside the local population. Mar. Ecol. Prog. Ser. 236: 219–232.

Meadows, D. 2001. Centre-edge differences in behaviour, territory size and fitness in clusters of territorial damselfish: patterns, causes and consequences. Behaviour 138: 1085–1116.

Miller-Sims, V.C., G. Gerlach, M.J. Kingsford and J. Atema. 2008. Dispersal in the spiny damselfish, *Acanthochromis polyacanthus*, a coral reef fish species without a larval pelagic stage. Mol. Ecol. 17: 5036–5048.

Mizushima, N., Y. Nakashima and T. Kuwamura. 2000. Semilunar spawning cycle of the humbug damselfish *Dascyllus aruanus*. J. Ethology 18: 105–108.

Montgomery, J.C., A. Jes, S.D. Simpson, M. Meekan and C. Tindle. 2006. Sound as an orientation cue for the pelagic larvae of reef fishes and decapod crustaceans. Adv. Mar. Biol. 51: 143–196.

Moyer, J.T. 1975. Reproductive behavior of the damselfish *Pomacentrus nagasakiensis* at Miyake-jima, Japan. Japanese J. Ichth. 22: 151–163.

Moyer, J.T. and L.J. Bell. 1976. Reproductive behavior of the anemonefish *Amphiprion clarkii* at Miyaki-Jima, Japan. Japanese J. Ichth. 23: 23–32.

Munday, P.L., J.M. Leis, J.M. Lough, C.B. Paris, M.J. Kingsford, M.L. Berumen and J. Lambrechts. 2009a. Climate change and coral reef connectivity. Coral Reefs 28: 379–395.

Munday, P.L., J.M. Donelson, D.L. Dixson and G.G.K. Endo. 2009b. Effects of ocean acidification on the early life history of a tropical marine fish. Proc. R. Soc. B-Biol. Sci. 276: 3275–3283.

Myrberg, A.A. and L. Fuiman. 2002. The sensory world of coral reef fishes. pp. 187–227. *In*: P.F. Sale (ed.). Coral Reef Fishes: Dynamics and Diversity in a Complex Ecosystem. Academic Press, San Diego.

Ohman, M.C., P.L. Munday, G.P. Jones and M.J. Caley. 1998. Settlement strategies and distribution patterns of coral reef fishes. J. Exp. Mar. Biol. Ecol. 225: 219–238.

Paris, C.B. and R.K. Cowen. 2004. Direct evidence of a biophysical retention mechanism for coral reef fish larvae. Limn. Ocean. 49: 1964–1979.

Parmentier, E., D. Lecchini, F. Lagardère and P. Vandewalle. 2004. Ontogenic and ecological control of metamorphosis onset in a carapid fish, *Carapus homei*: experimental evidence from vertebra and otolith comparisons. J. Exp. Biol. 301: 617–628.

Pini, J., S. Planes, E. Rochel, D. Lecchini and C. Fauvelot. 2011. Genetic diversity loss associated to high mortality and environmental stress during the recruitment stage of a coral reef fish. Coral Reefs 30: 399–404.

Pinsky, M.L., H.R. Montes and S.R. Palumbi. 2010. Using isolation by distance and effective density to estimate dispersal scales in anemonefish. Evolution 64: 2688–2700.

Planes, S. and G. Lecaillon. 2001. Caging experiment to examine mortality during metamorphosis of coral reef fish larvae. Coral Reefs 20: 211–218.

Planes, S., G.P. Jones and S.R. Thorrold. 2009. Larval dispersal connects fish populations in a network of marine protected areas. Proc. Nat. Acad. Sci. USA 106: 5693–5697.

Pressley, P.H. 1980. Lunar periodicity in the spawning of the yellowtail damselfish, *Microspathodon chrysurus*. Env. Biol. Fishes 5: 153–159.

Radford, C.A., J.A. Stanley, S.D. Simpson and A.G. Jes. 2011a. Juvenile coral reef fish use sound to locate habitats. Coral Reefs 30: 295–305.

Radford, C.A., C.T. Tindle, J.C. Montgomery and A.G. Jeffs. 2011b. Modeling a reef as an extended sound source increases the predicted range at which reef noise may be heard by fish larvae. Mar. Ecol. Prog. Ser. 438: 167–174.

Rasa, O. 1969. Territoriality and the establishment of dominance by means of visual cues in *Pomacentrus jenkinsi* (Pisces: Pomacentridae). Z. Tierpsychol. 26: 825–845.

Richardson, D., P. Harrison and V. Harriott. 1997. Timing of spawning and fecundity of a tropical and subtropical anemonefish (Pomacentridae: *Amphiprion*) on the east Coast of Australia. Mar. Ecol. Prog. Ser. 156: 175–181.

Robertson, D.R. 1973. Field observations on the reproductive behaviour of a pomacentrid fish, *Acanthochromis polyacanthus*. Z. Tierpsychol. 32: 319–324.

Robertson, D.R. 1990. Differences in the seasonalities of spawning and recruitment of some small neotropical reef fishes. J. Exp. Mar. Biol. Ecol. 144: 44–62.

Robertson, D.R., C.W. Petersen and J.D. Brawn. 1990. Lunar reproductive cycles of benthic-brooding reef fishes: reflections of larval biology or adult biology? Ecol. Monogr. 60: 311–329.

Ross, R.M. 1978. Reproductive behavior of the anemonefish *Amphiprion melanopus* on Guam. Copeia 1978: 103–107.

Saenz-Agudelo, P., G.P. Jones, S.R. Thorrold and S. Planes. 2009. Estimating connectivity in marine populations: an empirical evaluation of assignment tests and parentage analysis under different gene flow scenarios. Mol. Ecol. 18: 1765–1776.

Sale, P.F. 1972. Influence of corals in the dispersion of the pomacentrid fish, *Dascyllus aruanus*. Ecology 53: 741–744.

Siebeck, U.E. 2004. Communication in coral reef fish: the role of ultraviolet colour patterns in damselfish territorial behaviour. Anim. Behav. 68: 273–282.

Siebeck, U.E. and N.J. Marshall. 2001. Ocular media transmission of coral reef fish-can coral reef fish see ultraviolet light? Vision Research 41: 133–149.

Sikkel, P., S. Herzlieb and D. Kramer. 2005. Compensatory cleaner-seeking behavior following spawning in female yellowtail damselfish. Mar. Ecol. Prog. Ser. 296: 1–11.

Simpson, S.D., M.G. Meekan, N.J. Larsen, R.D. McCauley and A. Jeffs. 2010. Behavioral plasticity in larval reef fish: orientation is influenced by recent acoustic experiences. Behav. Ecol. 21: 1098–1105.

Stobutzki, I.C. and D.R. Bellwood. 1997. Sustained swimming abilities of the late pelagic stages of coral reef fishes. Mar. Ecol. Prog. Ser. 149: 35–41.

Thresher, R.E. 1983. Habitat effects on reproductive success in the coral reef fish, *Acanthochromis polyacanthus* (Pomacentridae). Ecology 64: 1184–1199.

Thresher, R.E. 1985. Distribution, abundance, and reproductive success in the coral reef fish *Acanthochromis polyacanthus*. Ecology 66: 1139–1150.

Thresher, R.E. and J.T. Moyer. 1983. Male success, courtship complexity and patterns of sexual selection in three congeneric species of sexually monochromatic and dichromatic damselfishes (Pisces: Pomacentridae). Anim. Behav. 31: 113–127.

Tolimieri, N., O. Haine, A. Jeffs, R. McCauley and J. Montgomery. 2004. Directional orientation of pomacentrid larvae to ambient reef sound. Coral Reefs 23: 184–191.

Tzioumis, V. and M.J. Kingsford. 1995. Periodicity of spawning of two temperate damselfishes: *Parma microlepis* and *Chromis dispilus*. Bull. Mar. Sci. 57: 596–609.

Victor, B.C. 1986. Duration of the planktonic larval stage of one hundred species of Pacific and Atlantic wrasses (family Labridae). Mar. Biol. 90: 317–326.

Vigliola, L. and M.G. Meekan. 2002. Size at hatching and planktonic growth determines post-settlement survivorship of a coral reef fish. Oecologia 131: 89–93.

Wellington, G.M. and B.C. Victor. 1989. Planktonic larval duration of one hundred species of Pacific and Atlantic damselfishes (Pomacentridae). Mar. Biol. 101: 557–567.

Whiteman, E.A. and I.M. Côté. 2004. Monogamy in marine fishes. Biol. Rev. Cambridge Phil. Soc. 79: 351–375.

Wilkinson, C. 2004. Status of Coral Reefs of the World. Australian Institute of Marine Science, Townsville, Australia.

Wilson, D.T. and M.I. McCormick. 1997. Spatial and temporal validation of settlement marks in the otoliths of tropical reef fishes. Mar. Ecol. Prog. Ser. 153: 259–271.

Wright, K.J., D.M. Higgs and A.J. Belanger. 2005. Auditory and olfactory abilities of pre-settlement larvae and post-settlement juveniles of a coral reef damselfish (Pisces: Pomacentridae). Mar. Biol. 147: 1425–1434.

Wright, K.J., D.M. Higgs, A.J. Belanger and J.M. Leis. 2008. Auditory and olfactory abilities of larvae of the Indo-Pacific coral trout *Plectropomus leopardus* (Lacepede) at settlement. J. Fish Biol. 72: 2543–56.

Wright, K.J., D.M. Higgs, D.H. Cato and J.M. Leis. 2010. Auditory sensitivity in settlement-stage larvae of coral reef fishes. Coral Reefs 29: 235–243.

Sex Change Strategies and Group Structure of Damselfishes

Akihisa Hattori[1],* and Margarida Casadevall[2]

Introduction

The evolution of sex change strategies in fishes is closely related to mating systems (Warner 1984, 1988a,b). Field studies have revealed that fish species utilize a variety of strategies, including sex change, in order to increase their reproductive potential within the social context of each species (Warner 1988a,b, Munday et al. 2006). Although almost all species of damselfishes are non-sex changers, some species exhibit sex changes. For example, species in the genera *Amphiprion* and *Premnas* include protandrous (male-to-female) sex changers and those within the genus *Dascyllus* include protogynous (female-to-male) sex changers (Table 1). These genera have unique mating systems that are quite different from those of other species, although their reproductive and parental care behavior is similar to that of other damselfishes. Thus, they offer excellent models for examining the evolution of mating systems and sex change in fishes. Sex-changing species inhabit particular refuges (i.e., giant sea anemones for *Amphiprion* and *Premnas*, and staghorn branching corals for *Dascyllus*), usually forming small social groups with a size-ordered dominance hierarchy. In contrast, non-sex changing species are solitary, aggregational (or colonial) or gregarious (see below), and they usually utilize ubiquitous shelters such as narrow gaps within

[1] Faculty of Liberal Arts and Education, Shiga University, 2-5-1 Hiratsu, Otsu, Shiga 520-0862, Japan.
[2] Environmental Sciences Department, Sciences Faculty, Girona University, 17071 Girona, Spain.
  Email: margarida.casadevall@udg.edu
* Corresponding author: hattori@edu.shiga-u.ac.jp

Table 1. Definitions of some important terms on sex change and mating systems in this chapter (see also Thresher 1984, Warner 1984, 1988a,b).

Protandrous sex changers	Individuals that have undergone sex change from *functional* male to *functional* female
Protogynous sex changers	Individuals that have undergone sex change from *functional* female to *functional* male
Pre-maturational sex changers	Functional males that have not been functionally female in protogynous species, or functional females that have not been functionally male in protandrous species
Primary males	Functional males that have not been functionally female in protogynous species (They are considered to be genetically determined pure males)
Diandric species	Protogynous species that have both primary males and secondary males that have been functionally female unlike primary males
Non-functional hermaphroditism	A proportion of individuals exhibit both testicular and ovarian tissues but only reproduce as either male or female during their lives
Functional hermaphroditism	A proportion of individuals function as both sexes at some time during their lives
Functional gonochorism	Functional males and females have never experienced sex change even if an individual has both ovary and testis in the gonads
Monogamy	A functional male and a functional female keep a long term pair-bond to reproduce
Polygyny	A functional male reproduces with more than one or more females
Harems (Resource defense polygyny)	A functional male defends an all-purpose territory within which more than one functional female has home range and reproduces with the male
Lek-like spawning aggregations	During the reproductive period, functional males temporally defend mating and nesting territories at a spawning site which functional females visit. If a female spawns only once at a nest of one male, this system can be recognized as polygyny
Promiscuity (polygynandry, i.e., multi-male-multi-female polygamy)	Functional males and females defend territories, where each male courts any females that pass by and each female spawns with several males during the reproductive period (It is hard to confirm whether their mating is random or not)
Multi-male polygyny (a transitional state from harem polygyny to promiscuity)	Several functional males live in a large habitat keeping their own harems but their control of females is not strict so that some females can undergo sex change
Random mating systems (random pair formation or large group spawning)	Functional males and females pair randomly to reproduce (i.e., random pair formation), or they temporally aggregate at a site to release eggs and sperms at the same time (i.e., large group spawning)
Asynchronous development of the follicles	Oocytes at all the stages of maturation present
Atretic oocytes	Degenerating oocytes
Postovulatory Follicles (POF)	Follicular layers that remain in the ovary after the release of the ovum during spawning
Ambisexual gonads	Coexistence of ovaric and testicular tissue in the same gonad (ovotestis)
Rudimentary gonads	Without histological sex differentiation

rocks or rubble, and cracks and coral heads on reefs. In this chapter, we first explain the key concepts underlying sex change strategies. Next, we classify damselfish species into three categories (i.e., solitary or aggregational; gregarious or schooling; social) before summarizing the relationship between these categories and sex change. Since non-sex changers are common in some sex-changing species, we describe the flexible sex determination mechanisms in these species. Finally, the relationship between growth, inter-group movement and sex change strategies among damselfishes will be discussed in relation to shelter use patterns.

Size Advantage as an Explanation of Flexible Sex Change Strategies

Many teleost fish species utilize sex change strategies. If expected fertility (the number of viable offspring) is related to body size, an individual who can change sex at the proper size can have more offspring than one remaining exclusively male or female (Ghiselin 1969). This size-advantage hypothesis (SAH) predicts that the direction of sex change in a species depends on its mating system. Sex change from female to male (protogyny) is found in species with polygyny because larger males monopolize mating opportunities. In this strategy, the expected reproductive success of males will expand rapidly with increasing body size (Charnov 1982, Warner 1984, 1988a,b, Kuwamura and Nakashima 1998). Sex change from male to female (protandry) is expected in species with mating systems in which the expected reproductive success of females increases with increasing body size, whereas that of males is less sensitive to body size (Ghiselin 1969, Warner 1975, 1988a,b, Charnov 1982) because they would be unable to monopolize mating opportunities (Warner 1984).

At intraspecific levels, the SAH predicts the optimal size at which sex change occurs in the population of a species having sex reversal ability. However, the timing of sex change does not always correspond to predictions based on expected fertility because many fish species have more complex life-history pathways in relation to sex change (Warner 1984, Munday et al. 2006). In protogynous species with harem polygyny, for instance, the disappearance of the male from a harem usually causes the remaining largest female to undergo a sex change, but some small individuals may become solitary in order to take over the dominant status in a harem after changing into bachelor males growing faster than females (Warner 1984). The wide variations in the timing of sex change are related to different factors describing reproductive value (Warner 1988a, Munday et al. 2006). This reproductive value is defined as the expected reproductive success of an individual, which takes into account growth rate, mortality and fertility. Growth and mortality, as well as fertility, of individuals are largely influenced by the social context (Munday et al. 2006). Accordingly, SAH based on sex-specific reproductive values indicates that an individual should change sex when the reproductive value of the other sex exceeds that of its current sexual status. Thus, individuals from the same population are expected to change sex flexibly (at different sizes or ages) due to the trade-offs between sex-specific growth, mortality and fertility of each individual (Munday et al. 2006).

Group Types, Mating Systems and Sex Change

Feeding around ubiquitous shelters or a particular refuge

Damselfishes are mainly grazers, plankton feeders or they feed on a variety of small invertebrates (see Chapter VII). Being typically small fishes, seldom exceeding 10–15 cm in length, they feed close to shelters such as narrow gaps, cracks, or coral heads on a reef, to which they can escape from predators (Lowe-McConnell 1987). On Ishigaki Island, for example (Table 2), 24 species of damselfishes were found on 84 very small patch reefs (involving coral heads and outcrops; total area = 166 m²) of

Table 2. Relationship among group types, sex change strategies and other characteristics of 24 damselfish species found on 84 very small patch reefs of the 0.28-hectare study site of Shiraho Reef, Ishigaki Island, Okinawa, Japan (Hattori and Shibuno 2010).

Species name	Group type	Sex change	TL (cm)	No.	Shelter or refuge	Feeding and spawning sites for adults
*Dascyllus aruanus**	Social[t]	♀→♂	5	545	Staghorn coral head	Almost same
*Amphiprion frenatus**	Social[t]	♂→♀	10	32	Giant sea anemone	Almost same
Amphiprion ocellaris	Social	♂→♀	6	3	Giant sea anemone	Almost same
*Chromis viridis**	Large school	None	6	321	*Acropora* coral head	Apparently different
*Chrysiptera cyanea**	Large shoal	None	4	716	Gaps and cracks	Almost same?
*Pomacentrus moluccensis**	Small shoal	None	5	151	Gaps and cracks	?
*Pomacentrus amboinensis**	Small shoal	None	7	19	Gaps and cracks	?
*Amblyglyphidodon curacao**	Small shoal	None	8	150	Gaps and cracks	Apparently different
*Abudefduf sexfasciatus**	Small school	None	8	27	Gaps and cracks	Apparently different
Abudefduf vaigiensis	Small school	None	10	1	Gaps and cracks	Apparently different
Amblyglyphidodon leucogaster	Small school	None	1.5[+]	7	Gaps and cracks	?
*Stegastes nigricans**	Solitary[t,a]	None	10	134	Gaps and cracks	Same (male); Different (female)
*Stegastes lividus**	Solitary[t,a]	None	14	71	Gaps and cracks	Same (male); Different (female)
*Cheiloprion labiatus**	Solitary[t,a]	None	4	41	*Acropora* coral head	Almost same
*Hemiglyphidodon plagiometopon**	Solitary[t] (single)	None	15	10	Gaps near reef base	Same (male); Different (female)?

Table 2. contd....

Table 2. contd.

Species name	Group type	Sex change	TL (cm)	No.	Shelter or refuge	Feeding and spawning sites for adults
Dischistodus prosopotaenia *	Solitary[t] (single)	None	14	19	Gaps within rocks	Same (male); Different (female)
Chrysiptera biocellata	Solitary[t] (single)	None	6	1	Gaps and cracks	Same (male); Different (female)
Neoglyphidodon melas	Solitary (single)	None	13	3	Soft coral	Apparently different
Neoglyphidodon nigroris	Solitary (single)	None	4[+]	3	Gaps and cracks	?
Pomacentrus chrysurus *	Solitary[t] (single)	None	4	10	Gaps and cracks	Same (male); Different (female)
Pomacentrus adelus *	Solitary (single)	None	7	264	Gaps and cracks	Same (male); Different (female)
Pomacentrus bankanensis	Solitary (single)	None	4[+]	1	Gaps and cracks	?
Chrysiptera rex	Solitary (single)	None	4[+]	2	Gaps and cracks	?
Plectroglyphidodon dickii	Solitary (single)	None	4[+]	1	*Pocillopora; Acropora*	?

*: Core species on the 84 patch reefs (the others are casual species). TL: maximum total length (cm) of individuals found at the site, +: only juveniles were found, No.: average number of individuals found in six censuses on the patch reefs (see Hattori and Shibuno 2010), t: territorial, a: aggregational. Information on sex change is based on Fish Base. Other information is based on only personal observations (AH) on fish behavior at the study site.

the shallow back reef of a fringing coral reef (< 2.5 m in depth, Hattori and Shibuno 2010, 2013). While many of these species live around ubiquitous shelters (e.g., narrow gaps within rocks or rubble or cracks or coral heads on the reef) and have similar food sources (e.g., filamentous algae, zooplankton or benthic invertebrates), several species have particular food resources (e.g., polyps of *Acropora* coral) and/or particular refuges (e.g., staghorn coral heads or giant sea anemones).

Group types and shelter use

Damselfishes can be classified into three categories based on group structure (Lowe-McConnell 1987): (1) solitary (single or aggregational), (2) gregarious (school or shoal making) and (3) the intermediate (social species). Individuals from solitary species, such as for example *Stegastes nigricans* and *S. lividus* (Table 2), maintain single permanent territories (i.e., all-purpose territories for males and feeding territories for females), protecting their shelter and/or food sources against conspecifics and other competitors (Hattori and Shibuno 2013). Although solitary species often form conspecific aggregations with each individual maintaining adjacent territories, they never form social groups that involve a hierarchy. In this chapter, we do not call the aggregation of individuals with contiguous territories a "colony" because these aggregations do not seem to have an obvious function or constitute a social unit as a whole, unlike nest colonies. While many solitary species living in ubiquitous shelters mainly feed on filamentous algae, some species can also use other living organisms for both food sources and shelter. For example, *Cheiloprion labiatus* uses *Acropora* branching coral and *Neoglyphidodon melas* uses soft coral (Allen 1991, Myers 1991).

Gregarious species form shoals or schools. Shoals are defined as aggregations of individuals that can swim in any direction; individual territories are not maintained, and these species have no dominance hierarchy. For example, *Amblyglyphidodon curacao* and *Chrysiptera cyanea* form shoals. Schooling species form groups that stay in a home range within and around ubiquitous shelters, such as cracks and coral heads on reefs. These fish hover and swim synchronously (Lowe-McConnell 1987). These schools do not show a dominance hierarchy. This characteristic allows the number of individuals within a school to grow, as shown, for example, by *Chromis viridis* which often form very large schools (Lecchini et al. 2006). Thus, gregarious species live around ubiquitous shelters usually feeding on zooplankton (Allen 1991). In gregarious species, group size is not determined by shelter size. Large group size may function as security against predators.

The intermediate species often form small groups with a size-ordered dominance hierarchy, when they use spatially discrete particular refuges such as branched coral heads or giant sea anemones. These social species sometimes defend their group territories against other conspecifics, when they use continuous refuges (see below). When group size becomes very large, the size-ordered hierarchy disappears. For example, *Dascyllus aruanus* and *Amphiprion frenatus* are social species inhabiting a particular refuge, with group size dependent on refuge size.

Group types, mating systems and sex change strategies

Spawning and parental care behavior is similar among species of damselfishes (Allen 1991, see Chapter III). During the reproductive period, adult males defend nesting sites on substrata. Simultaneously, they can also perform specific courtship displays to adult females. Adult females lay demersal eggs on the nests of the selected mates who fertilize, protect and ventilate the egg masses. Exceptionally, however, the species of a monotypic genus, *Acanthochromis polyacanthus*, show prolonged parental care behavior after hatching, defending the fly against predators (Robertson 1973).

Mating systems of damselfishes can be classified into four categories (Thresher 1984): (1) monogamy, (2) harems (resource defense polygyny), (3) lek-like spawning aggregations and (4) promiscuity (Table 1). Because it is rather difficult to follow the reproductive behavior of individually recognized fish, only a few studies have confirmed whether their mating is polygynous, random, or size-assortative in the last two systems.

Gregarious species basically form lek-like spawning aggregations during the reproductive period (Thresher 1984), in which each adult male temporally defends a mating and nesting site around the original shelter or on a different substratum, and adult females visit them to *select* mates. Consequently, females usually reproduce with two or more males, and single males usually fertilize the eggs of two or more females (Ochi 1986). In lek-like spawning aggregations, larger males and females seem to enjoy higher reproductive success because larger females have higher fecundity and they do not prefer to small males' nesting sites (Ochi 1985). Accordingly, sex change strategies have never been found in these species, as predicted by the SAH.

Promiscuous mating systems are present in solitary species, including those with aggregations such as *Stegastes nigricans* and those without aggregations such as *Dischistodus prosopotaenia*. Each male has an all-purpose territory and courts any female that temporally goes out of her territory and passes by his territory and each female spawns with several males. Even if their territories are contiguous to each other being confined to a very small patch reef, they do not show harem polygyny (Karino and Nakazono 1993). Because many solitary species use algae as food sources, each individual (irrespective of sex) may need a single non-overlapped large feeding territory, preventing individuals from forming harems. Thus, dominance hierarchies have not been found in these species. Accordingly, growth of individuals is likely not influenced by social contexts. Food availability or productivity within feeding territory may be an important determinant of an individual's reproductive value. Furthermore, surprisingly, female mate choice in "promiscuous" damselfishes has often been observed. Because low survival rates of the demersal eggs guarded by males would largely decrease the reproductive success of females and males (Knapp and Kovach 1991), females usually prefer males with a large body size (e.g., Schmale 1981) or vigorous courting behavior that indicate high ability of male parental care (e.g., Knapp and Warner 1991). In addition, larger females have higher fecundity. Therefore, sex changers have never been found in these species, as predicted by the SAH. The term "promiscuity" should be replaced by "polygynandry" (with non-random mating systems). Promiscuity must be characterized by random mating systems, in which protandrous sex change is usually favored according to SAH (Ghiselin 1969, Warner 1975, 1988a,b, Charnov 1982). Because protogynous and protandrous sex-changing damselfishes are not found in lek-like spawning aggregations and promiscuous mating systems, respectively, size-assortative mating would be observed more frequently than expected in these systems.

In contrast, mating systems of the social species differ considerably from those of other damselfish species. They live in unique refuges, such as *Dascyllus* in branched coral heads (with complicated hard structure) and *Amphiprion* and *Premnas* in giant sea anemones (without complicated hard structure). Because of this obligate anthozoan-dwelling nature, the survival rates of these species must be higher than those of other damselfish species (Karplus 2014). By inhabiting isolated refuges, these fish are confined to a narrow space, allowing the growth rates of subordinate individuals to be influenced by the dominant individuals. Thus, by dwelling within a small refuge, these fish can form small groups with a dominance hierarchy: feeding and breeding sites are restricted within narrow limits.

Many species living in small groups have the environmental potential for polygyny (EPP, Emlen and Oring 1977): the dominant individual in the group can monopolize mating opportunities. In a social species, such as *Dascyllus aruanus*, small groups show harem polygyny with protogynous sex change (Fig. 1), although temporal monogamy

Fig. 1. Small social group of *Dascyllus aruanus*, which inhabit particular refuge (Staghorn coral head).

will be displayed in very small refuges (Fricke and Holzberg 1974, Fricke 1977). When living in larger groups, however, these same species have no EPP. *Dascyllus aruanus* shows multi-male-multi-female polygamy without sex change (see below). However, small groups of another social species, *Amphiprion clarkii* do not show polygyny; rather, they are basically monogamous and show protandrous sex change. While living in habitats containing a high density of host sea anemones, monogamous pairs of *A. clarkii* can sometimes aggregate with temporal polyandry and few sex changes can be observed (Yanagisawa and Ochi 1986, Ochi 1989a,b, Hattori and Yanagisawa 1991).

We have to make a special reference to the unique damselfish, *Acanthochromis polyacanthus*. It seems to be a gregarious species but exceptionally does not form lek-like spawning aggregations during the reproductive period: adult males and females form breeding pairs defending small caves of dead corals for spawning and nesting sites in order to keep the pair territories, while they live in large groups during the non-reproductive period (Robertson 1973). This species shows neither social group nor sex change. This fact clearly indicates that monogamy is not directly related to sex change ability. Social groups with size-ordered hierarchy and the degree of body size differences between adjacently ranked group members are closely related to protogynous and protandrous sex change strategies, as shown below. Table 3 summarizes the relationship between group type, mating systems and sex change with reference to habitat type.

Table 3. Relationship between group type, mating systems and sex change in damselfishes. Habitat type (shelter or refuge) is also shown.

Group type	Mating systems	Sex change	Habitat type
Solitary or aggregational	Polygynandrous (with female mate choice), close to size-assortative mating	None	Ubiquitous shelter
Gregarious	Lek-like spawning aggregations (with female mate choice), close to size-assortative mating	None	Ubiquitous shelter; special sites for spawning and nesting
	Monogamy (only *Acanthochromis polyacanthus*)	None	Ubiquitous shelter; small caves for spawning and nesting sites
The intermediate (Social)	Harem polygyny	♀→♂	Unique refuge (with complicated hard structure)
	Monogamy	♂→♀	Unique refuge (without complicated hard structure)

Sex Change Strategies in Small Social Groups of *Dascyllus*, *Premnas* and *Amphiprion* spp.

The flexible protogynous strategy of Dascyllus *includes pre-maturational sex changers and non-sex changers*

The genus *Dascyllus* includes 10 mainly zooplanktivorous species, although it was recently shown that small species can also actively feed on small benthic crustaceans (Frédérich et al. 2010). Based on differences in morphology, biogeography and body coloration, they have been grouped into three species complexes: the aruanus, reticulatus and trimaculatus (Godwin 1995, Bernardi and Crane 1999, McCafferty et al. 2002). Individuals from the first two complexes are small-bodied (except for one large *D. flavicaudus* in the reticulatus complex), dwell in and around live branching coral colonies, and often form small social groups. Phylogenetic studies revealed that the ancestral *Dascyllus* species was small-bodied and closely associated with branched corals, and that the trimaculatus complex evolved recently, being less associated with corals (Bernardi and Crane 1999, McCafferty et al. 2002, Frédérich and Sheets 2010). The aruanus and reticulatus complexes contain protogynous species (Godwin 1995, McCafferty et al. 2002). For example, *Dascyllus aruanus*, the most studied species in the genus, shows a good correspondence between mating systems and sexuality pattern, as predicted by the SAH. In areas of scattered small branching corals, this species forms small, spatially discrete groups, each of which consists of one dominant adult male and several small adult females with a size-ordered hierarchy (Fricke 1977, Coates 1982). Sex change is socially controlled within group: disappearance of the male from the group causes the remaining largest female to undergo a sex change and become the dominant male (Fricke and Holzberg 1974, Coates 1982, Shpigel and Fishelson 1986). Juveniles may form pairs within very small corals, in which the larger individual is an adult male, not undergoing functional sex change, and the smaller becomes an adult female (Fricke and Holzberg 1974). When living in large branching coral colonies, in which a 1:1 functional sex ratio is found, this species often consists of primary males or pre-maturational sex changers (Table 1) and adult females (Cole 2002). The complicated branching structure of the coral may provide members with a type of screen, and dense coral cover maybe able to limit the amount of aggressive interactions from dominants to subordinates. In areas of continuous large coral colonies, *D. aruanus* shows *functional* gonochorism (Asoh 2003). Similar phenomena have also been observed in *D. melanurus*, which belongs to the aruanus complex (Asoh and Beaupre 2005) and in *D. flavicaudus*, which is the exceptionally large species in the reticulatus complex (Godwin 1995). The trimaculatus complex consists of large-bodied species that have higher mobility (Godwin 1995). As adults, each of these species often forms large feeding aggregates over ubiquitous shelters, like gregarious species, whereas the juveniles of these species always form groups within particular refuges (e.g., branching coral heads or giant anemones). This ontogenetic habitat shift occurs after they grow up (Booth 1995). All species of the complex are considered to be *functionally* gonochoristic (Godwin 1995, Bernardi and Crane 1999, McCafferty et al. 2002). In *D. albisella*, which belongs to the trimaculatus complex, monogamous and occasional polygynandrous mating systems (Table 1) without sex changers are described (Oliver and Lobel 2013).

In haremic species, the male's reproductive value increases according to its capacity for defending resources and sheltering sites required by the females, whereas the female's reproductive value increases with body size (Munday et al. 2006). The size of a group would increase as the size of its refuge increases, if group members can coexist without large body size differences between adjacently ranked individuals. A switch to multi-male polygyny (Table 1), as in *D. marginatus* (Fricke 1980), may be due to the increased number of individuals in the refuge, since the refuge may be too large to be defended by a single male. Experiments on the basslet *Pseudanthias squamipinnis* under field conditions indicate that the females of protogynous species can be induced to change sex by recruitment of a threshold number of additional adult females into the group, even in the presence of the resident male (Shapiro and Lubbock 1980). Multi-male polygyny has also been induced in the saddleback wrasses *Thalassoma duperrey* by field experiments (Ross et al. 1990).

Flexible protandrous strategy of anemonefishes, including pre-maturational sex changers and non-sex changers

Species in the genera *Amphiprion* and *Premnas* are usually called anemonefish (Chapter XII). *Premnas* includes only one species (*Premnas biaculeatus*) and *Amphiprion* consists of 29 spp., which are zooplanktivorous or omnivorous species (Allen 1975, Fautin and Allen 1992, Allen et al. 2008, 2010, Chapter XII). All of them are symbiotically associated with giant sea anemones. All these species are confirmed or inferred to have protandry with a monogamous mating system (Fig. 2; Allen 1975, Fautin and Allen 1992). These fishes inhabit isolated single or aggregated host sea anemones, which are essential resources for their refuge. Their hosts comprise of only 10 species in the field. Although each of the 10 anemone species has a single form,

Fig. 2. Small social group of *Amphiprion ocellaris*, which inhabit particular refuges (giant sea anemone).

two of these host species, *Entacmaea quadricolor* and *Heteractis magnifica*, often appear in aggregational form (Allen 1975, Fautin and Allen 1992). A breeding pair of anemonefish needs a sufficiently large host, allowing the female to lay a clutch of eggs on the hard substratum near the edge of the oral disc of the anemone, and the male takes care of the demersal eggs (Fautin and Allen 1992, Arvedlund et al. 2000).

Within a single isolated host, small groups of anemonefish form a size-ordered hierarchy (Fig. 2; Allen 1975, Fautin and Allen 1992). Socially controlled sex change within an isolated small group is well studied in three species: *Amphiprion percula* living in *H. magnifica* (Buston 2003a,b, 2004a,b), *A. ocellaris* living in *Stichodactyla gigantea* (Mitchell 2003, 2005, Hattori 2012) and *A. frenatus* living in *E. quadricolor* (Hattori 1991, 2005). Each small social group consists of an adult female, an adult male (with the female being much larger than the male) and zero to four non-breeders (i.e., subadults and juveniles). The female disappearance induces the male to change sex and the third-ranked (subadult) fish becomes the breeder male. When the adult male disappears, the third-ranked fish inherits the male breeding position. Some non-breeders may form groups on a very small host; when the host grows sufficiently large, the largest fish becomes female (i.e., without sex change) and the second member becomes male. Similar patterns of socially controlled sex change in small groups have been described in *Amphiprion bicinctus* (Fricke and Fricke 1977) and *Premnas biaculeatus* (Wood 1981, 1987), although information on host distributions is unavailable, and also in case of *A. melanopus* in isolated small aggregations of *E. quadricolor* (Ross 1978a,b), although no information on group isolation is available.

According to the SAH, protandrous sex changers are expected to occur in species having mating systems in which the expected reproductive value of males is less sensitive to their body size, whereas that of females increases with increasing body size (Ghiselin 1969, Charnov 1982, Warner 1988a,b). In random mating systems, pairing would occur randomly with respect to body size, so large males would not have a mating advantage. Pair formation in anemonefishes was often thought to occur randomly with respect to body size (Fricke and Fricke 1977, Kuwamura and Nakashima 1998). Larval settlement on a sparsely distributed host anemone was regarded as occurring randomly, with post-settlement movements between isolated single hosts thought to be very difficult (Allen 1975, Fricke and Fricke 1977). Moreover, due to the small carrying capacity of the anemone, only two adults were able to inhabit a single host (Fricke and Fricke 1977, Fautin and Allen 1992, Hattori and Yamamura 1995, Kuwamura and Nakashima 1998). However, recent studies have demonstrated that larval settlement onto a single host is not random (Elliott et al. 1995, Schmitt and Holbrook 1999, Elliott and Mariscal 2001, Buston 2003b, Hattori 2005, Mitchell 2005) and individuals often move between hosts (>40 m) in order to obtain larger mates or hosts if the opportunity arises (Hattori 1994, 2005, Hirose 1995). It was suggested that protandry in anemonefishes secures mates without the risk of adult movement between sparse hosts and ensures that the larger member of a breeding pair is always female, thus enhancing the fecundity of both the members of a pair (Fricke and Fricke 1977, Kuwamura and Nakashima 1998, Munday et al. 2006). However, this does not explain why anemonefishes are monogamous, since species living in small groups usually have EPP and in anemonefishes third-ranked fish within a group are sometimes larger than the smallest breeders in other groups (Hattori 1995, Buston 2004a,b, Mitchell 2005).

The summed body lengths (or total biomass) of a group of anemonefish are well known to be determined by the size of the host anemone (Fig. 3; Allen 1975, Ross 1978a, Hattori 1991, 2005, Mitchell and Dill 2005, Buston 2003a). Recent studies have revealed that large body size differences (or ratios) between group members adjacent in rank are constant within a population, which is necessary for subordinates to avoid *fatal eviction* from the refuge when single hosts are isolated (Buston 2003a,b, Hattori 2012). Accordingly, given the summed body lengths, a constant body size difference (or ratio) would determine the body size composition of the group members (Hattori 2012). In other words, the body size difference between the pair members would be

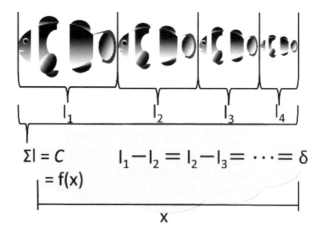

$$\Sigma l = C \qquad l_1 - l_2 = l_2 - l_3 = \cdots = \delta$$
$$= f(x)$$

Fig. 3. Summed body length of group members in *Amphiprion* can be considered to be a function of host anemone size (x). Thus, Σl is the constant for a given host size (x). In addition, body size differences between group members adjacent in rank are considered constant within a population.

predictable where the anemonefish live in an isolated single host, of which the size was given. This suggests that pair formation does not occur randomly with respect to body size.

In the limited shelter space of a host anemone, the growth of the dominant fish retards the growth of subordinate fishes (Allen 1975, Hattori 1991, 1995, Buston 2003b). Indeed, the dominant fish can use most of the resources, such as food and space (Allen 1975, Buston 2003b). In a social system where limited shelter space determines the summed body length of group members, the expected reproductive success via male function will not increase with increasing body size because the growth of a male retards the growth of females and the reproductive success of the male depends on the summed body length of females in the group (Hattori 2005). In contrast, the expected reproductive success via female function will increase with increasing body size (Hattori 2012). If anemonefish lived in haremic groups with dominant males, the growth of males would retard the growth of females. Thus, the total fecundity of all the small females within a group may be lower than the potential female fecundity of the dominant fish. Moreover, the fecundity of females may increase exponentially with increase in body size. The relationship between female fecundity and body size directly influences the fecundity of the dominant fish in haremic groups (Muñoz and Warner 2004). Although no study has yet assessed the relationship between female fecundity and gonad weight in anemonefishes, gonad weight is linearly associated with female fecundity in some fishes (Rhodes and Sadovy 2002, Sivakumaran et al. 2003). If the fecundity of an anemonefish female is linearly proportional to its gonad weight, then female fecundity will increase exponentially with body size (Fig. 4; Hattori 2012). When the dominant fish is female, accordingly, this large female can

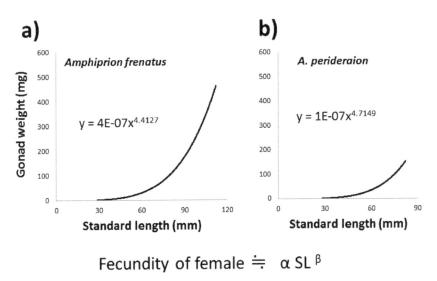

Fig. 4. Gonad weight increases exponentially with body size in *Amphiprion*: (a) *A. frenatus*, (b) *A. perideraion*. Fecundity of female, accordingly, can be assumed to increase exponentially with body size.

enjoy high fecundity. If the dominant fish was male, however, the total fecundity of small females might not be high because his growth retards the growth of females in a limited shelter space. Thus, the combination of exponential female fecundity functions and large body size differences between group members adjacent in rank can explain the function of protandry and monogamy under limited shelter space (Hattori 2012). Moreover, larger and older females, in addition to being more fecund, often produce high-quality offspring that grow faster. For example, the larger size of *Amphiprion chrysopterus* females was associated with a disproportionate increase in population replenishment (Beldade et al. 2012). Thus, when space is limited and the resulting group size is small, one large female would have the highest reproductive value in this small group meaning that a protandrous strategy with a monogamous mating system should be expected. The body size composition model with non linear female fecundity does not contradict the SAH and explains the function of monogamy and protandry when anemonefishes form small groups within small refuges.

On aggregational hosts, where shelter space is *not* limited, adult pairs hold territories (including several hosts) that are almost contiguous with each other, with subadults and juveniles usually having home ranges on the outskirts of or in the interstices between the pairs' territories in order to avoid aggressive behavior from adult pairs (Fricke 1979, Hattori and Yanagisawa 1991, Kobayashi and Hattori 2006). *Amphiprion akallopisos* living within the aggregational form of *H. magnifica* (Fricke 1979), *A. frenatus* living within the aggregational form of *E. quadricolor* (Kobayashi and Hattori 2006) and *A. clarkii* living within the densely distributed single form of *E. quadricolor* (Moyer 1980, Ochi 1989a,b, Hattori and Yanagisawa 1991) are typical examples. After the disappearance of an adult female from a breeding pair, the remaining mate rarely changes sex because a neighboring adult female usually immigrates to pair with the male in the habitats of high host density. Similar phenomenon is observed in *A. melanopus*, although no information on host anemone density is available (Godwin 1994a). In *A. clarkii*, the subsequently vacant female breeding post is frequently occupied by one of the subadults, which lives solitarily preparing to be a female. This type of subadult, which are named as subadult females, can be distinguished from other subadults based on the caudal fin coloration and behavior (Hattori and Yanagisawa 1991, Hattori and Yamamura 1995). When an adult male disappears, one of the subadults (except the subadult females) becomes the male breeder. This type of subadults, which are named as subadult males, may pair with each other and, when their host grows sufficiently large, the larger member becomes an adult female through the subadult female phase. Because the individuals, except for the small juveniles, are *not* confined to single hosts, these individuals can select mates with pair formation occurring in a size-assortative manner (Ochi 1989a,b, Hattori and Yanagisawa 1991, Hattori and Yamamura 1995). Similar phenomenon is described for *A. frenatus* in a habitat of high host density (Kobayashi and Hattori 2006). If they were polygynous, the fecundity of the male may not be high due to the high cost of territorial defense. One example is the monogamous longnose filefish *Oxymonacanthus longirostris*, in which the males defend territories so as not to decrease the females' feeding rates and fecundity (Kokita and Nakazono 1999). Polygynous territories are too large to be defended by males and consequently the cost of defense does not pay for the polygynous males. This explanation is applicable to anemonefishes in habitats of high host density.

Sex Change Mechanism and Gonad Development in Different Group Structure

The process of sex inversion involves a reorganization of the reproductive system, including replacement of gonadal cell types, reorganization of duct systems and behavioral changes. Changes can begin immediately upon a shift in the social status of an individual, and, depending on conditions, can be completed quickly in a few weeks, as it has be shown in *A. melanopus* (Godwin 1994a). The following section describes the most remarkable features of these gonad changes.

Sex changers, pre-maturational sex changers and non-sex changers of protogynous Dascyllus

Some species of *Dascyllus* have been reported to be non sex-changers, since protogyny might be demonstrated only via observation of individuals undergoing sex transition. In fact, in a study of *D. flavicaudus* (Parmentier et al. 2010), which is the exceptionally large species in the reticulatus complex (Godwin 1995), we have not found any mixed stage gonad in our samples, so we would have supposed that it was a gonochoric species. Actually, *D. flavicaudus* was reported as a diandric species (i.e., males can develop directly from the juvenile phase as well as by sex change from functional females, Table 1) since secondary testes were already observed (Asoh 2004). Godwin (1995) had also observed one individual of *D. flavicaudus*, the largest of its group, with atretic oocytes together with proliferating spermatogenic tissue, a clue of a recent sex change.

The testes of primary males develop directly from the juvenile state, which had not yet developed an ovarian cavity. However, the testes of secondary males develop from gonads with an ovarian lumen (Shapiro and Rasotto 1993). Species certainly reported with this secondary structure are *D. aruanus* (Cole 2002, Asoh 2003), *D. reticulatus* (Schwarz and Smith 1990) and *D. carneus* (Asoh and Yoshikawa 2003).

Secondary testes have also been observed in *D. albisella* (Asoh et al. 2001) and *D. trimaculatus* (Asoh and Kasuya 2002), previously considered to be gonochoric species (Yogo 1987, Godwin 1995). In fact, they both show a protogynous pattern of gonadal development, although they are non-functional hermaphrodites in the current ecological context (Asoh and Kasuya 2002). In non-functional hermaphroditism, a ratio of the population may exhibit both testicular and ovarian tissues but only reproduce as male or female (Sadovy and Liu 2008). This is the case of the *trimaculatus* complex species, considered to be *functionally* gonochoristic (Godwin 1995, Bernardi and Crane 1999, McCafferty et al. 2002).

Asoh (2004) suggests that diandric protogyny may be a common trait of *Dascyllus* species, meaning that, in a given situation, some females can switch sex and become secondary males. The trigger for the change could be a variation on the number of females in the area, for example, or perhaps changes in the environmental conditions (such as the size of the available resource patch), as proposed by Loft (1991).

It is known that a single species may have different mating systems under different ecological conditions (Ross 1990). *Dascyllus marginatus*, for example, switches from multi-male polygyny, to harem polygyny to monogamy (Fricke 1980). In the same way, some flexibility in sex change would also increase the reproductive success of a mating group member. Thus, the onset of secondary males would be the simplest strategy to change from a single male harem polygyny to a multi-male polygyny.

Dascyllus *gonadal development*

According to Strüssmann and Nakamura (2002), the rudimentary gonads of teleosts undergo first a period of somatic growth with no apparent histological differentiation. During the next period, the development of cellular or histological characteristics, such as the onset of intensive germ cell proliferation, signal the beginning of sex differentiation into ovaries or testes in some species. However, for other species, it marks the beginning of an all-female or intersex phase prior to the terminal differentiation into ovaries and testes.

Rudimentary or undifferentiated gonads of *Dascyllus* contain only primordial germ cells and somatic cells, and have no ovarian lumen, medial sperm ducts or sperm sinuses (Asoh and Yoshikawa 2003, Asoh and Kasuya 2002).

Primary testes develop from rudimentary gonads without a cavity or lumen and, in some species, it has been observed that the sperm ducts run as simple tubes in the medial hilar regions of testes throughout their length (Asoh 2003).

The first ovarian stage, which also develops from rudimentary gonads, already shows an ovarian lumen, and contains only primordial germs cells and somatic cells (Asoh and Kasuya 2002). The second stage begins with the development of at least one oocyte. Finally, the mature ovary contains oocytes from the cortical-alveolus stage to different vitellogenesis stages, and does not have testicular tissue. In this kind of asynchronous development of secondary growth follicles (Wallace and Selman 1981), oocytes at all stages of maturation are present, without a dominant population (Fig. 5a), and batches of eggs develop synchronously for each spawning event.

Female-to-male sex change of protogynous Dascyllus

In some species, the transition from female to male seems to occur both before and after the individuals have developed cortical-alveolus oocyte stage in their gonads, the former being pre-maturational sex changers. Though, in *D. aruanus* (Asoh 2003), developing spermatogenic tissue occurred only in gonads having pre-cortical alveolus stage oocytes suggesting that functional female-to-male sex change is rare or absent in this population of *D. aruanus*.

According to Asoh and Kasuya (2002), during the early reversal stage of the non-functionally protogynous *D. trimaculatus* gonads, degenerating oocytes in the primary growth stage are simultaneously observed with one or more clusters of spermatocytes, spermatids or spermatozoa. Oocytes at the mid-reversal stage, both in and beyond the primary growth, degenerate. Spermatogonia proliferate and spermatogenic cysts with

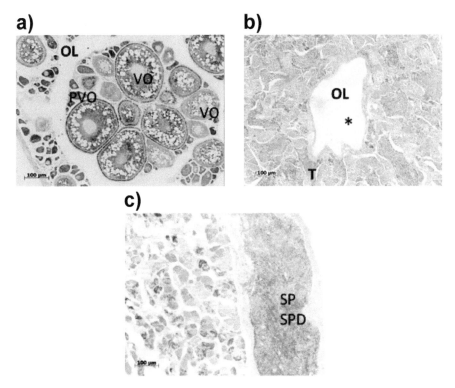

Fig. 5. *Dascyllus flavicaudus*: (a) Oocytes at different stages of development. VO: vitellogenic oocytes; PVO: previtellogenic oocytes; (b) Testicular tissue (T) showing different stages of spermatogenesis and a possible remnant of the ovarian lumen (*); (c) Male gonad with the spermduct (SPD) full of spermatozoa (SP).

cells in early stages of development predominate. They are distributed throughout the entire gonad amidst the stromal-like cells. Finally, at late reversal stage, the gonad testis of *D. trimaculatus* consists entirely of spermatogenic tissue, except for remnants of the prior ovarian phase (oocytes and the ovarian lumen) (Asoh and Kasuya 2002).

Dimorphism in sperm duct configuration, as a criterion to distinguish between primary and secondary males (Reinboth 1962, 1970), is not clear in the genera *Dascyllus*. According to Shapiro and Rasotto (1993), in the testes of the secondary males of *Thalassoma bifasciatum* (with a lumen), the sperm duct runs within the testis wall in the form of sperm sinuses. On the other hand, the sperm ducts of primary males seem to run as simple tubes in the medial region of the gonads (Asoh 2003). The testes of the diandric protogynous species *D. flavicaudus* develop from the ovaries and exhibit an ovarian lumen (Asoh 2004). We have also seen a probable remnant of the ovarian lumen in some *D. flavicaudus* testes (Fig. 5b).

Sex changers, pre-maturational sex changers and non-sex changers of protandrous Amphiprion

In protandric species, the development of male cells is both preceded and followed by ovarian development (Shapiro 1992). It would mean that female development is primary and that male development is a temporary phase initiated by a masculinizing mechanism and terminated by its cessation. This primary ovarian differentiation has been observed at least in *A. clarkii, A. frenatus, A. perideraion, A. melanopus* and *A. akallopisos* (Miura et al. 2003, Hattori and Yanagisawa 1991, Hattori 1991, 1994, 2000, Godwin 1994b, Casadevall et al. 2010). In these species, immature gonads show predominant ovarian tissue with primary (early and perinucleolus) oocytes, while spermatogenic germ cells are not distinguishable. After the immature stage, ambisexual gonads appeared in juveniles: oocytes appear early and are a prominent part of the gonad in both juveniles and functional males (Elofsson et al. 1997). Francis (1992) suggests that female development is the "default" pattern for teleosts, with gonads developing along a female pathway unless male development is somehow induced.

In *A. polymnus*, the first sexually differentiated sex cells were hermaphrodite, having clones both of oogonia and spermatogonia (Rattanayuvakorn et al. 2006). Nonetheless, in the very young gonads of *A. ocellaris*, we observed a bigger proportion of ovarian tissue (M.C. unpublished data; Fig. 6a).

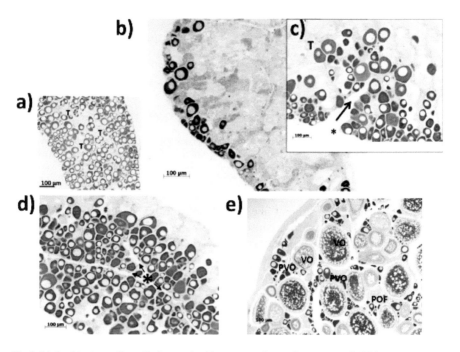

Fig. 6. (a) *Amphiprion ocellaris.* Early gonad, with not many clones of spermatogonia (T) between oocytes; (b) *A. akallopisos* male gonad. Gonad consisting mainly of testicular tissue, with some primary growth phase oocytes at the periphery; (c) *A. akallopisos.* Early reversal stage. Invagination of the ovarian tissue (*) in lateral view; (d) *A. akallopisos.* Early reversal stage. Invagination of the ovarian tissue (*) in frontal view; (e) *A. akallopisos* female gonad. Oocytes in different stages of development, previtellogenic (PVO) and vitellogenic as well (VO). Post ovulatory follicles are also observed (POF).

Amphiprion *gonadal development*

Juveniles are non-breeders with ambisexual gonads. Abol-Munafi et al. (2011) observed that non-breeders of *A. ocellaris* possess an intermixed ovotestis without boundaries between the testicular and ovarian regions. According to Miura et al. (2003, 2008), the ovarian development of *A. clarkii* begins two months after hatching, but testicular tissue does not begin its differentiation until the fifth month.

In *A. akallopisos*, the subadult male gonads show different stages of development of male germinal cells, including some spermatozoa (but not in the seminiferous tubules or in the sperm duct) and primary growth phase oocytes (Casadevall et al. 2010). In those males, testicular tissue gradually dominates the gonad and there is no ovarian cavity (Fig. 6b). The same has been observed in subadult males of *A. ocellaris* (M.C. unpublished data). In both species, the female region is peripheral whereas there is a larger proportion of mature testicular tissue with seminiferous tubules. In these tubules, cells in all the different stages of development are organized in cysts and show synchronous development. Upon completion of spermiogenesis, cysts open up and the spermatozoa are discharged into the lobular lumen, from where they head to the sperm duct. Only oocytes in the primary growth phase are observed in the ovarian tissue, peripherally located: oogonia, chromatin-nucleolar and perinucleolar stages. Oogenic activity of the ovotestis does not proceed beyond the perinucleolus stage either in *A. polymnus* (Rattanayuvakorn et al. 2006), *A. frenatus* (Bruslé-Sicard and Reinboth 1990, Hattori 1991), *A. clarkii* (Hattori and Yanagisawa 1991, Hattori 1994) and *A. perideraion* (Hattori 2000).

The transition to a functional male involves proliferation of spermatogenic tissue. The male tissue is composed of sexual cells in all the different stages and includes large amounts of spermatozoa. *Amphiprion akallopisos* testicular tissue has a remarkable increase in size and the spermiduct is full of ripe sperm, with clearly visible blood vessels close to this area (Casadevall et al. 2010).

The central gonadal lumen or ovarian cavity is not observed in ambisexual or male gonads, neither in *A. akallopisos* nor in *A. ocellaris*.

Male-to-female sex change of protandrous Amphiprion

In some species, sex differentiation follows a gradient from the anterior to the posterior areas of the gonads regardless of sex (Strüssmann and Ito 2005). Sex reversal in *A. akallopisos* and *A. ocellaris* also occurs in a cephalocaudal gradient (Casadevall et al. 2010, Abol-Munafi et al. 2011), with the formation of an ovarian cavity or lumen. The cavity formation has also been related to the sex reversal process in *A. frenatus* (Hattori 1991), *A. perideraion* (Hattori 2000), and *A. clarkii* (Hattori and Yanagisawa 1991).

Miura et al. (2003, 2008) observed that the initial ovarian cavity formation was indicated by the presence of two elongated aggregations of somatic cells in the basal and distal portions of the gonads. However, in *A. akallopisos*, the gonad lumen formation (Fig. 6c,d) appears to be due to an invagination of the ovarian tissue (Casadevall et al. 2010), in a very similar manner to the process described by Godwin (1994b) for *A. melanopus*; the spermatogenic tissue being rejected to the periphery of the gonad. The whole process of sex reversal in *A. akallopisos* (transverse section) is illustrated in Fig. 7.

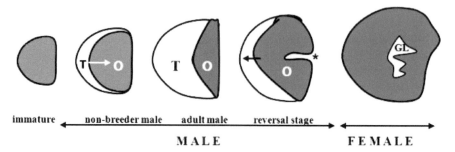

Fig. 7. *Amphiprion akallopisos.* Process of sex reversal from immature stage to female stage (in a transverse section). GL indicates the gonadal lumen formation from the ovarian tissue invagination.

During the lumen development of the latter, the ovarian tissue is mainly composed of oocytes in the primary growth phase (oogonia, chromatin-nucleolar and perinucleolar stages) but an additional stage of cortical alveoli of some oocytes, with lipid drops in their cytoplasm, is observed, indicating that vitellogenesis process is ready to start. It is possible to observe all the spermatogenic stages (from spermatogonia to spermatozoa) in the testicular tissue.

The mid-reversal stage ends with the complete delimitation of the gonad lumen. At the late reversal stage of *A. akallopisos*, the degeneration of testicular tissue is observed. Simultaneously, the ovarian tissue develops numerous oocytes in the first cortical alveoli development stage.

When oocytes begin to mature, the testicular parts of the intersexual gonads disappear in *A. akallopisos*. This was also observed in *A. clarkii*, *A. polymnus*, *A. perideraion*, *A. sandaracinos* and *A. ocellaris* by Moyer and Nakazono (1978).

In the mature female gonad (Fig. 6e), ovarian tissue is well developed around the straight luminal spaces. All the oocyte stages are found: chromatin-nucleolar, perinucleolar, cortical alveoli, vitellogenic stages, mature and atretic oocytes. The ovarian development is clearly asynchronous, with oocytes of all stages, without dominant populations (Wallace and Selman 1981).

Postovulatory follicles (POF) were also observed in active females, indicating a recent spawning event (Casadevall et al. 2010). The spawning periodicity is variable in clownfishes, but it has been related to lunar cycles in some tropical species. In *A. melanopus*, for instance, spawning peaks coincide approximately with the first and third quarters of the moon (Ross 1978a). Although the relationship between spawning and lunar cycle is not clear for all the species (Gordon and Bok 2001), it means that a period of *ca.* 15 days minimum is required between the development of the clusters of cortical alveoli oocytes to the next spawning event.

Neuroendocrine Aspects and Social Control of Sex Change

Signals used by anemonefish for social control of sex change are diverse. Such signals could be attributed to behavioral interaction between sexes, relative size, sexual proportion, and possible stimuli such as pheromones, visual cues or sound communication (Loft 1991, Guerrero-Estévez and Moreno-Mendoza 2010, Fricke 1980, Fricke and Fricke 1977, Colleye et al. 2009, Colleye and Parmentier 2012).

In *Amphiprion* species, behavioral interaction between sexes seems to be stronger and more hierarchical. According to Ross (1990), protandrous suppression refers to suppression of sex change in protandrous species, and occurs when one or more members of a social group prevent sex change in a candidate fish, normally through aggressive dominance. Sex change is initiated or completed only when the suppression condition is removed (disinhibition), normally when the single dominant terminal-sex individual dies or vacates the social group. This is the case of *Amphiprion* species, and induction by other specific stimuli does not seem necessary.

In *A. ocellaris*, sex differentiation of the upper-ranked individuals is gradually determined by long term social interactions (Iwata et al. 2008). However, Fricke and Fricke (1977) had observed that sex reversal in *A. akallopisos* also occurred in the absence of intragroup social pressure. In all those cases, females were not present, so that they concluded that females control the emergence of other females. It is also possible that some non-breeders become females without passing through a functional male state when the opportunity arises, as in *A. clarkii* (Ochi and Yanagisawa 1987, Ochi 1989b, Hattori and Yanagisawa 1991) and *A. frenatus* (Hattori 1991).

Actually, behavioral sex change can be faster than gonadal changes, and sometimes occurs between minutes and hours (Robertson 1972, Warner and Swearer 1991). According to Francis (1992), for species with environmentally labile sex determination that respond to social cues, the brain is the likely site of recognition of social events, and hypothalamic GnRH (gonadotropin releasing hormone) is the key regulator. As stated by Godwin (2009), social control of sex change in fishes must be initiated through changes in neural signalling, but the understanding of the mechanisms is still very basic.

In a recent study of *A. melanopus*, findings by Kim et al. (2012) support the hypothesis that GnRH plays an important role in the regulation of gonadal development and its sex change. It is suggested that GnRH peptides are probably involved in the regulation of gonadal function as autocrine or paracrine regulators.

Hypothalamic GnRH neurons are the control centre of the hypothalamo- pituitary-gonadal axis and reproductive function. In fact, studies of hermaphroditic fish suggest that brain GnRH and arginine vasotocin (AVT) are involved in the sex change process (Foran and Bass 1999, Godwin 2009, 2010). More specifically, AVT has been closely linked with the display of aggressive social dominance, either increasing or decreasing this behavior (Godwin 2010).

Munakata and Kobayashi (2010) propose a sexually bipotential brain for some teleost fishes: when a protogynous hermaphroditic fish is in a female phase, the female portion of the brain is active and the male portion is quiescent; at the time the individual starts to behave as a male, the male portion of the brain is activated and the female portion becomes quiescent.

According to Elofsson et al. (1997), the male sex is associated with greater numbers of preoptic GnRH hypothalamic cells; increase in GnRH cell number correlates with protogynous sex change and teleost sexual maturation, and the decrease of the GnRH cells with protandrous sex change.

Thus, the complete gonad sex change probably involves the brain GnRH, and pituitary together with gonadotropins gonad steroid hormones (androgens and estrogens). Indeed, anemonefishes show higher levels of estradiol in females, while 11-ketotestosterone levels are higher in males (Godwin and Thomas 1993).

However, despite recent insights, the molecular mechanism underlying gonadal sex change in hermaphroditic fish is not yet well understood.

Growth, Inter-group Movement and Sex Change Strategies in Relation to Types of Refuge

We have reviewed the sexuality patterns of the genera *Dascyllus*, *Amphiprion* and *Premnas*. Sex-changing species in damselfishes are all habitat-specialists. When inhabiting small discrete refuges, sex-changing species form small groups with a size-ordered dominance hierarchy. However, the body size composition of group members largely differs between *Dascyllus* and anemonefish (*Amphiprion* and *Premnas*) species.

In anemonefishes living in isolated small groups, body size differences (or ratios) between group members adjacent in rank are large and constant, groups are rather stable, and individual inter-group movements rarely occur. When their hosts are the single form, the home ranges of subordinates are often confined to single hosts. However, aggressive interactions among group members are rarely observed (Hattori 1991, 1995, 2000, 2005, Buston and Cant 2006). When they inhabit aggregational hosts or densely distributed single hosts, where inter-host movements frequently occur, aggressive interactions between group members (or neighbors) are intense and body size differences between males and females become small (Fricke and Fricke 1977, Ross 1978b, Fricke 1979, Hattori and Yanagisawa 1991). Godwin (1994a) reported highly aggressive interactions between the members of small groups in *A. melanopus* at Madang, Papua New Guinea, but information on host anemones are unavailable: these groups may be formed at isolated small aggregations of anemones and the body size differences are small, like at Guam (Ross 1978a). Thus, the large body size differences may minimize costly conflicts among individuals that live in the isolated small groups (Buston et al. 2007, Buston and Zink 2009). In *Dascyllus*, even those inhabiting discrete branching corals, the body size differences between adjacent ranked members are small or trivial, groups are unstable, and individual inter-group movements often occur (Fricke 1977, Asoh 2003). The large body size differences must be unnecessary for *Dascyllus* group members to coexist, even those inhabiting discrete branching coral heads. We speculate that positive or negative effects of the changes in neural signalling caused by visual cues on the production of GnRH may be a crucial determinant in the social control of sex change.

The differences between *Dascyllus* and anemonefish species may be due to differences in refuge structure (Fig. 8) and distribution pattern. Giant sea anemones, in which *Amphiprion* is present, have a tentacle-grown soft basin structure, in which

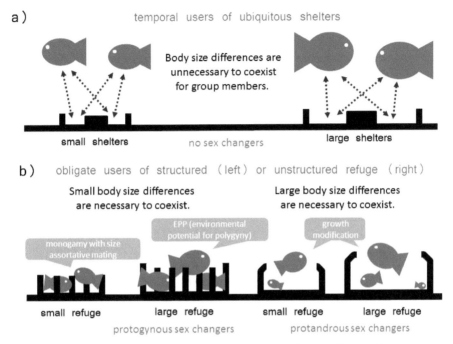

Fig. 8. Model of coexistence of group members in a damselfish species in relation to habitat use pattern. (a) Aggregational and gregarious species that temporally use ubiquitous shelters. Body size differences are unnecessary for group members to coexist in a site. (b) Social species that always use structured or unstructured refuges. Large body size differences are necessary for group members to coexist in an unstructured refuge, otherwise subordinates would receive fatal eviction by dominant fish. Small body size differences are necessary for group members to coexist in a structured refuge, although the complicated structure of refuge may provide the members with a type of screen, which may weaken the social stress by visual cues that may be able to retard the growth of subordinates.

individuals of similar body size could hardly coexist while interacting aggressively, as frequently seen in a small aquarium with open space. Consequently, subordinates adopt a growth modification strategy to prevent *fatal eviction* by dominant fish outside the anemone in the field (Buston 2003a,b). In short, because of the unstructured refuge, low average density of anemones and a high risk of mobility, subordinates must adopt a mating strategy that involves waiting for vacated breeding posts (Hattori and Yamamura 1995, Hattori 2000). In contrast, live branching corals have a 3-D hard framework structure. This may look like a small aquarium in which a structured object is centered. Subordinate individuals of a group can grow gradually in one side in order to achieve dominant status in a new group because the structured refuge, the high density of the refuge and their low risk of mobility (i.e., no fatal eviction by dominant fish) allow them to escape from social stress exerted by dominants in the field. Similar reproductive tactics are described in other protogynous fishes (Munday et al. 2006). The difference in body size composition of group members between *Dascyllus* and anemonefish species is attributable to the difference in their growth and inter-group movement strategies while using different types of refuges.

Conclusion and Prospects

Comparing sex changers and non-sex changers among damselfishes, we have found that sex changers are all obligate refuge-dwelling species, while the others are basically temporal users of ubiquitous shelters. In addition, we can conclude that frequent sex changes can be observed where social species utilize isolated small patches. Relatively large and/or structurally complicated refuges may be able to weaken the social stress among subordinates because excessive space or complicated hard structure would allow them to escape from behavioral stress that influences the social control of sex change (Fig. 8). Accordingly, large body size differences between adjacently ranked group members must be necessary when they are confined to small and/or unstructured refuges. Probably, this body size difference results from their growth modification strategies to coexist with dominants and to avoid fatal eviction. In a structured refuge, as an intermediate state, only small body size differences may be necessary.

Bidirectional sex change has been recently found in some protogynous coral reef fishes (e.g., Kuwamura et al. 1994, Munday 2002, Munday et al. 2006). In fact, some species and clades have evolved the capacity to change sex in either direction because their local, social or environmental conditions can change dramatically and unpredictably in a way that favors repeated changes of sex (Erisman et al. 2013). Among demersal spawners, bidirectional sex changers are always found in small habitat specialists, such as polygynous *Trimma* and monogamous *Gobiodon* and *Paragobiodon* gobies (Kuwamura and Nakashima 1998, Wong and Buston 2013, Karplus 2014). Accordingly, one may expect that polygynous *Dascyllus* and monogamous anemonefishes also have the ability of bidirectional sex change. Cole (2002) suggests the presence of bidirectional sex changers in *Dascyllus* based on observations of the gonad structure, and Kuwamura and Nakashima (1998) conducted several experiments to find out whether *Amphiprion* have this ability. However, it has not been confirmed in the damselfishes (Wong and Buston 2013, Karplus 2014). Both-ways sex changers may need very small "structured" refuges, where EPP is unclear and pairs of similar body sizes can be formed (Fig. 8). In the bidirectional sex-changing gobies, because of the smaller body size, the burden for males in parental care must be severer than in case of the social species of damselfishes and consequently subordinate females can grow faster than males, unlike in case of the damselfishes. In addition, they mate with the first partner they encounter in a discrete refuge avoiding further risky movements (Kuwamura and Nakashima 1998, Wong and Buston 2013, Karplus 2014). Accordingly, the alternative sex-changing strategies can be observed in the gobies under various social statuses of individuals.

For anemonefishes, however, the necessity of large body size differences between pair members for coexistence may prevent bidirectional sex change. Indeed, forcible pair formation of females with a small body size difference always causes the death or emigration of one partner (Fricke and Fricke 1977, Kuwamura and Nakashima 1998). After a sex change from male to female, the rest of the previous testicular structure in the gonads disappears in *Amphiprion* (Godwin 1994b, Casadevall et al. 2010, Abol-Munafi et al. 2011). Thus, the reverse sex change does not seem to be an option. Further studies are required to confirm whether *Dascyllus* have the ability of bidirectional sex change or not.

Acknowledgements

We are grateful to E. Parmentier, B. Frédérich, S.P. Clack, D. Price, and an anonymous reviewer, for their valuable advice regarding the manuscript. F. Iwasaki provided valuable information on references. This work was supported in part by a grant-in-aid for Science Research from the Japan Ministry of Education, Science, Culture, and Sports (No. 23570022).

References

Abol-Munafi, A.B., N.H. Lorazmi-Lokman, N.A. Asma, S. Sarmiza and N.Y. Abduh. 2011. Histological study of the protandrous anemonefish (*Amphiprion ocellaris*). J. Animal Vet. Adv. 10 (22): 3031–3036.
Allen, G.R. 1975. Anemonefishes: Their Classification and Biology, 2nd ed. TFH Publications, Neptune City, New Jersey.
Allen, G.R. 1991. Damselfishes of the World. Mergus Publisher, Melle, Germany.
Allen, G.R., J. Drew and L. Kaufman. 2008. *Amphiprion barberi*, a new species of anemonefish (Pomacentridae) from Fiji, Tonga, and Samoa. Aqua. 14: 105–114.
Allen, G.R., J. Drew and D. Fenner. 2010. *Amphiprion pacificus*, a new species of anemonefish (Pomacentridae) from Fiji, Tonga, Samoa, and Wallis Island. Aqua. 16: 129–138.
Arvedlund, M., I. Bundgaard and L.E. Nielsen. 2000. Host imprinting in anemonefishes (Pisces: Pomacentridae): does it dictate spawning site preferences? Environ. Biol. Fish. 58: 203–213.
Asoh, K. 2003. Gonadal development and infrequent sex change in a population of the humbug damselfish, *Dascyllus aruanus*, in continuous coral-cover habitat. Mar. Biol. 142: 1207–1218.
Asoh, K. 2004. Gonadal development in the coral reef damselfish *Dascyllus flavicaudus* from Moorea, French Polynesia. Mar. Biol. 146: 167–179.
Asoh, K. and M. Kasuya. 2002. Gonadal development and mode of sexuality in a coral-reef damselfish, *Dascyllus trimaculatus*. J. Zool. (Lond.) 256: 301–310.
Asoh, K. and T. Yoshikawa. 2003. Gonadal development and non-functional protogyny in the Indian damselfish, *Dascyllus carneus*. J. Zool. (Lond.) 260: 23–39.
Asoh, K. and S.J. Beaupre. 2005. Frequency of functional sex change in two populations of *Dascyllus melanurus* conforms to a prediction from sex allocation theory. Copeia 2005: 732–743.
Asoh, K., T. Yoshikawa and M. Kasuya. 2001. Gonadal development and non-functional protogyny in a coral reef damselfish *Dascyllus albisella* Gill. J. Fish Biol. 58: 1601–1616.
Beldade, R., S.J. Holbrook, R.J. Schmitt, S. Planes, D. Malone and G. Bernardi. 2012. Larger female fish contribute disproportionately more to self-replenishment. Proc. R. Soc. B-Biol. Sci. 279: 2116–2121.
Bernardi, G. and N.L. Crane. 1999. Molecular phylogeny of the humbug damselfishes inferred from mtDNA sequences. J. Fish Biol. 54: 1210–1217.
Booth, D.J. 1995. Juvenile groups in a coral reef damselfish: density-dependent effects on individual fitness and population demography. Ecology 76: 91–106.
Bruslé-Sicard, S. and R. Reinboth. 1990. Protandric hermaphrodite peculiarities in *Amphiprion frenatus* Brevoort (Teleostei, Pomacentridae). J. Fish Biol. 36: 383–390.
Buston, P.M. 2003a. Forcible eviction and prevention of recruitment in the clown anemonefish. Behav. Ecol. 14: 576–582.
Buston, P.M. 2003b. Size and growth modification in clownfish: sex change is not the only way these fish achieve dominance: they grow into the role. Nature 424: 145–146.
Buston, P.M. 2004a. Does the presence of non-breeders enhance the fitness of breeders? An experimental analysis in the clown anemonefish. Behav. Ecol. Sociobiol. 57: 23–31.
Buston, P.M. 2004b. Territory inheritance in the clown anemonefish. Proc. R. Soc. B-Biol. Sci. 271: S252–S254.
Buston, P.M. and M.A. Cant. 2006. A new perspective on size hierarchies in nature: patterns, causes, and consequences. Oecologia 149: 362–372.
Buston, P.M. and A.G. Zink. 2009. Reproductive skew and the evolution of conflict resolution: a synthesis of transactional and tug-of-war models. Behav. Ecol. 20: 672–684.

Buston, P.M., H.K. Reeve, M.A. Cant, S.L. Vehrencamp and S.T. Emlen. 2007. Reproductive skew and the evolution of group dissolution tactics: a synthesis of concession and restraint models. Anim. Behav. 74: 1643–1654.

Casadevall, M., E. Delgado, O. Colleye, S.B. Monserrat and E. Parmentier. 2010. Histological study of the sex-change in the Skunk Clownfish *Amphiprion akallopisos*. The Open Fish Science Journal 2: 25–31.

Charnov, E.L. 1982. The Theory of Sex Allocation. Princeton University Press, Princeton, New Jersey.

Coates, D. 1982. Some observations on the sexuality of humbug samselfish, *Dascyllus aruanus* (Pisces, Pomacentridae) in the field. Z. Tierpsychol. 59: 7–18.

Cole, K.S. 2002. Gonad morphology, sexual development, and colony composition in the obligate coral-dwelling damselfish *Dascyllus aruanus*. Mar. Biol. 140: 151–163.

Colleye, O. and E. Parmentier. 2012. Overview on the diversity of sounds produced by clownfishes (Pomacentridae): importance of acoustic signals in their peculiar way of life. PloS ONE 7(11): e49179.

Colleye, O., B. Frederich, P. Vandewalle, M. Casadevall and E. Parmentier. 2009. Agonistic sounds in the skunk clownfish *Amphiprion akallopisos*: size-related variation in acoustic features. J. Fish Biol. 75: 908–916.

Elliott, J.K. and R.N. Mariscal. 2001. Coexistence of nine anemonefish species: differential host and habitat utilization, size and recruitment. Mar. Biol. 138: 23–36.

Elliott, J.K., J.M. Elliott and R.N. Mariscal. 1995. Host selection, location, and association behaviors of anemonefishes in field settlement experiment. Mar. Biol. 122: 377–389.

Elofsson, U., S. Winberg and R.C. Francis. 1997. Number of preoptic GnRH-immunoreactive cells correlates with sexual phase in a protandrously hermaphroditic fish, the dusky anemonefish (*Amphiprion melanopus*). J. Comp. Physiol. A 181: 484–492.

Emlen, S.T. and L.W. Oring. 1977. Ecology, sexual selection and the evolution of mating systems. Science 197: 215–223.

Erisman, B.E., C.W. Petersen, P.A. Hastings and R.R. Warner. 2013. Phylogenetic perspectives on the evolution of functional hermaphroditism in teleost fishes. Integr. Comp. Biol. 53: 736–754.

Fautin, D.G. and G.R. Allen. 1992. Anemonefishes and their Host Sea Anemones. Western Australian Museum, Perth (Australia).

Foran, C.M. and A.H. Bass. 1999. Preoptic GnRH and AVT: axes for sexual plasticity in teleost fish. Gen. Comp. Endocrinol. 116: 141–152.

Francis, R.C. 1992. Sexual lability in teleosts: developmental factors. Q. Rev. Biol. 67: 1–18.

Frédérich, B. and H.D. Sheets. 2010. Evolution of ontogenetic allometry shaping giant species: a case study from the damselfish genus *Dascyllus* (Pomacentridae). Biol. J. Linn. Soc. 99: 99–117.

Frédérich, B., O. Lehanse, P. Vandewalle and G. Lepoint. 2010. Trophic niche width, shift, and specialization of *Dascyllus aruanus* in Toliara Lagoon, Madagascar. Copeia 2010: 218–226.

Fricke, H.W. 1977. Community structure, social organization and ecological requirements of coral reef fish (Pomacentridae). Helgolander wiss. Meeresunters 30: 412–426.

Fricke, H.W. 1979. Mating system, resource defence and sex change in the anemonefish *Amphiprion akallopisos*. Z. Tierpsychol. 50: 313–326.

Fricke, H.W. 1980. Control of different mating systems in a coral reef fish by one environmental factor. Anim. Behav. 28: 561–569.

Fricke, H.W. and S. Holzberg. 1974. Social units and hermaphroditism in a pomacentrid fish. Naturwissenschaften 61: 367–368.

Fricke, H.W. and S. Fricke. 1977. Monogamy and sex change by aggressive dominance in coral reef fish. Nature 266: 830–832.

Ghiselin, M.T. 1969. The evolution of hermaphroditism among animals. Q. Rev. Biol. 44: 189–208.

Godwin, J. 1994a. Behavioural aspects of protandrous sex change in the anemonefish, *Amphiprion melanopus*, and endocrine correlates. Anim. Behav. 48: 551–567.

Godwin, J. 1994b. Histological aspects of protandrous sex change in the anemonefish *Amphiprion melanopus* (Pomacentridae, Teleostei). J. Zool. (Lond.) 232: 199–213.

Godwin, J. 1995. Phylogenetic and habitat influences on mating system structure in the humbug damselfishes (Dascyllus, Pomacentridae). Bull. Mar. Sci. 57: 637–652.

Godwin, J. 2009. Social determination of sex in reef fishes. Semin. Cell Dev. Biol. 20: 264–270.

Godwin, J. 2010. Neuroendocrinology of sexual plasticity in teleost fishes. Front. Neuroendocrinol. 31: 203–216.

Godwin, J. and P. Thomas. 1993. Sex change and steroid profiles in the protandrous anemonefish *Amphiprion melanopus* (Pomacentridae, Teleostei). Gen. Comp. Endocrinol. 91: 144–157.

Gordon, A.K. and A.W. Bok. 2001. Frequency and periodicity of spawning in the clownfish *Amphiprion akallopisos* under aquarium conditions. Aquarium Sci. Conserv. 3: 307–313.

Guerrero-Estévez, S. and N. Moreno-Mendoza. 2010. Sexual determination and differentiation in teleost fish. Rev. Fish. Biol. Fish. 20: 101–121.

Hattori, A. 1991. Socially controlled growth and size-dependent sex change in the anemonefish *Amphiprion frenatus* in Okinawa, Japan. Jpn. J. Ichthyol. 38: 165–177.

Hattori, A. 1994. Inter-group movement and mate acquisition tactics of the protandrous anemonefish, *Amphiprion clarkii*, on a coral reef, Okinawa. Jpn. J. Ichthyol. 41: 159–165.

Hattori, A. 1995. Coexistence of two anemonefish, *Amphiprion clarkii* and *A. perideraion*, which utilize the same host sea anemone. Environ. Biol. Fish. 42: 345–353.

Hattori, A. 2000. Social and mating systems of the protandrous anemonefish *Amphiprion perideraion* under the influence of a larger congener. Aust. Ecol. 25: 187–192.

Hattori, A. 2005. High mobility of the protandrous anemonefish *Amphiprion frenatus*: nonrandom pair formation in limited shelter space. Ichthyol. Res. 52: 57–63.

Hattori, A. 2012. Determinants of body size compositions in limited shelter space: why are anemonefishes protandrous? Behav. Ecol. 23: 512–520.

Hattori, A. and Y. Yanagisawa. 1991. Life-history pathways in relation to gonadal sex differentiation in the anemonefish, *Amphiprion clarkii*, in temperate waters of Japan. Environ. Biol. Fish. 31: 139–155.

Hattori, A. and N. Yamamura. 1995. Co-existence of subadult males and females as alternative tactics of breeding post acquisition in a monogamous and protandrous anemonefish. Evol. Ecol. 9: 292–303.

Hattori, A. and T. Shibuno. 2010. The effect of patch reef size on fish species richness in a shallow coral reef shore zone where territorial herbivores are abundant. Ecol. Res. 25: 457–468.

Hattori, A. and T. Shibuno. 2013. Habitat use and coexistence of three territorial herbivorous damselfishes on different-size patch reefs. J. Mar. Biol. Assoc. UK 93: 2265–2272.

Hirose, Y. 1995. Pattern of pair formation in protandrous anemonefishes, *Amphiprion clarkii*, *A. frenatus* and *A. perideraion*, on coral reefs of Okinawa, Japan. Environ. Biol. Fish. 43: 153–161.

Iwata, E., Y. Nagai, M. Hyoudou and H. Sasaki. 2008. Social environment and sex differentiation in the false clown anemonefish, *Amphiprion ocellaris*. Zool. Sci. 25: 123–128.

Karino, K. and A. Nakazono. 1993. Reproductive behavior of the territorial herbivore *Stegastes nigricans* (Pisces: Pomacentridae) in relation to colony formation. J. Ethol. 11: 99–110.

Karplus, I. 2014. Symbiosis in Fishes: The Biology of Interspecific Partnerships. Wiley-Blackwell, Hoboken, New Jersey.

Kim, N.N., S.S. Hyun, R.H. Hamid, J. Lee and Ch.Y. Choi. 2012. Expression profiles of three types of GnRH during sex-change in the protandrous cinnamon clownfish, *Amphiprion melanopus*: Effects of exogenous GnRHs. Comp. Biochem. Physiol. B: Biochem. Mol. Biol. 161: 124–133.

Knapp, R.A. and J.T. Kovach. 1991. Courtship as an honest indicator of male parental quality in the bicolor damselfish, *Stegastes partitus*. Behav. Ecol. 2: 295–300.

Knapp, R.A. and R. Warner. 1991. Male parental care and female choice in the bicolor damselfish, *Stegastes partitus*: bigger is not always better. Anim. Behav. 41: 747–756.

Kobayashi, M. and A. Hattori. 2006. Spacing pattern and body size composition of the protandrous anemonefish *Amphiprion frenatus* inhabiting colonial host sea anemones. Ichthyol. Res. 53: 1–6.

Kokita, T. and A. Nakazono. 1999. Pair territoriality in the longnose filefish, *Oxymonacanthus longirostris*. Ichthyol. Res. 46: 297–302.

Kuwamura, T. and Y. Nakashima. 1998. New aspects of sex change among reef fishes: recent studies in Japan. Environ. Biol. Fish. 52: 125–135.

Kuwamura, T., Y. Nakashima and Y. Yogo. 1994. Sex change in either direction by growth-rate advantage in the monogamous coral goby, *Paragobiodon echinocephalus*. Behav. Ecol. 5: 4345–438.

Lecchini, D., Y. Nakamura, J. Grignon and M. Tsuchiya. 2006. Evidence of density-independent mortality in a settling coral reef damselfish, *Chromis viridis*. Ichthyol. Res. 53: 298–300.

Loft, D.F. 1991. Intraspecific Variation in the Social Systems of Wild Vertebrates. Cambridge Studies in Behavioural Biology. Cambridge University Press, Cambridge.

Lowe-McConnell, R.H. 1987. Ecological Studies in Tropical Fish Communities. Cambridge University Press, Cambridge.

McCafferty, S., E. Bermingham, B. Quenouille, S. Planes, G. Hoelzer and K. Asoh. 2002. Historical biogeography and molecular systematics of the Indo-Pacific genus *Dascyllus* (Teleostei: Pomacentridae). Mol. Ecol. 11: 1377–1392.

Mitchell, J.S. 2003. Social correlates of reproductive success in false clown anemonefish: subordinate group members do not pay-to-stay. Evol. Ecol. Res. 5: 89–104.

Mitchell, J.S. 2005. Queue selection and switching by false clown anemonefish, *Amphiprion ocellaris*. Anim. Behav. 69: 643–652.

Mitchell, J.S. and L.M. Dill. 2005. Why is group size correlated with the size of the host anemone in the false clown anemonefish? Can. J. Zool. 83: 372–376.

Miura, S., T. Komatsu, M. Higa, R.K. Bhandari, S. Nakamura and M. Nakamura. 2003. Gonadal sex differentiation in protandrous anemone fish, *Amphiprion clarkii*. Fish Physiol. Biochem. 28: 165–166.

Miura, S., S. Nakamura, Y. Kobayashi, F. Piferrer and M. Nakamura. 2008. Differentiation of ambisexual gonads and immunohistochemical localization of P450 cholesterol side-chain cleavage enzyme during gonadal sex differentiation in the protandrous anemonefish, *Amphiprion clarkii*. Comp. Biochem. Physiol. B: Biochem. Mol. Biol. 149: 29–37.

Moyer, J.T. 1980. Influence of temperate waters on behaviour of the tropical anemonefish *Amphiprion clarkii* at Miyake-jima, Japan. Bull. Mar. Sci. 30: 261–272.

Moyer, J.T. and A. Nakazono. 1978. Protandrous hermaphroditism in six species of the anemonefish genus *Amphiprion* in Japan. Jpn. J. Ichthyol. 25: 101–106.

Munakata, A. and M. Kobayashi. 2010. Endocrine control of sexual behavior in teleost fish. Gen. Comp. Endocrinol. 165: 456–468.

Munday, P.L. 2002. Bi-directional sex change: testing the growth-rate advantage model. Behav. Ecol. Sociobiol. 52: 247–254.

Munday, P.L., P.M. Buston and R.R. Warner. 2006. Diversity and flexibility of sex-change strategies in animals. Trends Ecol. Evol. 21: 89–95.

Muñoz, R.C. and R.R. Warner. 2004. Testing a new version of the size-advantage hypothesis for sex change: sperm competition and size-skew effects in the bucktooth parrotfish, *Sparisoma radians*. Behav. Ecol. 15: 129–136.

Myers, R.F. 1991. Micronesian Reef Fishes. Coral Graphics, Barrigada, Guam.

Ochi, H. 1985. Termination of parental care due to small clutch size in the temperate damselfish, *Chromis notata*. Environ. Biol. Fish. 12: 155–160.

Ochi, H. 1986. Breeding synchrony and spawning intervals in the temperate damselfish *Chromis notata*. Environ. Biol. Fish. 17: 117–124.

Ochi, H. 1989a. Acquisition of breeding space by nonbreeders in the anemonefish *Amphiprion clarkii* in temperate waters of southern Japan. Ethology 83: 279–294.

Ochi, H. 1989b. Mating behavior and sex change of the anemonefish *Amphiprion clarkii* in the temperate waters of southern Japan. Environ. Biol. Fish. 26: 257–275.

Ochi, H. and Y. Yanagisawa. 1987. Sex change and social structure in the anemonefish in temperate waters. pp. 517–543. *In*: Y. Ito, J.L. Brown and J. Kikkawa (eds.). Animal Societies: Theories and Facts. Japan Scientific Societies Press, Tokyo.

Oliver, S.J. and P.S. Lobel. 2013. Direct mate choice for simultaneous acoustic and visual courtship displays in the damselfish, *Dascyllus albisella* (Pomacentridae). Environ. Biol. Fish. 96: 447–457.

Parmentier, E., L. Kéver, M. Casadevall and D. Lecchini. 2010. Diversity and complexity in the acoustic behaviour of *Dascyllus flavicaudus* (Pomacentridae). Mar. Biol. 157: 2317–2327.

Rattanayuvakorn, S., P. Mungkornkarn, A. Thongpan and K. Chatchavalvanich. 2006. Gonadal development and sex inversion in saddleback anemonefish *Amphiprion polymnus* Linnaeus (1758). Kasetsart J. (Nat. Sci.) 40: 196–203.

Reinboth, R. 1962. Morphologische und funktionelle Zweigeschlechtlichkeit bei marinen Teleostiern (Serranidae, Sparidae, Centracanthidae, Labridae). Zool. Jahrb. Physiol. 69: 405–543.

Reinboth, R. 1970. Intersexuality in fishes. Mem. Soc. Endocrinol. 18: 515–543.

Rhodes, K.L. and Y. Sadovy. 2002. Reproduction in the camouflage grouper (Pisces: Serranidae) in Pohnpei, Federated States Micronesia. Bull. Mar. Sci. 70: 851–869.

Robertson, D.R. 1972. Social control of sex reversal in a coral-reef fish. Science 177: 1007–1009.

Robertson, D.R. 1973. Field observations on reproductive behaviour of a pomacentrid fish, *Acanthochromis polyacanthus*. Z. Tierpsychol. 32: 319–324.

Ross, R.M. 1978a. Reproductive behavior of the anemonefish *Amphiprion melanopus* on Guam. Copeia 1978: 103–107.

Ross, R.M. 1978b. Territorial behaviour and ecology of the anemonefish *Amphiprion melanopus* on Guam. Z. Tierpsychol. 36: 71–83.

Ross, R.M. 1990. The evolution of sex-change mechanisms in fishes. Environ. Biol. Fish. 29: 81–93.

Ross, R.M., T.F. Hourigan, M.M.F. Lutnesky and I. Singh. 1990. Multiple simultaneous sex changes in social groups of a coral-reef fish. Copeia 1990: 427–433.

Sadovy de Mitcheson, Y. and M. Liu. 2008. Functional hermaphroditism in teleosts. Fish. Fish. 9: 1–43.

Schmale, M. 1981. Sexual selection and reproductive success in males of the bicolor damselfish, *Eupomacentrus partitus* (Pisces: Pomacentridae). Anim. Behav. 29: 1172–1184.

Schmitt, R.J. and S.J. Holbrook. 1999. Settlement and recruitment of three damselfish species: larval delivery and competition for shelter space. Oecologia 118: 76–86.

Schwarz, A.L. and C.L. Smith. 1990. Sex change in the damselfish *Dascyllus reticulatus* (Richardson) (Perciformes: Pomacentridae). Bull. Mar. Sci. 46: 790–798.

Shapiro, D.Y. 1992. Plasticity of gonadal development and protandry in fishes. J. Exp. Zool. Part A 261: 194–203.

Shapiro, D.Y. and R. Lubbock. 1980. Group sex ratio and sex reversal. J. Theor. Biol. 82: 411–426.

Shapiro, D.Y. and M.B. Rasotto. 1993. Sex differentiation and gonadal development in the diandric, protogynous wrasse, *Thalassoma bifasciatum* (Pisces, Labridae). J. Zool. (Lond.) 230: 231–245.

Shpigel, M. and L. Fishelson. 1986. Behavior and physiology of coexistence in two species of *Dascyllus* (Pomacentridae, Teleostei). Environ. Biol. Fish. 17: 253–265.

Sivakumaran, K.P., P. Brown, D. Stoessel and A. Giles. 2003. Maturation and reproductive biology of female wild carp, *Cyprinus carpio*, in Victoria, Australia. Environ. Biol. Fish. 68: 321–332.

Strüssmann, C.A. and M. Nakamura. 2002. Morphology, endocrinology and environmental modulation of gonadal sex differentiation in teleost fishes. Fish. Physiol. Biochem. 26: 13–29.

Strüssmann, C.A. and L.S. Ito. 2005. Where does gonadal sex differentiation begin? Gradient of histological sex differentiation in the gonads of pejerrey, *Odontesthes bonariensis* (Pisces, Atherinidae). J. Morph. 265: 190–196.

Thresher, R.E. 1984. Reproduction in Reef Fishes. TFH Publications, Neptune City, New Jersey.

Wallace, R.A. and K. Selman. 1981. Cellular and dynamic aspects of oocyte growth in Teleosts. Amer. Zool. 21: 325–343.

Warner, R.R. 1975. The adaptive significance of sequential hermaphroditism in animals. Amer. Nat. 109: 61–82.

Warner, R.R. 1984. Mating behaviour and hermaphroditism in coral reef fishes: the diverse forms of sexuality found among tropical marine fishes can be viewed as adaptations to their equally diverse mating systems. Am. Scientist 72: 128–136.

Warner, R.R. 1988a. Sex change and size-advantage model. Trends Ecol. Evol. 3: 133–136.

Warner, R.R. 1988b. Sex change in fishes: hypotheses, evidence, and objections. Environ. Biol. Fish. 22: 81–90.

Warner, R.R. and S.E. Swearer. 1991. Social-control of sex change in the bluehead wrasse, *Thalassoma bifasciatum* (Pisces, Labridae). Biol. Bull. 181: 199–204.

Wong, M.Y.L. and P.M. Buston. 2013. Social systems in habitat-specialist reef fishes: key concepts in evolutionary ecology. BioScience 63: 453–463.

Wood, E.M. 1981. Sex change and other social strategies in anemonefish. Prog. Underwater Sci. 6: 61–64.

Wood, E.M. 1987. Behaviour and social organization in anemonefish. Prog. Underwater Sci. 12: 53–60.

Yanagisawa, Y. and H. Ochi. 1986. Step-fathering in the anemonefish *Amphiprion clarkii*: a removal study. Anim. Behav. 35: 1769–1780.

Yogo, Y. 1987. Hermaphroditism and evolutionary aspects of its occurrences in fishes. pp. 1–47. *In*: A. Nakazono and T. Kuwamura (eds.). Sex Change in Fishes. Tokai University Press, Tokyo.

Habitat-use and Specialisation among Coral Reef Damselfishes

Morgan S. Pratchett,[1,] Andrew S. Hoey,[1] Shaun K. Wilson,[2] Jean-Paul A. Hobbs[3] and Gerald R. Allen[4]*

Introduction

The high diversity of fishes found on coral reefs is traditionally attributed to the high complexity and large number of distinct microhabitats available within reef habitats (e.g., Ross 1986). Habitat complexity is important in facilitating the co-existence of species across a broad range of size classes (Beukers and Jones 1998), largely by moderating predation. The diversity of microhabitats meanwhile, may lead to increased diversity of reef associated fishes (Messmer et al. 2011), especially in case of sympatric fishes that exhibit highly specialised and complementary patterns of habitat use. One striking example of this is the very high levels of habitat specialisation and clear partitioning of anemone species by anemonefishes (e.g., Allen 1972, Fautin and Allen 1997, Elliot and Mariscal 2001). It is apparent, however, that sympatric fishes

---

[1] ARC Centre of Excellence for Coral Reef Studies, James Cook University, Townsville, Australia.
  Email: andrew.hoey1@jcu.edu.au
[2] Marine Science Program, Department of Parks and Wildlife 7 Dick Perry Ave, Kensington, WA 6151, Australia.
  Email: shaun.wilson@dpaw.wa.gov.au
[3] Department of Environment and Agriculture, Curtin University, Perth, WA 6845, Australia.
  Email: jp.hobbs@curtin.edu.au
[4] Western Australian Museum, Locked Bag 49, Welshpool DC, Perth, Western Australia 6986, Australia.
  Email: gerald.allen@wa.gov.au
* Corresponding author: morgan.pratchett@jcu.edu.au

often do not exhibit clear resource partitioning (Sale 1977). Most coral reef fishes use a very wide and often broadly overlapping range of different resources (Roughgarden 1974, Pratchett 2005). Highly specialised reef fishes may also exhibit highly convergent habitat preferences and high resource overlap (e.g., Munday 2004, Pratchett et al. 2012). On the northern Great Barrier Reef, highly specialised coral-dwelling damselfishes (*Dascyllus aruanus, D. reticulatus*, and *Pomacentrus moluccensis*) all prefer the same coral host, *Pocillopora damicornis* (Pratchett et al. 2012).

The extent to which species are specialist or generalist in their use of resources has significant ramifications not only for their competitive coexistence, but is also expected to influence patterns of distribution, abundance and vulnerability to disturbance (MacNally 1995). Highly specialised species, for example, are expected to be much more constrained by the availability of resources and less able to cope with fluctuations in resource availability as compared to their generalist counterparts. This leads to greatly increased vulnerability to habitat loss, if not elevated risk of extinction (Vázquez and Simberloff 2002). Several studies on terrestrial animals have revealed strong, often negative correlations between specialisation and range-size (e.g., Eeley and Foley 1999), attributed to limitations in the range of locations that can support species with very specific resource requirements (Brown 1984, McKinney 1997). These relationships are assumed to apply equally for both marine and terrestrial species, and are increasingly being used in assessing extinction vulnerability among coral reef fishes (e.g., Graham et al. 2011). Graham et al. (2011) used three different measures of specialisation (diet, habitat and settlement preferences) to establish the vulnerability of coral reef fishes to climate-induced habitat degradation, and suggested that fishes with very specific settlement preferences were most vulnerable to sustained and ongoing climate change. There are, however, few studies that have effectively quantified resource specialisation among marine species, let alone specifically tested ecological ramifications of resource specialisation (but see Munday 2004, Pratchett et al. 2012). More generally, there is very limited data on the critical prey or habitat resources that are required by different reef fishes (Pratchett et al. 2012).

Habitat degradation has a devastating influence on the structure and dynamics of ecological assemblages (Vitousek 1997, Fahrig 2001) and is increasingly being recognised as the major contributor to global biodiversity loss (Brooks et al. 2002, Hoekstra et al. 2005). Habitat degradation results from damage to and/or loss of key habitat forming species (e.g., trees, kelp, corals) leading to declines in habitat-area and structural complexity, or increased habitat fragmentation (Caley et al. 2001, Allison 2004). The specific effects of disturbances on habitat forming species, as well as habitat-associated species, depend on the frequency, severity and selectivity of individual disturbances. Moderate disturbances may have highly selective effects, often impacting only a very limited suite of different species, but these moderate disturbances may nonetheless have very important influences on biodiversity and community structure (Connell 1978). Understanding species-specific responses to habitat degradation requires extensive knowledge of patterns of habitat-use, including measures of habitat specialisation, as well as knowledge of specific effects of disturbances on each habitat type (McKinney 1997). While habitat specialists are expected to be much more vulnerable to habitat degradation as compared to species with generalised habitat requirements (Brown 1984, McKinney 1997, Vázquez and

Simberloff 2002), it is also possible that specialist species may escape any effects from major disturbances because they use a relatively narrow range of resources. However, what is clear is that specialist species are less able to exploit alternative resources following the depletion of their preferred resources, and are thus likely to experience more severe fitness consequences as compared to generalist species that can generally withstand all but the most severe and comprehensive disturbances to resource availability.

Although patterns of habitat use are key to predicting the effects of habitat degradation on motile species (e.g., Wilson et al. 2008), there is surprisingly a limited amount of data available on the critical habitat requirements for most coral reef fishes. Exceptions can be found in case of the coral reef damselfishes (Pomacentridae), for which several studies have documented the preferential use of different coral microhabitats (e.g., Holbrook et al. 2000, Wilson et al. 2010, Pratchett et al. 2012) or anemone species (e.g., Elliot and Mariscal 2001, Litsios et al. 2012). The family Pomacentridae is one of the most speciose families of perciform fishes, with approximately 380 extant species described and validated (Chapter II). They can be found in temperate and tropical localities, in marine and estuarine habitats, but are particularly abundant on coral reefs, where they represent 6%–22% of biodiversity (Bellwood and Hughes 2001, Allen and Werner 2002). Coral reef damselfishes vary greatly in their habitat specialisation, ranging from obligate coral-dwelling damselfishes or anemone fishes that live exclusively within a limited range of specific host species to generalist species that can live in a broad range of environments (Elliot and Mariscal 2001, Wilson et al. 2008). Some coral reef damselfishes (e.g., *Chrysiptera rollandi*) specifically select rubble habitats (Hoey et al. 2007, Wilson et al. 2008), which are generally abundant habitats that provide shelter for small-bodied fishes.

The purpose of this chapter was to review the specific habitat requirements and relative levels of ecological specialisation for coral reef damselfishes, especially, coral-dwelling damselfishes and anemonefishes. These data will then be used to test whether specialist species have greater vulnerability to widespread and increasing disturbances to coral reef habitats, mostly linked to sustained and ongoing effects of global climate change, following Pratchett et al. (2008, 2011b). Moreover, the data collated are used to address widely held, but as yet untested, macroecological theories relating to ecological specialisation. More specifically, this chapter tests whether highly specialist coral reef damselfishes have lower abundance and generally narrower geographic ranges as compared to their generalist counterparts (*sensu* Brown 1984).

Resource Requirements and Habitat Diversity

Species within a range of ecosystems are being increasingly viewed in terms of their ecosystem function, as opposed to their taxonomic affinities or trophic guild (e.g., Blondel 2003, Graham et al. 2006, Fox et al. 2009). Despite the high diversity of coral reef fish assemblages and the assumption that there will be a high degree of functional redundancy among species, several recent studies have shown that coral reef fishes can have highly specialised diets (e.g., Wilson and Bellwood 1997, Pratchett 2005), and perform unique and critical functions (Bellwood et al. 2003, Hoey and Bellwood

2009). This is especially evident among herbivorous coral reef fishes where dietary and functional specialisation has been demonstrated to lead to a high degree of complementarity in ecosystem function (e.g., Burkepile and Hay 2008, Rasher et al. 2013). Consequently, the broad trophic groups traditionally used to classify reef fishes are now being refined and subdivided to reflect differences in ecosystem function (e.g., Hoey and Bellwood 2011, Pratchett et al. 2011b). The broadest functional distinction for coral reef damselfishes is between those species that feed predominantly in the water column (planktivores) versus those species that feed predominantly on benthic prey, whether they are live corals, algae, or other benthic food types (Table 1).

Table 1. Extant species of coral reef damselfishes (family Pomacentridae) showing the diversity in patterns of resource use, including key functional groups (Planktivores, Benthic – Benthic feeders (excluding algal farmers and coral feeders), Alg Farm – Algal farmers, which were further divided into Extensive (E), Intensive (I) and Indeterminate (Ind), and Coral - Coral feeders) and microhabitat requirements (Coral – Coral-dwelling species, further divided into those that are obligate (O) or facultative (F) coral dwellers, "?" indicates where it is not yet known if the use of corals as microhabitats is obligate or facultative, and Anemone – anemonefishes, which have an obligate dependence on anemones). Species were classified according to their predominant (though not necessarily exclusive) feeding habits and habitat associations.

Species	Depth (m)	Max length (mm)	Functional group	Microhabitat requirements
Abudefduf				
A. abdominalis	1–50	330	Planktivore	
A. bengalensis	1–6	85	Planktivore	Coral (F)
A. concolor	1–5	200	Planktivore	
A. conformis	0.5–12	170	Planktivore	
A. declivifrons	1–5	180	Planktivore	
A. margariteus	2–8	210	Planktivore	
A. natalensis	1–25	190	Planktivore	
A. notatus	1–12	190	Planktivore	
A. saxatilis	1–12	200	Planktivore	
A. septemfasciatus	0–3	210	Benthic	Coral (F)
A. sexfasciatus	1–15	190	Planktivore	Coral (F)
A. sordidus	0–3	220	Benthic	
A. sparoides	1–6	160	Planktivore	
A. taurus	0–4	220	Benthic	
A. troschelii	1–12	200	Planktivore	
A. vaigiensis	1–12	220	Benthic	
A. whitleyi	1–5	190	Planktivore	
Acanthochromis				
A. polyacanthus	1–65	150	Planktivore	Coral (F)

Table 1. contd....

Table 1. contd....

Species	Depth (m)	Max length (mm)	Functional group	Microhabitat requirements
Altrichthys				
A. azurelineatus	1–10	75	Planktivore	Coral (O)
A. curatus	1–10	65	Planktivore	Coral (O)
Amblyglyphidodon				
A. aureus	12–35	130	Planktivore	
A. batunai	2–10	85	Planktivore	Coral (O)
A. curacao	1–15	120	Planktivore	Coral (F)
A. flavilatus	12–20	100	Planktivore	Coral (O)
A. flavopurpureus	10–30	120	Planktivore	
A. indicus	1–15	120	Planktivore	Coral (O)
A. leucogaster	2–45	125	Planktivore	Coral (F)
A. melanopterus	1.5–15	150	Planktivore	
A. orbicularis	12–27	140	Planktivore	Coral (F)
A. silolona	1–10	125	Planktivore	Coral (O)
A. ternatensis	1–12	130	Planktivore	Coral (O)
Amblypomacentrus				
A. breviceps	2–35	70	Planktivore	Coral (F)
A. clarus	15–25	60	Planktivore	
Amphiprion				
A. akallopisos	3–25	110	Planktivore	Anemone
A. akindynos	1–25	120	Planktivore	Anemone
A. allardi	1–30	150	Planktivore	Anemone
A. barberi	2–10	125	Planktivore	Anemone
A. bicinctus	1–30	140	Planktivore	Anemone
A. chagosensis	10–25	110	Planktivore	Anemone
A. chrysogaster	2–40	150	Planktivore	Anemone
A. chrysopterus	1–20	170	Planktivore	Anemone
A. clarkii	1–55	130	Planktivore	Anemone
A. ephippium	2–15	140	Planktivore	Anemone
A. frenatus	1–2	140	Planktivore	Anemone
A. fuscocaudatus	5–30	140	Planktivore	Anemone

Table 1. contd....

Table 1. contd....

Species	Depth (m)	Max length (mm)	Functional group	Microhabitat requirements
Amphiprion				
A. latezonatus	10–45	140	Planktivore	Anemone
A. latifasciatus	1–15	130	Planktivore	Anemone
A. leucokranos	2–5	90	Planktivore	Anemone
A. mccullochi	2–45	120	Planktivore	Anemone
A. melanopus	1–10	120	Planktivore	Anemone
A. nigripes	2–15	110	Planktivore	Anemone
A. ocellaris	1–15	110	Planktivore	Anemone
A. omanensis	2–10	140	Planktivore	Anemone
A. pacificus	1–10	80	Planktivore	Anemone
A. percula	1–15	110	Planktivore	Anemone
A. perideraion	3–20	100	Planktivore	Anemone
A. polymnus	2–30	130	Planktivore	Anemone
A. rubrocinctus	1–8	120	Planktivore	Anemone
A. sandaracinos	3–20	140	Planktivore	Anemone
A. sebae	2–25	160	Planktivore	Anemone
A. tricinctus	3–40	130	Planktivore	Anemone
Cheiloprion				
C. labiatus	1–3	80	Coral (O)	Coral (O)
Chromis				
C. abrupta	1–22	70	Planktivore	
C. abyssicola	90–152	160	Planktivore	
C. abyssus	107–150	126	Planktivore	
C. acares	2–37	55	Planktivore	
C. agilis	3–56	100	Planktivore	
C. albicauda	25–70	175	Planktivore	
C. albomaculata	12–40	200	Planktivore	
C. alleni	11–30	90	Planktivore	
C. alpha	18–95	110	Planktivore	
C. amboinensis	5–65	80	Planktivore	
C. analis	10–70	140	Planktivore	

Table 1. contd....

Table 1. contd....

Species	Depth (m)	Max length (mm)	Functional group	Microhabitat requirements
Chromis				
C. athena	60–65	80	Planktivore	
C. atripectoralis	2–15	110	Planktivore	Coral (F)
C. atripes	10–35	70	Planktivore	
C. axillaris	60–80	130	Planktivore	
C. bami	12–40	80	Planktivore	
C. brevirostris	90–120	90	Planktivore	
C. caudalis	20–50	100	Planktivore	
C. chrysura	6–30	180	Planktivore	
C. cinerescens	3–15	130	Planktivore	
C. cyanea	3–55	130	Planktivore	
C. degruyi	85–120	110	Planktivore	
C. delta	10–80	65	Planktivore	
C. dimidiata	2–20	70	Planktivore	Coral (F)
C. durvillei	100–200	92	Planktivore	
C. earina	75–116	93	Planktivore	
C. elerae	12–70	70	Planktivore	
C. enchrysura	20–100	70	Planktivore	
C. fatuhivae	20–40	90	Planktivore	
C. fieldi	1–40	88	Planktivore	Coral (F)
C. flavaxilla	3–20	70	Planktivore	Coral (O)
C. flavicauda	50–61	110	Planktivore	
C. flavipectoralis	2–15	70	Planktivore	Coral (F)
C. fumea	3–25	130	Planktivore	
C. hanui	6–50	80	Planktivore	
C. insolata	20–100	160	Planktivore	
C. iomelas	3–35	85	Planktivore	Coral (F)
C. kennensis	6–40	120	Planktivore	
C. lepidolepis	2–20	90	Planktivore	
C. leucura	20–119	70	Planktivore	
C. lineata	2–10	55	Planktivore	
C. margaritifer	2–20	90	Planktivore	Coral (F)
C. mirationis	40–155	150	Planktivore	

Table 1. contd....

Table 1. contd....

Species	Depth (m)	Max length (mm)	Functional group	Microhabitat requirements
Chromis				
C. monochroma	40–55	65	Planktivore	
C. multilineata	2–40	150	Planktivore	
C. nigroanalis	20–40	120	Planktivore	
C. nigrura	1–30	55	Planktivore	
C. nitida	5–25	90	Planktivore	Coral (F)
C. okamurai	150–150	120	Planktivore	
C. onumai	51–90	162	Planktivore	
C. opercularis	10–40	160	Planktivore	
C. ovatiformes	10–40	100	Planktivore	Coral (F)
C. pamae	5–20	130	Planktivore	
C. pelloura	30–50	140	Planktivore	
C. pembae	15–50	120	Planktivore	
C. planesi	50–54	138	Planktivore	
C. pura	40–65	120	Planktivore	
C. retrofasciata	5–65	55	Planktivore	Coral (F)
C. scotochiloptera	5–20	160	Planktivore	
C. scotti	5–100	90	Planktivore	
C. struhsakeri	99–183	110	Planktivore	
C. ternatensis	2–15	90	Planktivore	Coral (O)
C. trialpha	3–50	60	Planktivore	
C. unipa	42–70	80	Planktivore	
C. vanderbilti	2–20	60	Planktivore	
C. viridis	1–12	90	Planktivore	Coral (O)
C. weberi	3–25	130	Planktivore	
C. westaustralis	2–75	110	Planktivore	
C. woodsi	50–175	110	Planktivore	
C. xanthochira	10–48	150	Planktivore	
C. xanthopterygia	2–20	110	Planktivore	
C. xanthura	3–40	160	Planktivore	
C. xouthos	12–50	135	Planktivore	
C. xutha	2–20	70	Planktivore	

Table 1. contd....

Table 1. contd....

Species	Depth (m)	Max length (mm)	Functional group	Microhabitat requirements
Chrysiptera				
C. albata	40–55	36	Planktivore	
C. annulata	0–2	70	Planktivore	
C. arnazae	3–20	50	Planktivore	Coral (O)
C. biocellata	0–5	110	Alg Farm (Ind)	
C. bleekeri	3–12	90	Planktivore	
C. brownriggi	0–2	80	Alg Farm (Ind)	Coral (F)
C. caeruleolineata	30–65	55	Planktivore	
C. chrysocephala	3–6	70	Planktivore	
C. cyanea	0–10	80	Planktivore	Coral (F)
C. cymatilis	3–20	55	Planktivore	Coral (O)
C. flavipinnis	3–38	80	Planktivore	Coral (F)
C. galba	1–30	90	Planktivore	
C. giti	3–20	55	Planktivore	Coral (O)
C. glauca	0–2	110	Alg Farm (Ind)	
C. hemicyanea	1–15	65	Planktivore	Coral (O)
C. kuiteri	15–30	55	Planktivore	
C. niger	0–2	65	Planktivore	Coral (O)
C. oxycephala	1–16	90	Planktivore	Coral (O)
C. parasema	1–16	65	Planktivore	Coral (O)
C. pricei	3–10	50	Planktivore	Coral (O)
C. rapanui	3–38	70	Planktivore	
C. rex	1–6	70	Benthic	
C. rollandi	2–35	55	Benthic	
C. sinclairi	1–15	65	Planktivore	Coral (O)
C. springeri	5–30	55	Planktivore	Coral (O)
C. starcki	25–52	90	Planktivore	
C. talboti	6–35	60	Planktivore	
C. taupou	0–5	80	Planktivore	
C. traceyi	5–30	60	Planktivore	
C. tricincta	10–38	60	Planktivore	
C. unimaculata	0–2	80	Alg Farm (Ind)	

Table 1. contd....

Table 1. contd....

Species	Depth (m)	Max length (mm)	Functional group	Microhabitat requirements
Dascyllus				
D. albisella	1–50	130	Planktivore	Coral (O)
D. aruanus	1–12	90	Planktivore	Coral (O)
D. auripinnis	1.5–30	150	Planktivore	
D. carneus	5–35	65	Planktivore	Coral (O)
D. flavicaudus	3–40	120	Planktivore	Coral (F)
D. marginatus	1–15	65	Planktivore	Coral (O)
D. melanurus	1–10	90	Planktivore	Coral (O)
D. reticulatus	1–50	90	Planktivore	Coral (O)
D. strasburgi	5–15	120	Planktivore	Coral (F)
D. trimaculatus	1–55	140	Planktivore	Coral (F)
Dischistodus				
D. chrysopoecilus	1–5	160	Alg Farm (I)	
D. darwiniensis	1–8	150	Alg Farm (E)	
D. fasciatus	1–8	150	Alg Farm (E)	
D. melanotus	1–10	170	Alg Farm (I)	
D. perspicillatus	1–10	210	Alg Farm (I)	Coral (?)
D. prosopotaenia	1–12	200	Alg Farm (E)	
D. pseudochrysopoecilus	1–5	180	Alg Farm (I)	
Hemiglyphidodon				
H. plagiometopon	1–20	200	Alg Farm (E)	Coral (F)
Lepidozygus				
L. tapeinosoma	5–25	90	Planktivore	Coral (F)
Microspathodon				
M. chrysurus	0–10	210	Alg Farm (E)	Coral (F)
Neoglyphidodon				
N. carlsoni	1–5	130	Benthic	
N. crossi	2–5	130	Benthic	
N. melas	1–12	170	Coral (F)	Coral (F)
N. mitratus	10–45	135	Benthic	Coral (F)
N. nigroris	2–23	120	Alg Farm (Ind)	
N. oxyodon	0–4	160	Benthic	

Table 1. contd....

Table 1. contd....

Species	Depth (m)	Max length (mm)	Functional group	Microhabitat requirements
Neoglyphidodon				
N. polyacantus	2–30	160	Benthic	
N. thoracotaeniatus	15–45	110	Benthic	Coral (F)
Neopomacentrus				
N. anabatoides	2–15	110	Planktivore	Coral (O)
N. azysron	1–12	80	Planktivore	Coral (F)
N. bankieri	3–12	65	Planktivore	Coral (F)
N. cyanomos	5–18	90	Planktivore	Coral (F)
N. filamentosus	5–12	80	Planktivore	
N. fuliginosus	1–10	90	Planktivore	
N. metallicus	2–10	80	Planktivore	
N. miryae	2–25	110	Planktivore	Coral (F)
N. nemurus	1–10	70	Planktivore	Coral (F)
N. sidensis	1–10	80	Planktivore	
N. sororius	1–12	85	Planktivore	
N. taeniurus	0–3	110	Planktivore	
N. violascens	5–25	65	Planktivore	
N. xanthurus	1–15	65	Planktivore	Coral (?)
Parma				
P. oligolepis	2–20	220	Alg Farm (Ind)	
P. polylepis	1–40	230	Alg Farm (Ind)	
Plectroglyphidodon				
P. dickii	1–12	110	Benthic	Coral (O)
P. flaviventris	5–12	90	Benthic	Coral (O)
P. imparipennis	0–3	60	Benthic	
P. johnstonianus	2–12	90	Coral (O)	Coral (O)
P. lacrymatus	2–12	110	Alg Farm (E)	
P. leucozonus	0–2	110	Alg Farm (Ind)	
P. phoenixensis	0–8	90	Alg Farm (Ind)	Coral (O)
P. randalli	1–4	90	Benthic	
P. sagmarius	0–3	64	Benthic	
P. sindonis	0–3	130	Benthic	

Table 1. contd....

Table 1. contd....

Species	Depth (m)	Max length (mm)	Functional group	Microhabitat requirements
Pomacentrus				
P. adelus	0–8	90	Alg Farm (Ind)	
P. agassizi	1–4	110	Planktivore	Coral (F)
P. albicaudatus	1–12	65	Planktivore	
P. albimaculatus	10–20	90	Planktivore	
P. alexanderae	5–30	90	Planktivore	
P. alleni	3–15	65	Planktivore	
P. ambionensis	2–40	110	Planktivore	Coral (F)
P. aquilus	0–15	110	Alg Farm (Ind)	
P. arabicus	1–6	150	Planktivore	
P. armillatus	5–20	80	Planktivore	
P. atriaxillaris	12–27	82	Planktivore	
P. aurifrons	2–14	75	Planktivore	Coral (F)
P. auriventris	2–12	90	Planktivore	
P. australis	5–35	90	Planktivore	
P. azuremaculatus	5–30	110	Planktivore	
P. baenschi	1–10	100	Planktivore	
P. bankanensis	0–12	90	Alg Farm (Ind)	Coral (F)
P. bintanensis	1–5	86	Planktivore	
P. bipunctatus	5–43	80	Planktivore	
P. brachialis	6–40	110	Planktivore	Coral (F)
P. burroughi	2–16	90	Alg Farm (Ind)	
P. caeruleopunctatus	5–15	95	Planktivore	
P. caeruleus	1–10	90	Planktivore	
P. callainus	1–12	95	Planktivore	
P. cheraphilus	10–18	75	Planktivore	
P. chrysurus	0–3	90	Alg Farm (Ind)	Coral (F)
P. coelestis	1–12	90	Planktivore	
P. colini	10–18	90	Planktivore	
P. cuneatus	1–6	100	Planktivore	
P. fakfakensis	2–8	74	Planktivore	
P. geminospilos	5–20	80	Planktivore	

Table 1. contd....

Table 1. contd....

Species	Depth (m)	Max length (mm)	Functional group	Microhabitat requirements
Pomacentrus				
P. grammorhynchus	2–12	120	Alg Farm (Ind)	Coral (F)
P. imitator	2–15	110	Planktivore	
P. indicus	1–15	120	Planktivore	
P. javanicus	10–18	90	Planktivore	
P. lepidogenys	1–12	90	Planktivore	Coral (F)
P. leptus	1–10	70	Planktivore	
P. limosus	5–10	62	Planktivore	
P. littoralis	0–5	110	Planktivore	
P. maafu	2–15	86	Planktivore	Coral (O)
P. melanochir	1–8	70	Planktivore	
P. micronesicus	1–12	90	Planktivore	
P. microspilos	2–30	95	Planktivore	
P. milleri	1–6	100	Planktivore	
P. moluccensis	1–14	70	Planktivore	Coral (O)
P. nagasakiensis	5–30	120	Planktivore	Coral (F)
P. nigromanus	6–60	90	Planktivore	Coral (F)
P. nigromarginatus	20–50	90	Planktivore	
P. opisthostigma	6–16	90	Planktivore	
P. pavo	1–16	110	Planktivore	Coral (F)
P. philippinus	1–12	110	Planktivore	
P. pikei	1–6	110	Planktivore	
P. polyspinus	3–10	90	Planktivore	
P. proteus	2–10	100	Planktivore	
P. reidi	12–70	120	Planktivore	
P. rodriguesensis	9–18	105	Planktivore	
P. saksonoi	10–15	100	Planktivore	
P. similis	2–12	90	Planktivore	
P. simsiang	0–10	90	Planktivore	
P. smithi	2–14	70	Planktivore	Coral (F)
P. spilotoceps	2–12	78	Planktivore	
P. stigma	2–10	130	Planktivore	

Table 1. contd....

Table 1. contd....

Species	Depth (m)	Max length (mm)	Functional group	Microhabitat requirements
Pomacentrus				
P. sulfureus	1–5	110	Benthic	Coral (O)
P. trichourus	1–43	110	Planktivore	
P. trilineatus	0–4	100	Planktivore	
P. tripunctatus	0–3	100	Alg Farm (Ind)	
P. vaiuli	3–45	90	Alg Farm (Ind)	
P. wardi	1–20	110	Alg Farm (E)	Coral (F)
P. xanthosternus	1–8	100	Planktivore	
P. yoshii	1.5–22	95	Planktivore	
Pomachromis				
P. exilis	8–12	65	Planktivore	
P. fuscidorsalis	1–18	80	Planktivore	
P. guamensis	3–33	60	Planktivore	
P. richardsoni	10–20	80	Planktivore	
Premnas				
P. biaculeatus	1–6	170	Planktivore	Anemone
Pristotis				
P. cyanostigma	5–10	110	Planktivore	
P. obtusirostris	5–80	150	Planktivore	
Stegastes				
S. adustus	0–3	130	Alg Farm (I)	Coral (F)
S. albifasciatus	0–2	120	Alg Farm (I)	
S. apicalis	1–5	150	Alg Farm (I)	Coral (F)
S. arcifrons	1–20	170	Alg Farm (I)	
S. aureus	1–5	110	Alg Farm (I)	
S. baldwini	1–5	120	Alg Farm (I)	
S. diencaeus	2–5	110	Alg Farm (E)	Coral (F)
S. emeryi	1–18	100	Alg Farm (I)	
S. fasciolatus	0–5	170	Alg Farm (E)	Coral (F)
S. fuscus	3–18	120	Alg Farm (E)	Coral (F)
S. gascoynei	2–30	160	Alg Farm (I)	
S. insularis	1–3	110	Alg Farm (I)	

Table 1. contd....

Table 1. contd....

Species	Depth (m)	Max length (mm)	Functional group	Microhabitat requirements
Stegastes				
S. leucostictus	0–5	90	Alg Farm (E)	Coral (F)
S. limbatus	1–2	160	Alg Farm (I)	
S. luteobrunneus	0–5	150	Alg Farm (I)	
S. nigricans	1–12	150	Alg Farm (I)	Coral (F)
S. obreptus	2–6	160	Alg Farm (E)	
S. partitus	0–45	90	Alg Farm (I)	Coral (F)
S. pictus	3–20	120	Alg Farm (I)	
S. planifrons	1–30	120	Alg Farm (E)	Coral (F)
S. punctatus	1–5	170	Alg Farm (I)	Coral (F)
S. variabilis	0–30	110	Alg Farm (I)	Coral (F)
Teixeirichthys				
T. jordani	10–20	140	Planktivore	

Among the best-studied group of damselfishes are the territorial algal farmers, which are an ecologically important group of benthic feeders (Chapter VI). These damselfishes are important in maintaining algal communities that differ in taxonomic composition (Ceccarelli 2007, Hoey and Bellwood 2010), have a greater biomass and greater productivity, than those on adjacent substrata (reviewed by Ceccarelli et al. 2001). Algal-farming damselfishes can also influence the settlement and survival of corals within their territories (Wellington 1982, Arnold et al. 2010, Gochfeld 2010). However, territorial algal farming damselfishes are not a uniform group. Recent studies (e.g., Hata and Kato 2004, Ceccarelli 2007, Emslie et al. 2012) have further divided them into intensive, extensive, and indeterminate farmers based on the intensity of farming and territory defence, and the resultant size and algal composition of their territories. Intensive farmers have relatively small territories containing low diversity algal turfs that they weed intensively and defend aggressively, whereas extensive farmers have larger territories with diverse algal assemblages that they weed and defend less intensively (Hata and Kato 2004). Indeterminate farmers defend and weed less intensively than the former two groups, and only have subtle effects on the algal composition within their territories (Emslie et al. 2012). While it was traditionally assumed that territorial farming damselfish were herbivorous, cultivating the algal communities as a direct source of food, it is becoming increasingly apparent that they contain a mix of both herbivorous and detritivorous species (Wilson and Bellwood 1997, Ceccarelli 2007). For example, the extensive farming *Hemiglyphidodon plagiometopon* is a specialised detritivore, actively avoiding ingesting algae within its territories (Wilson and Bellwood 1997). In contrast, the intensive farming *Stegastes apicalis* and extensive farming *Pomacentrus wardi* appear to be specialised herbivores with diets dominated by filamentous algae such as *Polysiphonia* (Ceccarelli 2007).

While territorial algal-farming damselfishes can have significant localised effects on habitat structure (Lobel 1980, Ceccareli et al. 2001), they do not have the same functional importance ascribed to them as many larger bodied herbivorous fishes do (Bellwood et al. 2004). These larger bodied herbivores are considered critical for preventing and/or reversing phase shifts from coral- to macroalgal dominance of benthic reef habitats. Notably, the degradation of coral reef ecosystems (manifest as declines in live coral cover) may actually lead to increased abundance of algal farming damselfishes (Ceccarelli et al. 2006, Emslie et al. 2012) and other species with generalist habitat requirements (Bellwood et al. 2006b, 2012), at the cost of specialist coral-dwelling damselfishes and anemonefishes.

Coral-dwelling damselfishes

Among the many microhabitats available to damselfishes on coral reefs, both adults and juveniles are often associated with live coral. Previous studies suggest that 20%–40% of pomacentrids living on coral reefs are dependent on live corals (Munday et al. 2008, Wilson et al. 2008), and may be adversely affected by extensive coral loss. Of the 322 damselfishes that are reported from coral reefs, 73 species (22.7%) have been recorded to use live coral as habitat (Table 1, see also Coker et al. 2014). Although many coral reef damselfishes associate with live corals and/or structurally complex habitats, other species display a preference for structurally simple areas dominated by sand or rubble. For example, both juvenile and adult *Chrysiptera rollandi* have been shown to select sand and rubble habitats and actively avoid habitats with a high cover of corals (Hoey et al. 2007, Wilson et al. 2008). While there has not been any rigorous phylogenetic analysis of microhabitat use and specialisation for coral-dwelling damselfishes, it appears that the use of live corals has arisen independently within several distinct genera (Table 1). It is also apparent that while some damselfishes are obligately dependent upon their live coral hosts, other species (facultative coral-dwelling damselfishes) have much more generalist microhabitat requirements and use corals infrequently or opportunistically (Table 1). Habitat preferences can also change with ontogeny, where adults of some species may be habitat generalists, but juveniles of the same species have specific habitat requirements (Wilson et al. 2008). Jones et al. (2004) showed that up to 65% of reef fishes (not just damselfishes) preferentially settle in live corals, while only a fraction of these species maintain close association with corals as adults. The key to understanding the vulnerability of damselfishes to sustained and ongoing coral loss is to establish the extent to which these fishes rely on live corals either at settlement or throughout their lives (Wilson et al. 2008).

The distinction between obligate and facultative coral-dwelling fishes is important in understanding how fishes may respond to coral depletion (e.g., Pratchett et al. 2008, Coker et al. 2014), and is generally quantified based on the frequency with which different species associate with live coral hosts, as opposed to dead corals or other structural habitats. Reliance on live corals is expected to be much greater among those fishes that have a strong and permanent association with specific coral hosts (Munday et al. 1997, Holbrook et al. 2000) as opposed to those species that only sometimes occupy specific host colonies, but may nonetheless derive significant benefit from living in habitats with rich coral growth (Jones et al. 2004, Wilson et al. 2006). Where quantitative data on coral use are available (e.g., Wilson et al. 2008, Pratchett et al. 2012) obligate coral-dwelling fishes are defined as those species for which >80% of individuals are found living within live coral hosts. Of the 73 damselfishes that are recorded to live in live coral colonies, at least 24 species are obligately dependent upon live corals, living predominantly within specific live coral colonies as juveniles and adults (Table 1). In contrast, a relatively small proportion of individuals of facultative coral-dwelling damselfishes (<30%; Wilson et al. 2008) are found sheltering within live coral colonies, or individual fishes only occasionally use coral habitats and do not exhibit strong fidelity for specific coral hosts (e.g., Pratchett et al. 2012). Facultative coral-dwelling damselfishes include a large number of territorial algal farming species that tend to establish their territories (at least initially) among the branches of live arborescent *Acropora* corals, though there is often very limited live coral available within the area of well-established territories (Ceccarelli 2007). Other facultative coral-dwelling damselfishes (e.g., *Dascyllus trimaculatus*) utilise specific coral hosts in much the same way as obligate coral-dwelling damselfishes, with distinct social groups established in specific coral colonies (Asoh and Kasuya 2002). However, *D. trimaculatus* typically settles in anemones, associating with corals as adults, and is not restricted to coral reefs. On subtropical reefs, *D. trimaculatus* establishes social groups within close proximity of crevices or many other non-coral structural habitats. The final group of facultative coral-dwelling fishes are those fishes that opportunistically shelter within live corals, but have low fidelity to anyone coral colony (e.g., *Chromis atripectoralis* and *Pomacentrus amboinensis*) and do not appear to be strictly reliant on habitat forming corals (Pratchett et al. 2008).

Obligate coral-dwelling damselfishes are predominantly planktivores, which live in close association with one or more specific host colonies, feeding in the water column directly above their host corals and sheltering within the corals during the night or at any approach of danger (Forrester 1990, Allen 1991; but see Frédérich et al. 2010). Of the obligate coral-dwelling damselfishes only two species depend upon host corals for food (Table 1). Primarily, live corals are important in providing shelter and/or settlement habitat for damselfishes (Feary et al. 2007, Pratchett et al. 2012, Coker et al. 2012, 2014), whereby it is the structural complexity of coral skeletons that is important in reducing predation (e.g., Hixon and Beets 1993, Holbrook and Schmitt 2002). Accordingly, Coker et al. (2015) found no change in the physiological condition of *Dascyllus aruanus* (based on hepatocyte vacuolation) between fish living on bleached versus healthy coral colonies. However, live, healthy colonies are more effective predator refuges than bleached or dead colonies that are structurally intact (Coker et al. 2009). This may be because the live coral tissues better obscure visual and/or olfactory cues of resident fishes, or because the biological defences (e.g., nematocysts) of live corals aid in deterring potential predators (Coker et al. 2009).

Corals colonies vary, both within and among species, in their suitability as settlement and adult habitats (Noonan et al. 2012, Coker et al. 2014), mainly due to overall differences in colony morphology. Virtually all of the corals that are inhabited by coral-dwelling damselfishes have a branching morphology (Fig. 1; see also Wilson et al. 2008). However, coral-dwelling damselfishes exhibit habitat specificity well beyond the level of broadly defined coral morphologies, either using only a very restricted subset of available coral taxa (genera or species) or using different corals

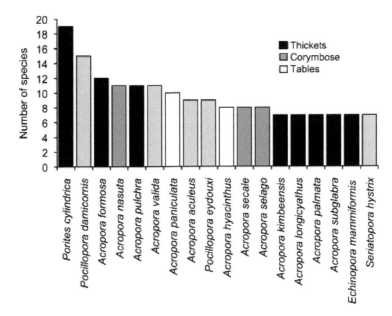

Fig. 1. Number of damselfish species that associate with each of the most commonly used host coral species. All coral species shown are branching, but range from sprawling thickets (black bars), to distinct corymbose colonies (grey bars), or tables with relative short branches (white bars). Redrawn from data presented in Coker et al. (2014).

disproportionately to their availability (e.g., Pratchett et al. 2008, Wilson et al. 2010). On Indo-Pacific reefs the most important coral species are *Porites cylindrica* (19 spp.), *Pocillopora damicornis* (15 spp.) and *Acropora formosa* (12 spp.). These branching corals are common and often dominate coral assemblages in shallow back reef habitats (Veron 2000). *Porites cylindrica* and *A. formosa* form large sprawling thickets, which provide large areas of habitat for fishes of varying sizes, and enable coexistence of a very high diversity of different fishes (Johnson et al. 2011). *Pocillopora damicornis* is a corymbose coral that mostly occurs as small, distinct colonies, but very high densities of colonies in lagoonal and back reef habitats lead to large areas of almost continuous habitat (Pratchett et al. 2008).

A large overlap in the coral species used by sympatric species of coral-dwelling damselfishes (e.g., Shpigel 1982, Holbrook and Schmitt 2002, Bonin et al. 2009, Pratchett et al. 2012) suggests that there are relatively few different distinct coral types (e.g., *Acropora*, *Pocillopora*, *Stylophora* and/or *Seriatopora* spp.) that maximise fitness (growth, survival and/or reproductive success). Consistent with this hypothesis, Jones (1988) and Beukers and Jones (1998) showed that the survivorship of damselfishes (specifically *D. aruanus*, *P. amboinensis*, and *P. moluccensis*) was much higher in *P. damicornis* compared with other branching corals, such as *Acropora nobilis*. The corals that are preferred by coral-dwelling damselfishes also overlap with those used by other coral-dwelling fishes (e.g., Kuwamurra et al. 1994, Munday et al. 2004) as well as crustaceans that live exclusively within live corals (Knudsen 1967, Pratchett

2001). Such similarities in preferred corals are presumably because it is these corals that represent the most effective refuges from potential predators.

Strong habitat associations of coral-dwelling damselfishes may become established at settlement (Wellington 1992, Schmitt and Holbrook 1999), but are also likely to be reinforced by variation in post-settlement growth and survivorship (Wilson et al. 2010). Many obligate coral-dwelling damselfishes recruit directly to preferred adult habitats (Ohman et al. 1998, Feary et al. 2007), often in direct response to the presence or odour of conspecifics (Sweatman 1985, Booth 1992, Lecchini et al. 2007). However, some species settle in a much wider range of habitats than those which appear to be used by adult conspecifics (Roberston 1996, Bay et al. 2001, McCormick and Hoey 2004, Wilson et al. 2010), suggesting that these fishes settle indiscriminately, but only persist in a restricted range of microhabitats (Lecchini and Galzin 2005). For example, *P. amboinensis* settles among rubble, dead coral, and live branching coral, but individuals associated with rubble habitats experience much higher mortality in the 24 h post-settlement period (McCormick and Hoey 2004).

Coral-dwelling damselfishes often exhibit highly convergent habitat preferences (using mostly *Pocillopora* or *Acropora* corals), but vary in the extent to which they are more or less specialised in their use of different microhabitats (e.g., Wilson et al. 2006, Pratchett et al. 2008). At Lizard Island, on the Great Barrier Reef, the most specialised species was *Dascyllus reticulatus*, which was found living in only 8 different coral species from 5 different genera (Pratchett et al. 2008). Moreover, the majority (77%) of individuals were found in *P. damicornis*. *Pocillopora damicornis* was also the predominant coral used by *Dascyllus aruanus* and *Pomacentrus moluccensis*, though these fishes used a greater range of different coral habitats (10 and 31 coral species, respectively). In order to further explore interspecific differences in ecological specialisation among coral-dwelling damselfishes, it is important to assess patterns of coral use across a wide range of different locations throughout their geographic range (e.g., Pratchett 2013). This is because seemingly high levels of ecological specialisation at just one location, or in just one habitat, may be due to local constraints on the availability or accessibility of different resource (prey or habitat) types. For highly specialised species, patterns of resource use would be expected to be invariant among different locations (Devictor et al. 2010). Conversely, generalist species may use only a few different resources at each location, but the specific resources used vary in relation to availability, or strong constraints due to environmental factors or biological interactions (Fox and Morrow 1981). For this reason, we attempted to compile data on the complete range of different coral species that are used by obligate coral-dwelling damselfishes from a range of studies conducted in different geographic locations.

Data on the specific host corals used by coral-dwelling damselfishes was surprisingly limited; most studies of habitat use by these fishes do not identify host corals to species, but use either morphological categories (e.g., Wilson et al. 2008) or only identify corals to genus or family. While this may facilitate effective comparisons of relative specialisation for a given suite of species at a single location (e.g., Wilson et al. 2008), it precludes meaningful comparisons among locations and among studies. Overall, species-level data of coral use was available for only 7 (out of 24) obligate coral-dwelling damselfishes (Fig. 2). The most specialised of these fishes was *Dascyllus marginatus*, which has been recorded from only 8 different

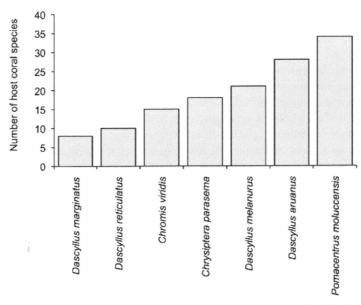

Fig. 2. Variation in the number of coral species used by each of seven obligate coral-dwelling damselfishes. Only species for which there have been studies of habitat-use in at least two distinct geographic locations are shown. Species are ordered based on the total number of coral species used, which is a proxy for ecological specialisation.

coral species, while the most generalist species, *Pomacentrus moluccensis*, has been reported to use 34 different coral species. All but the most specialised damselfish (*D. marginatus*) inhabit *Pocillopora damicornis*, and all but *D. marginatus* and *D. reticulatus* use *Porites cylindrica*.

Interspecific differences in ecological specialisation may be reflective of differences in the competitive abilities of different coral-dwelling damselfishes (Holbrook and Schmitt 2002), whereby the most specialised species monopolise preferred host corals (or at least limit access to preferred microhabitats), forcing subordinate species to use other suboptimal coral hosts. The competitive hierarchy of these obligate coral-dwelling damselfishes is not known, but other studies of coral reef fishes have confirmed that specialist species tend to be competitively dominant (Blowes et al. 2013) because the relative value of preferred resources is much higher for specialist species (Genner et al. 1999). However, interspecific differences in body size, shape, and behaviour may add to differences in the range of corals used by different coral-dwelling damselfishes. Most notably, *P. moluccensis* uses branching coral species that have both much tighter and more open branching as compared to the corals used by *Dascyllus* spp. Small and juvenile *P. moluccensis* are often found in the small spaces between tightly packed branches of complex branching *Seriatopora hystrix*. However, adult *P. moluccensis* maybe found in large coral colonies with wider branching, such as *Echinopora* spp. Moreover, juvenile *Chromis retrofasciata* select colonies of *Seriatopora hystrix* based on branch spacing, preferentially settling in and experiencing higher survival in colonies with intermediate, as opposed to broad or narrow, branch spacing (Noonan et al. 2012).

Anemonefishes

The strong association between some damselfishes (mostly *Amphiprion* spp.) and a few distinct species of "large" sea anemones is often regarded as one of the most specialised habitat associations among coral reef fishes (Jones et al. 2002). In the wild, these fishes are only very rarely seen, and cannot reproduce, without their specific host anemones (Fautin 1991). The abundance and distribution of anemonefishes are therefore limited by that of their host anemones. There are a total of 28 species of damselfishes (commonly referred to as anemonefishes, Fautin 1991, Chapter XII) that have an obligate and symbiotic association with sea anemones (Table 1). This includes several recently described species, for which the full range of host anemones (*Amphiprion barberi*, *A. chagosesnsis* and *A. pacificus*) are not yet known, but excludes the formerly well studied species, *Amphiprion leucokranos* and *A. thiellei*, that are probably hybrids (Litsios and Salamin 2014). The true anemonefishes are a monophyletic group, comprising of two currently recognised genera (*Premnas* and *Amphiprion*) within the subfamily Pomacentrinae, tribe Amphiprionini (Cooper et al. 2009, Litsios et al. 2012). However, *Premnas* occurs within, rather than outside, the *Amphiprion* clade (Santini and Polacco 2006, Litsios et al. 2012). Other damselfishes (e.g., *Dascyllus trimaculatus*) may also be found within anemones, but these are not considered to be true anemonefishes because the association is not obligate and is most apparent among juvenile fishes (Fautin and Allen 1997).

Despite the diversity of anemones that occur on tropical coral reefs, anemonefishes have so far been recorded to use just ten different species of anemones in the wild (Fautin 1991). Anemones that host anemonefishes tend to be large and most abundant in warm shallow waters, but are otherwise very diverse, belonging to five genera and three different families (Fauntin 1991). The number of host anemones used by each species of anemonefish ranges from 1 to 10 (Fauntin 1991, Fautin and Allen 1997), revealing an obvious gradient in ecological specialisation (Fig. 3). Seven (out of 24) species of anemonefishes have been recorded with just one anemone (Fig. 2). This includes the relatively widespread species, *Amphiprion sebae*, which occurs throughout the northern Indian Ocean from Yemen to Sumatra, but is always found living within *Stichdactyla haddoni* (Fauntin and Allen 1997). The most generalist species, *A. clarkii*, has been recorded with each of the ten different host anemones, including two species of anemone (*Cryptodendrum adhaesivum* and *Heteractis malu*) that are not used by any other anemonefishes (Fautin and Allen 1997). The most frequently used species of host anemone is *Heteractis crispa*, which occurs in a range of habitats and is very widespread, and is used by 13 species of anemonefishes (Fig. 2). Other important host anemones are *Entacmaea quadricolor*, *Heteractis magnifica*, *Stichodactyla mertensii*, each of which hosts at least 11 different anemonefishes, including at least one specialist species (Fig. 2).

Variation among anemonefishes in their degree of habitat specialisation may be linked to competitive ability. Experimental manipulations of anemone occupation by adult anemonefishes show that the relative abundance and host specificity of sympatric species is positively correlated to their competitive ability (Fautin 1986). For example, *Premnas biaculeatus* is only found on *Entacmaea quadricolor* at Lizard Island and was the dominant anemonefish found on this anemone (Fautin 1986).

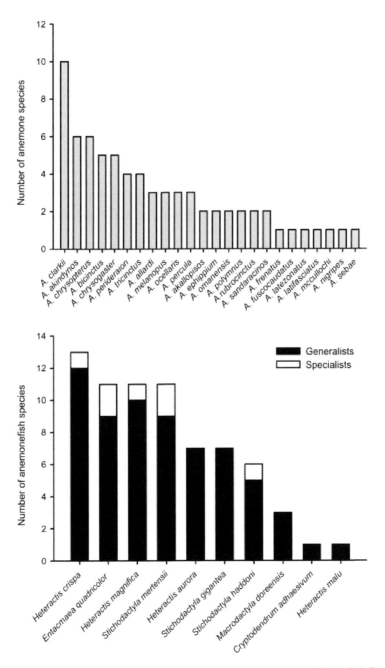

Fig. 3. Associations between anemonefishes (24 species) and their host anemones (10 species). Top panel shows the number of host anemones (out of 10) used by each of the anemonefishes, ordered from the most generalist species (*A. clarkii*), which uses all ten host anemones to the most specialised species, that only use a single species of host anemone. Bottom panel shows the total number of anemonefishes that use each of the different host anemones, including both specialist anemonefishes (white bars) that only use a single species of anemone, and more generalist anemonefishes (black bars) that use at least 2 different host anemones.

Even so, *P. biaculeatus* was not able to completely exclude other more generalist anemonefishes (*A. melanopus* and *A. akindynos*) from occupying or co-occurring on *E. quadricolor* (Fautin 1986) partly because the outcomes of interspecific competition were size dependent. Hattori (2002) also found that the generalist *A. clarkii* was an inferior competitor to the specialist *A. perideraion*. However, following a severe bleaching event that resulted in the mass mortality of the host anemone dominated by *A. perideraion*, the generalist *A. clarkii* persisted by moving on to other host anemone species, and the specialist *A. perideraion* went locally extinct (Hattori 2002). Further research is required to assess the generalities of these findings, however, it does appear that ecological specialists may be superior competitors but they may also have a higher risk of extinction.

Habitat Specialisation and Vulnerability to Disturbance

Scientific interest in the evolution of ecological specialisation (Futuyma and Moreno 1988) is largely centred on the specific conditions (e.g., the relative incidence and intensity of habitat perturbations) that promote specialisation. For species with specific adaptations that constrain resource use, these adaptations are expected to increase efficiency in the capture or assimilation of prey resources (e.g., Futuyma and Moreno 1988), or enable utilisation of resources that cannot be accessed by generalist counterparts. Generally, specialist species are expected to outperform generalist species when and where their primary resources are available and abundant (e.g., Caley and Munday 2003). If however, resource specialisation is due to strong selectivity in resource use by species that are otherwise able to exploit a wide range of resources, then preferred prey resources are expected to be those that provide the greatest energetic return or fitness benefits (Hughes 1980). In these instances, the most specialist species would be expected to be competitively superior and to monopolise access to preferred resources (e.g., Fautin 1986, Hattori 2002). Both ecological and evolutionary specialisation would typically arise (or persist) only where preferred resources were abundant or widespread. However, the extent to which fishes are obligately dependent upon specific resource types will fundamentally affect their vulnerability to disturbance and resource depletion. Vulnerability to disturbance will also be further enhanced if specialist species also have smaller population sizes and/or restricted geographic ranges as compared to generalist species (Brown 1984).

Vulnerability to disturbance

Environmental disturbances are a normal part of the ecology and dynamics of coastal and marine ecosystems (Hoegh-Guldberg and Bruno 2010, Pratchett et al. 2012), often resulting in localised depletion of key habitat forming organisms, such as kelp (Dayton 1985, Steneck et al. 2002), seagrasses (Duarte 2002), mangroves (Alongi 2002) and scleractinian corals (Bruno and Selig 2007). However, global climate change and other more direct anthropogenic disturbances are compounding upon natural and pre-existing disturbances, to greatly increase the frequency, severity and geographic extent of habitat degradation and loss across a wide range of marine ecosystems

(Pratchett et al. 2011a). On coral reefs, the effects of climate change are largely manifest as acute episodes of mass bleaching and widespread mortality of zooxanthellate organisms (e.g., corals and anemones; Fig. 4), which result from sustained increases in ocean temperatures (Hoegh-Guldberg 1999, Donner et al. 2005). The degradation and loss of these important habitat forming organisms will in turn, have significant ramifications for the large number of fishes that use these habitats (Pratchett et al. 2008), leading to marked declines in the abundance of fishes (Pratchett et al. 2011b), significant declines in local diversity (Wilson et al. 2006), and/or directional shifts in the composition of fish assemblages (Bellwood et al. 2006b). Bellwood et al. (2006b)

Fig. 4. Damselfishes residing in bleached coral hosts. Top panel: *Dascyllus aruanus* in a recently bleached *Acropora* colony in Moorea, French Polynesia in 2010. Photo by Morgan Pratchett. Bottom panel: *Amphiprion clarkii* in a bleached *Cryptodendrum adhaesivum* during a bleaching event at Christmas Island in 2010. Photo by Justin Gilligan.

reported no change in the abundance and diversity of crytpobenthic fishes (including damselfishes) following localised bleaching and coral loss in 1998 on an inshore reef of the Great Barrier Reef. However, this disturbance had strong selective effects on the abundance of individual species, with marked declines in the abundance of specialist coral-dwelling (e.g., *P. moluccensis*) and coral feeding fishes (e.g., *Gobiodon* spp.), but concomitant increases in the abundance of species (*Neopomacentrus bankieri* and *Istogobius* spp.) with much more generalist food and habitat requirements. Moreover, the pre-disturbance structure of fish assemblage is yet to be restored >12 years after the bleaching and despite recovery of benthic assemblages (Bellwood et al. 2012).

Specialisation is increasingly being considered alongside population size and geographical range as one of the key predictors of a species' extinction risk (e.g., McKinney 1997, Graham et al. 2011). In general, highly specialised species are expected to be constrained by the availability of resources and are less able to cope with fluctuations in resource availability as compared to their generalist counterparts, making them more susceptible to population collapse or extinction following major disturbances that reduce the local availability of essential resources ("the specialisation-disturbance hypothesis", Vázquez and Simberloff 2002). In support of this hypothesis, butterflyfishes that consume a smaller range of different coral prey tend to exhibit much more pronounced declines in abundance during extensive coral loss as compared to those species (e.g., *C. lunulatus*) that consume a wider range of different corals (Pratchett et al. 2008, Wilson et al. 2013). The specialisation-disturbance hypothesis presupposes, however, that specialist species are actually constrained to feed on a restricted set of prey types (i.e., "obligate specialists", Cornell 1967), whereas the dietary breadth of some species ("facultative specialists", Glasser 1982) may vary in response to changes in prey availability (e.g., Liem 1980, Robinson and Wilson 1998). Importantly, facultative specialists may withstand changes in the availability of preferred resources by expanding the range of resources that they use following major disturbances (e.g., Pratchett et al. 2004, Berumen et al. 2005).

Vulnerability and responses of damselfishes to coral loss

Although many coral reef damselfishes occupy habitats (e.g., corals and anemones) that are highly vulnerable to climate change and other largescale disturbances (e.g., Bellwood et al. 2006b, Jones et al. 2008, Pratchett et al. 2008, Saenz-Agudelo et al. 2011, Hobbs et al. 2013), there is little information on whether these species are facultative or obligate specialists, and how individual species may respond to localised habitat loss. In order to explore the responses of damselfishes to coral loss, we consulted data compiled from 26 studies at 75 reef locations around the world (Pratchett et al. 2011b), each of which documents changes in the abundance of fishes during acute disturbance events (mostly, climate induced coral bleaching) that caused >25% decline in live coral cover. In order to standardise responses according to the severity of coral loss, proportional changes in the individual abundance of each species are expressed as a ratio of the proportional loss of coral cover, following Wilson et al. (2006) and Pratchett et al. (2008, 2011b). Standardised responses are then averaged across all studies that consider each fish species (Fig. 5). Based on these data, it is expected that specialist coral-dwelling damselfishes would exhibit much more

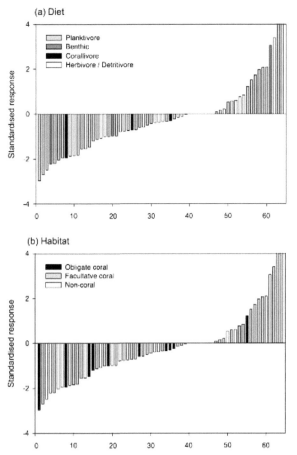

Fig. 5. Standardised responses of 64 species of damselfishes to declines in coral cover, following Wilson et al. (2006). Responses are calculated based on the proportional changes in the local abundance of each species relative to declines in total live coral cover, averaged across up to 24 distinct studies. Species are coded according to their (a) diet and (b) habitat requirements, with darker shading indicating increased reliance on live corals. List of species: 1 *Dascyllus reticulatus,* 2 *Chromis margaritifer,* 3 *Chromis lepidolepis,* 4 *Chrysiptera rex,* 5 *Neopomacentrus azysron,* 6 *Abudefduf saxatilis,* 7 *Chromis weberi,* 8 *Cheiloprion labiatus,* 9 *Dischistodus prosopotaenia,* 10 *Microspathodon chrysurus,* 11 *Chromis nitida,* 12 *Pomacentrus nagasakiensis,* 13 *Dischistodus* sp., 14 *Pomacentrus moluccensis,* 15 *Chromis viridis,* 16 *Abudefduf sexfasciatus,* 17 *Pomacentrus vaiuli,* 18 *Stegastes lividus,* 19 *Dascyllus melanurus,* 20 *Chrysiptera rollandi,* 21 *Pomacentrus amboinensis,* 22 *Plectroglyphidodon lacrymatus,* 23 *Neoglyphidodon nigroris,* 24 *Pomacentrus philippinus,* 25 *Neoglyphidodon melas,* 26 *Chrysiptera flavipinnis,* 27 *Dascyllus carneus,* 28 *Stegastes nigricans,* 29 *Chrysiptera cyanea,* 30 *Chromis cyanea,* 31 *Dischistodus melannotus,* 32 *Pomacentrus wardi,* 33 *Pomacentrus chrysurus,* 34 *Amblyglyphidodon ternatensis,* 35 *Plectroglyphidodon dickii,* 36 *Dascyllus aruanus,* 37 *Stegastes fasciolatus,* 38 *Pomacentrus bankanensis,* 39 *Stegastes apicalis,* 40 *Hemiglyphidodon plagiometopon,* 41 *Neopomacentrus cyanomos,* 42 *Pomacentrus pavo,* 43 *Dischistodus perspicillatus,* 44 *Dischistodus pseudochrysopoecilus,* 45 *Chrysiptera talboti,* 46 *Pomachromis richardsoni,* 47 *Amblyglyphidodon curacao,* 48 *Pomacentrus lepidogenys,* 49 *Stegastes acapulcoensis,* 50 *Chromis multilineata,* 51 *Pomacentrus grammorhynchus,* 52 *Pomacentrus adelus,* 53 *Stegastes partitus,* 54 *Stegastes planifrons,* 55 *Chromis ternatensis,* 56 *Dascyllus trimaculatus,* 57 *Chromis atripectoralis,* 58 *Amblyglyphidodon leucogaster,* 59 *Neopomacentrus bankieri,* 60 *Pomacentrus coelestis,* 61 *Abudefduf vaigiensis,* 62 *Pomacentrus brachialis,* 63 *Acanthochromis polyacanthus,* 64 *Stegastes leucostictus.*

pronounced responses than generalist counterparts. However, no account is taken of the selectivity of different coral disturbances, such that extensive coral loss may occur without actually affecting the narrow range of corals used by specialist species. It is also difficult to separate the effects of coral loss from associated declines in structural complexity, which could have strong negative effects on fishes with only minimal reliance on live coral (Pratchett et al. 2008). The extent to which coral-dwelling damselfishes are impacted by localised coral loss will depend upon (i) whether their specific host coral was affected, (ii) the overall extent of coral loss, (iii) changes in the relative and absolute abundance of different coral species, and (iv) the extent to which structural complexity is also affected.

Of the 64 damselfishes considered in these analyses, 17 species exhibited declines in abundance that were disproportionate (i.e., <–1) to the loss of live coral (Fig. 5). This included many species that depend on corals for food or habitat. For example, all corallivorous damselfishes (both obligate and facultative) and all obligate coral-dwelling damselfishes exhibited declines in abundance following extensive coral loss. However, declines in abundance (including some very marked declines) were also recorded for some damselfishes with no apparent dependency on live corals. Benthic reef assemblages are highly diverse, creating an array of different habitats, which fosters high species richness within fish assemblages (Messmer et al. 2011). In this diverse environment the availability of shelter may be especially important for damselfishes, since most are small bodied (TL < 10 cm), and are frequently the prey of larger fishes (Hiatt and Strasburg 1960, Kingsford 1992, Farmer and Wilson 2011). High cover and diversity of live corals is likely to greatly increase the availability of predator refuges, as well as moderate interspecific competition among coral reef damselfishes (Robertson 1996, Bonin et al. 2009) which will benefit small bodied fishes regardless of their specific affinities for live coral habitats. Despite the apparent lack of reliance on live coral or structural complexity, *C. rollandi* experienced declines in abundance that are approximately proportional to the loss of live coral (standardised response = –0.97; Fig. 5). Together with *C. rollandi*, another 16 damselfish species that display no apparent reliance on live coral declined in abundance following declines in coral cover (Fig. 5). The underlying mechanisms for such declines are not readily apparent, but highlight the complexities of the relationships between live corals and coral reef damselfishes.

Marked declines in the local abundance of coral-dwelling damselfishes following coral loss caused by coral-bleaching (Fig. 5) suggest that at least some species are obligate specialists and have a limited capacity to utilise alternative coral habitats. However, the responses of individual fishes to the loss of their specific host corals (or anemone hosts) will depend not only on the local abundance of alternative habitats, but also on the capacity of fishes to move from established habitats and successfully recolonise new habitats that are often already occupied by conspecifics and/or congenerics (Coker et al. 2013). For coral reef damselfishes, 60% (39 out of 64) of species exhibit declines in abundance following acute disturbances that cause extensive loss (25%–60%) of corals (Fig. 5). The species that exhibited the most pronounced decline in abundance is the habitat specialist *Dacyllus reticulatus* (Fig. 5). Moreover, among obligate coral-dwelling damselfish, those that associate with fewer coral species exhibited greater declines in abundance than generalist coral-dwellers that associate with a larger number of coral species (Fig. 6).

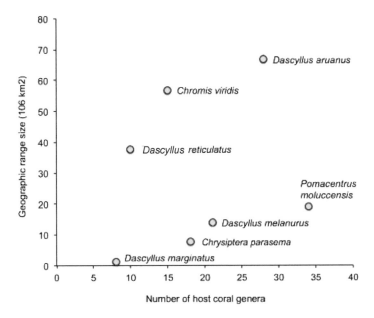

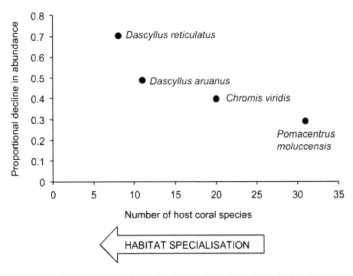

Fig. 6. Consequences of ecological specialisation for coral-dwelling damselfishes. Top panel shows the relationship between habitat specialisation (the number of coral genera reported to be used by each of 7 different damselfishes) and their geographic range. Bottom panel shows proportional declines in the abundance of coral-dwelling damselfishes following outbreaks of crown-of-thorns starfish at Lizard Island, northern GBR relative to the number of different coral species actually used. Apparent relationships are based on very limited data and will need to be tested across a much wider range of species and locations.

The majority of coral-dwelling damselfishes inhabit corals (e.g., *Pocillopora damicornis*) that are highly susceptible to climate induced bleaching and mortality (Pratchett et al. 2008). Any bleaching event, therefore, is likely to markedly affect the availability of the majority of preferred coral habitats. Consequently, we may expect a marked decline in coral-dwelling damselfishes following widespread coral bleaching and/or coral mortality (Pratchett et al. 2008), with the recovery of populations being dependent on the supply of larvae. In support of this, Halford et al. (2004) reported marked declines in the abundance, diversity, and composition of damselfishes following storm damage in the southern GBR. While the overall abundance of damselfishes recovered relatively quickly, the diversity and composition did not (see also Bellwood et al. 2006a, 2012, Halford et al. 2004).

Similar patterns are also evident when comparing damselfishes that are more or less specialised in terms of diet (as opposed to habitat use). Corallivorous damselfishes (e.g., *Plectroglyphidodon johnstonianus*) almost invariably exhibit declines in abundance following widespread coral loss, which is to be expected (Pratchett et al. 2008). However, planktivorous damselfishes and other more generalist benthic feeders (e.g., herbivorous/detrivorous species) showed mixed responses (Fig. 5). Even within refined functional groups, responses to coral loss vary spatially, and often depend on the nature of the disturbance that caused coral depletion. For example, the abundance of extensive and indeterminate algal farming damselfishes increased following an outbreak of crown-of-thorns starfish on mid-shelf reefs on the GBR, but showed no change (extensive farmers) or decreased (indeterminate farmers) in response to coral bleaching on inshore reefs in the central GBR (Emslie et al. 2012).

Anemonefishes are critically reliant on their host anemones in the field, especially for reproduction, and therefore their future persistence is intrinsically linked to that of their host anemones. Host anemones are vulnerable to increasing sea temperatures, which can cause the anemones to bleach, shrink and die (Hobbs et al. 2013). These changes in the host anemones have negative impacts on anemonefishes including reduced reproductive output and local extinction (Hattori 2002, Saenz-Agudelo et al. 2011). The degree of impact differs between anemonefishes and it appears that species that are superior competitors, use a limited range of host anemones and have limited adult mobility are the most vulnerable to disturbance (Hattori 2002). Thus, the vulnerability of damselfishes is likely to be a consequence of interactions between several biological (mobility, body size, life history traits, habitat vulnerability) and ecological factors (competitive ability, degree of specialisation).

Habitat Specialisation: Consequences for Geographic Range and Population Size

Ecological theory predicts that specialists should have smaller geographic ranges than generalist species because the distributions of specialist species are more likely to be constrained by the distribution of a few key resources (Brown 1984). If this relationship holds, highly specialised species may face a 'double jeopardy' of extinction or even a 'triple jeopardy' where specialist species are also less common than their generalist counterparts (Hawkins et al. 2000, Munday 2004). The relative abundance of generalist versus specialist species is likely to depend upon the availability of preferred resources, such that specialist species might be more abundant in specific locations where preferred resources are abundant. However, given the inherent constraints in resource availability at larger scales, it is expected that generalist species will be found in a greater range of habitats and may therefore have larger overall populations (Fox and Morrow 1981, Brown 1984). Williams et al. (2006), however, revealed a strong negative relationship between dietary evenness and geographical range for frogs (*Cophixalus* spp.) in northern Australia, showing that the most specialised species actually had the largest ranges. Williams et al. (2006) suggested that this pattern is a product of extinction filtering, whereby restricted range and therefore extinction prone species are more likely to persist if they are diet generalists. Similarly for obligate coral feeding butterflyfishes, the most specialised species, *Chaetodon trifascialis* is distributed throughout the Indo-West Pacific and has one of the largest geographical ranges of all butterflyfishes. In contrast, several endemic corallivorous butterflyfishes (e.g., *C. rainfordi* and *C. tricinctus*) have very generalised diets (Pratchett 2013, Pratchett et al. 2014).

Among coral-dwelling damselfishes, it is apparent that one of the species with the smallest geographic range is among the most specialised; *Dascyllus marginatus* is restricted to the Red Sea and Gulf of Oman, and has been reported to use a very limited range of different coral hosts, mainly *Acropora*, *Seriatopora* and *Pocillopora* (Liberman et al. 1995, and references therein). Based on the limited data available on the range of coral genera used by different coral-dwelling damselfishes, it does appear that less specialised species (e.g., *Pomacentrus moluccensis* and *Dascyllus aruanus*) have generally larger geographic ranges (Fig. 6). However, relative levels of ecological specialisation among these fishes are based on the absolute number of coral genera that they use, which could be confounded by (i) differences in the amount of research on patterns of habitat use, where it is clear that those species (*P. moluccensis* and *D. aruanus*) reported to use the greatest range of host corals are also the best studied, and (ii) variation in the availability of alternative coral habitats over geographical scales, especially when considering the extremes of the Indo-Pacific gradients of coral diversity (e.g., Red Sea), where many coral groups are rare or absent. In the absence of rigorous and comparable data on ecological specialisation we have not undertaken rigorous statistical analyses to test whether the relationship between habitat use and geographic range is significant, but these data suggest that coral-dwelling damselfishes may support the predictions of Brown (1984).

Although specialisation data are lacking for most damselfishes, the habitat use of anemonefishes has been well documented (Fautin 1991). For anemonefishes, there is evidence of a positive relationship between range size and specialisation when measured as the number of host species of anemones (Fig. 7), but this is not significant ($r^2 = 0.06$, $F_{1,22} = 1.43$, $p = 0.25$). Similarly, there was no relationship between specialisation and depth range inhabited ($R^2 = 0.01$, $F_{1,22} = 0.25$, $p = 0.62$). That is, specialist species can have geographic ranges that are just as small, and just as large, as generalist counterparts. Anemonefishes do not conform to the relationships between specialisation and abundance and specialisation and geographic range size (Fig. 7), and therefore specialist anemonefishes are not expected to have the elevated

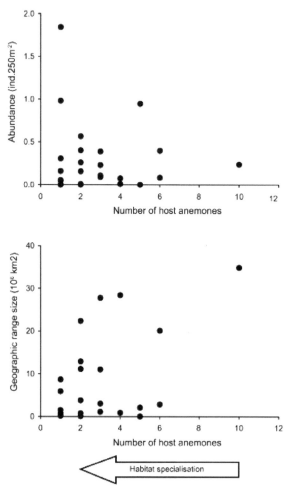

Fig. 7. Relationship between ecological specialisation (the number of anemone hosts used by each anemonefish) and mean abundance and geographic range. Neither relationship is significant (see text for statistical results).

risk of extinction that is predicted for other specialist species. However, other factors (such as specialising on a vulnerable habitat) are likely to be important in determining the vulnerability of anemonefishes to disturbances.

Abundance data recorded throughout the geographic ranges of most species (De Brauwer, unpublished data) provides the opportunity to test relationships between habitat specialisation (number of different host anemones used; Fautin 1991) and abundance. Examining this data for 24 species of anemonefishes failed to find a relationship between average density and specialisation as measured by either the number of different host anemone species used ($R^2 = 0.02$, $F_{1,22} = 0.37$, $p = 0.55$) or the breadth of the depth range inhabited ($R^2 = 0.06$, $F_{1,22} = 1.48$, $p = 0.24$). That is, the degree of habitat specialisation of an anemonefish does not predict its abundance. Similarly, in coral reef angelfishes, local abundance is not linked to the breadth of habitat used or diet (Hobbs et al. 2010). However, studies on other coral reef fishes have found that specialisation can be associated with low (Bean et al. 2002, Munday 2004) and high abundance (Jones et al. 2002, Pratchett et al. 2008). These contrasting results indicate that the relationship between specialisation and abundance is complex, varies between different taxonomic groups, and specialist species can be rare or common.

Conclusion

Coral reef damselfishes represent a broad range of different functional groups, with marked interspecific variation in their reliance on corals and other potentially vulnerable habitats (e.g., anemones) as well as differing levels of ecological specialisation. Widespread degradation of coral reef ecosystems and strong directional shifts in the structure of coral assemblages (Pandolfi et al. 2011, Hughes et al. 2012) are likely to lead to localised losses of some highly specialised coral reef damselfishes, but may benefit other damselfishes with more generalist habitat requirements. Indeed the local extinction of specialist anemonefishes has occurred following bleaching induced mass mortality of anemones, while generalist species were largely unaffected (Hattori 2002). This suggests that the composition of damselfishes on degraded reefs is likely to change (as for many other families of reef fishes), even if the overall abundance of damselfishes does not change (e.g., Bellwood et al. 2006b).

While ecological specialists are much more sensitive to the depletion of preferred resources (compared to generalist counterparts), this does not necessarily translate into an increased risk of extinction (Lawton et al. 2011). Importantly, ecological specialisation is one of a number of different factors that influence extinction risk and there is no evidence that specialist species of coral reef damselfishes necessarily have narrower ranges or smaller populations (see also Jones et al. 2002, Pratchett et al. 2008), which would otherwise enhance vulnerability to disturbance and the risk of extinction (Brown 1984). All coral reef damselfishes, however, may be susceptible to direct effects of increasing temperature (Johansen and Jones 2011) and ocean acidification (Munday et al. 2012). Johansen and Jones (2011) suggested that significant loss of damselfish species will occur if ocean temperatures increase $\geq 3°C$ above present and unless adaptation can occur. There is however, a definite need for increased research on patterns of habitat use, especially among coral-dwelling damselfishes, in order to better resolve apparent consequences of ecological specialisation.

References

Allen, G.R. 1972. The Anemonefishes: Their Classification and Biology. TFH Publications, Neptune City.
Allen, G.R. 1991. Damselfishes of the World. Mergus, Melle, Germany.
Allen, G.R. and T.B. Werner. 2002. Coral reef fish assessment in the 'coral triangle' of southeastern Asia. Environ. Biol. Fishes 65: 209–214.
Allison, G. 2004. The influence of species diversity and stress intensity on community resistance and resilience. Ecol. Monogr. 74: 117–134.
Alongi, D.M. 2002. Present state and future of the world's mangrove forests. Environ. Conserv. 29: 331–349.
Arnold, S.N., R.S. Steneck and P.J. Mumby. 2010. Running the gauntlet: inhibitory effects of algal turfs on the process of coral recruitment. Mar. Ecol.-Prog. Ser. 414: 91–105.
Asoh, K. and M. Kasuya. 2002. Gonadal development and mode of sexuality in a coral reef damselfish, *Dascyllus trimaculatus*. J. Zool. 256(3): 301–309.

Bay, L.K., G.P. Jones and M.I. McCormick. 2001. Habitat selection and aggression as determinants of spatial segregation among damselfish on a coral reef. Coral Reefs 20: 289–298.

Bean, K., G.P. Jones and M.J. Caley. 2002. Relationships among distribution, abundance and microhabitat specialisation in a guild of coral reef triggerfish (family Balistidae). Mar. Ecol.-Prog. Ser. 233: 263–272.

Bellwood, D.R. and T.P. Hughes. 2001. Regional-scale assembly rules and biodiversity of coral reefs. Science 292: 1532–1535.

Bellwood, D.R., A.S. Hoey and J.H. Choat. 2003. Limited functional redundancy in high diversity systems: resilience and ecosystem function on coral reefs. Ecol. Lett. 6: 281–285.

Bellwood, D.R., T.P. Hughes, C. Folke and M. Nyström. 2004. Confronting the coral reef crisis. Nature 429: 827–833.

Bellwood, D.R., T.P. Hughes and A.S. Hoey. 2006a. Sleeping functional group drives coral-reef recovery. Curr. Biol. 16: 2434–2439.

Bellwood, D.R., A.S. Hoey, J.L. Ackerman and M. Depczynski. 2006b. Coral bleaching, reef fish community phase shifts and the resilience of coral reefs. Glob. Change Biol. 12: 1587–1594.

Bellwood, D.R., S. Klanten, M.S. Pratchett, N. Konow and L. van Herwerden. 2009. Evolutionary history of the butterflyfishes (f: Chaetodontidae) and the rise of coral feeding fishes. J. Evol. Biol. 23: 335–349.

Bellwood, D.R., A.H. Baird, M. Depczynski, A. González-Cabello, A.S. Hoey, C.D. Lefèvre and J.K. Tanner. 2012. Coral recovery may not herald the return of fishes on damaged coral reefs. Oecologia 170: 567–573.

Berumen, M.L., M.S. Pratchett and M.I. McCormick. 2005. Within-reef differences in diet and body condition of coral-feeding butterflyfishes (Chaetodontidae). Mar. Ecol.-Prog. Ser. 287: 217–227.

Beukers, J.S. and G.P. Jones. 1998. Habitat complexity modifies the impact of piscivores on a coral reef fish population. Oecologia 114: 50–59.

Blondel, J. 2003. Guilds or functional groups: does it matter? Oikos 100: 223–231.

Blowes, S.A., M.S. Pratchett and S.R. Connolly. 2013. Heterospecific aggression and dominance in a guild of coral-feeding fishes: the roles of dietary ecology and phylogeny. Am. Nat. 182: 157–168.

Bonin, M.C., M. Srinivasan, G.R. Almany and G.P. Jones. 2009. Interactive effects of interspecific competition and microhabitat on early post-settlement survival in a coral reef fish. Coral Reefs 28: 265–274.

Booth, D.J. 1992. Larval settlement patterns and preferences by domino damselfish *Dascyllus albisella* Gill. J. Exp. Mar. Biol. Ecol. 155: 85–104.

Brooks, T.M., R.A. Mittermeier, C.G. Mittermeier, G.A. Da Fonseca, A.B. Rylands, W.R. Konstant, P. Flick, J. Pilgrim, S. Oldfield, G. Magin and C. Hilton-Taylor. 2002. Habitat loss and extinction in the hotspots of biodiversity. Conserv. Biol. 16(4): 909–923.

Brown, J.H. 1984. On the relationship between abundance and distribution of species. Am. Nat. 124: 255–279.

Bruno, J.F. and E.R. Selig. 2007. Regional decline of coral cover in the Indo-Pacific: timing, extent, and subregional comparisons. PLoS ONE 2: e711.

Burkepile, D.E. and M.E. Hay. 2008. Herbivore species richness and feeding complementarity affect community structure and function on a coral reef. Proc. Natl. Acad. Sci. USA 105: 16201–16206.

Caley, M.J. and P.L. Munday. 2003. Growth trades off with habitat specialisation. Proc. R. Soc. B-Biol. Sci. 270(Suppl. 2): S175–S177.

Caley, M.J., K.A. Buckley and G.P. Jones. 2001. Separating the effects of habitat fragmentation, degradation, and loss on coral commensals. Ecology 82: 3435–3448.

Ceccarelli, D.M. 2007. Modification of benthic communities by territorial damselfish: a multi-species comparison. Coral Reefs 26: 853–866.

Ceccarelli, D.M., G.P. Jones and L.J. McCook. 2001. Territorial damselfishes as determinants of the structure of benthic communities on coral reefs. Oceanogr. Mar. Biol. Annu. Rev. 39: 355–389.

Ceccarelli, D.M., T.P. Hughes and L.J. McCook. 2006. Impacts of simulated overfishing on the territoriality of coral reef damselfish. Mar. Ecol.-Prog. Ser. 309: 255–262.

Coker, D.J., M.S. Pratchett and P.L. Munday. 2009. Coral bleaching and habitat degradation increases susceptibility to predation for coral-dwelling fishes. Behav. Ecol. 20: 1204–1210.

Coker, D.J., N.A.J. Graham and M.S. Pratchett. 2012. Interactive effects of live coral and structural complexity on the recruitment of reef fishes. Coral Reefs 31: 919–927.

Coker, D.J., S.P.W. Walker, P.L. Munday and M.S. Pratchett. 2013. Social group entry rules may limit population resilience to patchy habitat disturbance. Mar. Ecol.-Prog. Ser. 493: 237–242.

Coker, D.J., S.K. Wilson and M.S. Pratchett. 2014. Importance of live coral habitat for reef fishes. Rev. Fish. Biol. Fish. 24: 89–126.

Coker, D.J., J.P. Nowicki and M.S. Pratchett. 2015. Body condition of the coral-dwelling fish *Dascyllus aruanus* (Linnaeus 1758) following host colony bleaching. Environ. Biol. Fishes 98: 691–695.

Connell, J.H. 1978. Diversity in tropical rain forests and coral reefs. Science 199: 1302–1310.

Cooper, W.J., L.L. Smith and M.W. Westneat. 2009. Exploring the radiation of a diverse reef fish family: Phylogenetics of the damselfishes (Pomacentridae), with new classifications based on molecular analyses of all genera. Mol. Phylogenet. Evol. 52: 1–16.

Cornell, H. 1967. Search strategies and the adaptive significance of switching in some general predators. Am. Nat. 110: 317–319.

Dayton, P.K. 1985. Ecology of kelp communities. Annu. Rev. Ecol. Evol. Syst. 16: 215–245.

Devictor, V., J. Clavel, R. Julliard, S. Lavergne, D. Mouillot, W. Thuiller, P. Venail, S. Villéger and N. Mouquet. 2010. Defining and measuring ecological specialisation. J. Appl. Ecol. 47: 15–25.

Donner, S.D., W.J. Skirving, C.M. Little, M. Oppenheimer and O. Hoegh-Guldberg. 2005. Global assessment of coral bleaching and required rates of adaptation under climate change. Glob. Change Biol. 11: 2251–2265.

Duarte, C.M. 2002. The future of seagrass meadows. Environ. Conserv. 29: 192–206.

Eeley, H.A.C. and R.A. Foley. 1999. Species richness, species range size and ecological specialisation among African primates: geographical patterns and conservation implications. Biodivers. Conserv. 8: 1033–1056.

Elliott, J.K. and R.N. Mariscal. 2001. Coexistence of nine anemonefish species: differential host and habitat utilization, size and recruitment. Mar. Biol. 138(1): 23–36.

Emslie, M.J., M. Logan, D.M. Ceccarelli, A.J. Cheal, A.S. Hoey, I. Miller and H.P.A. Sweatman. 2012. Regional-scale variation in the distribution and abundance of farming damselfishes on Australia's Great Barrier Reef. Mar. Biol. 159: 1293–1304.

Fahrig, L. 2001. How much habitat is enough? Biol. Conserv. 100: 65–74.

Farmer, B. and S.K. Wilson. 2011. Diet of targeted finfishes and the implications for trophic cascades. Environ. Biol. Fishes 91: 71–85.

Fautin, D.G. 1986. Why do anemonefishes inhabit only some host actinians? Environ. Biol. Fishes 15: 171–180.

Fautin, D.G. 1991. The anemonefish symbiosis: what is known and what is not. Symbiosis 10: 23–46.

Fautin, D.G. and G.R. Allen. 1997. Anemonefishes and their Host Sea Anemones. Western Australian Museum, Perth.

Feary, D.A., G.R. Almany, M.I. McCormick and G.P. Jones. 2007. Habitat choice, recruitment and the response of coral reef fishes to coral degradation. Oecologia 153: 727–737.

Forrester, G.E. 1990. Factors influencing the juvenile demography of a coral reef fish. Ecology 71(5): 1666–1681.

Fox, L.R. and P.A. Morrow. 1981. Specialisation: species property or local phenomenon. Science 211: 887–893.

Fox, R.J., T.L. Sunderland, A.S. Hoey and D.R. Bellwood. 2009. Estimating ecosystem function: contrasting roles of closely related herbivorous rabbitfishes (Siganidae) on coral reefs. Mar. Ecol.-Prog. Ser. 385: 261–269.

Frédérich, B., O. Lehanse, P. Vandewalle and G. Lepoint. 2010. Trophic niche width, shift, and specialisation of *Dascyllus aruanus* in Toliara lagoon, Madagascar. Copeia 2010: 218–226.

Futuyma, D.J. and G. Moreno. 1988. The evolution of ecological specialisation. Annu. Rev. Ecol. Evol. Syst. 19: 207–233.

Genner, M.J., G.F. Turner and S.J. Hawkins. 1999. Foraging of rocky habitat cichlid fishes in Lake Malawi: coexistence through niche partitioning? Oecologia 121: 283–292.

Glasser, J.W. 1982. A theory of trophic strategies: the evolution of facultative specialists. Am. Nat. 119: 250–262.

Gochfeld, D.J. 2010. Territorial damselfishes facilitate survival of corals by providing an associational defense against predators. Mar. Ecol.-Prog. Ser. 398: 137–148.

Graham, N.A.J., S.K. Wilson, S. Jennings, N.V.C. Polunin, J.P. Bijoux and J. Robinson. 2006. Dynamic fragility of oceanic coral reef ecosystems. Proc. Natl. Acad. Sci. USA 103: 8425–8429.

Graham, N.A.J., P. Chabanet, R.D. Evans, S. Jennings, Y. Letourneur, A.M. MacNeil, T.R. McClanahan, M.C. Öhman, N.V.C. Polunin and S.K. Wilson. 2011. Extinction vulnerability of coral reef fishes. Ecol. Lett. 14: 341–348.

Halford, A., A.J. Cheal, D. Ryan and D.M. Williams. 2004. Resilience to large-scale disturbance in coral and fish assemblages on the Great Barrier Reef. Ecology 85: 1892–1905.

Hata, H. and M. Kato. 2004. Monoculture and mixed-species algal farms on a coral reef are maintained through intensive and extensive management by damselfishes. J. Exp. Mar. Biol. Ecol. 313: 285–296.

Hattori, A. 2002. Small and large anemonefishes can coexist using the same patchy resources on a coral reef, before habitat destruction. J. Anim. Ecol. 71(5): 824–831.

Hattori, A. 2005. High mobility of the protandrous anemonefish *Amphiprion frenatus*: nonrandom pair formation in limited shelter space. Ichthyol. Res. 52(1): 57–63.

Hawkins, J.P., C.M. Roberts and V. Clark. 2000. The threatened status of restricted-range coral reef fish species. Anim. Conserv. 3: 81–88.

Hiatt, R.W. and D.W. Strasburg. 1960. Ecological relationships of the fish fauna on coral reefs of the Marshall Islands. Ecol. Monogr. 30: 65–127.

Hixon, M.A. and W.N. Brostoff. 1983. Damselfish as keystone species in reverse: intermediate disturbance and diversity of reef algae. Science 220: 511–513.

Hixon, M.A. and J.P. Beets. 1993. Predation, prey refuges, and the structure of coral-reef fish assemblages. Ecol. Monogr. 63: 77–101.

Hobbs, J.P.A., G.P. Jones and P.L. Munday. 2010. Rarity and extinction risk in coral reef angelfishes on isolated islands: interrelationships among abundance, geographic range size and specialisation. Coral Reefs 29: 1–11.

Hobbs, J.P.A., A.J. Frisch, B.M. Ford, M. Thums, P. Saenz-Agudelo, K.A. Furby and M.L. Berumen. 2013. Taxonomic, spatial and temporal patterns of bleaching in anemones inhabited by anemonefishes. PLoS ONE 8(8): e70966.

Hoekstra, J.M., T.M. Boucher, T.H. Ricketts and C. Roberts. 2005. Confronting a biome crisis: global disparities of habitat loss and protection. Ecol. Lett. 8: 23–29.

Hoegh-Guldberg, O. 1999. Climate change, coral bleaching and the future of the world's coral reefs. Mar. Freshw. Res. 50: 839–866.

Hoegh-Guldberg, O. and J.F. Bruno. 2010. The impact of climate change on the world's marine ecosystems. Science 328: 1523–1528.

Hoey, A.S. and M.I. McCormick. 2004. Selective predation for low body condition at the larval-juvenile transition of a coral reef fish. Oecologia 139: 23–29.

Hoey, A.S. and D.R. Bellwood. 2009. Limited functional redundancy in a high diversity system: single species dominates key ecological process on coral reefs. Ecosystems 12: 1316–1328.

Hoey, A.S. and D.R. Bellwood. 2010. Damselfish territories as a refuge for macroalgae on coral reefs. Coral Reefs 29: 107–118.

Hoey, A.S. and D.R. Bellwood. 2011. Suppression of herbivory by macroalgal density: a critical feedback on coral reefs? Ecol. Lett. 14: 267–273.

Hoey, J., M.I. McCormick and A.S. Hoey. 2007. Influence of depth on sex-specific energy allocation patterns in a tropical reef fish. Coral Reefs 26: 603–613.

Holbrook, S.J. and R.J. Schmitt. 2002. Competition for shelter space causes density-dependent predation mortality in damselfishes. Ecology 83: 2855–2868.

Holbrook, S.J., G.E. Forrester and R.J. Schmitt. 2000. Spatial patterns in abundance of a damselfish reflect availability of suitable habitat. Oecologia 122: 109–120.

Hughes, R.N. 1980. Optimal foraging theory in the marine context. Oceanogr. Mar. Biol. Annu. Rev. 18: 423–481.

Hughes, T.P., A.H. Baird, E.A. Dinsdale, N.A. Moltschaniwskyj, M.S. Pratchett, J.E. Tanner and B.L. Willis. 2012. Assembly rules of reef corals are flexible along a climatic gradient. Curr. Biol. 22: 736–741.

Jennings, S., A.S. Brierley and J.W. Walker. 1994. The inshore fish assemblages of the Galápagos Archipelago. Biol. Conserv. 70: 49–57.

Johansen, J.L. and G.P. Jones. 2011. Increasing ocean temperature reduces the metabolic performance and swimming ability of coral reef damselfishes. Glob. Change Biol. 17: 2971–2979.

Johnson, M., S. Holbrook, R. Schmitt and A. Brooks. 2011. Fish communities on staghorn coral: effects of habitat characteristics and resident farmerfishes. Environ. Biol. Fishes 91(4): 429–448.

Jones, A.M., S. Gardner and W. Sinclair. 2008. Losing 'Nemo': bleaching and collection appear to reduce inshore populations of anemonefishes. J. Fish Biol. 73(3): 753–761.

Jones, G.P. 1988. Experimental evaluation of the effects of habitat structure and competitive interactions on the juveniles of two coral reef fishes. J. Exp. Mar. Biol. Ecol. 123: 115–126.

Jones, G.P., M.J. Caley and P.L. Munday. 2002. Rarity in coral reef fish communities. pp. 81–101. *In*: P.F. Sale (ed.). Coral Reef Fishes: Dynamics and Diversity in a Complex Ecosystem. Academic Press, San Diego.

Jones, G.P., M.I. McCormick, M. Srinivasan and J.V. Eagle. 2004. Coral decline threatens fish biodiversity in marine reserves. Proc. Natl. Acad. Sci. USA 101: 8251–8253.

Kingsford, M.J. 1992. Spatial and temporal variation in predation on reef fishes by coral trout (*Plectropomus leopardus*, Serranidae). Coral Reefs 11: 193–198.

Knudsen, J.W. 1967. *Trapezia* and *Tetralia* (Decapoda, Brachyura, Xanthidae) as obligate ectoparasites of pocilloporid and acroporid corals. Pac. Sci. 21: 50–57.

Kuwamura, T., Y. Yogo and Y. Nakashima. 1994. Population-dynamics of goby *Paragobiodon echinocephalus* and host coral *Stylophora pistillata*. Mar. Ecol.-Prog. Ser. 103: 17–23.

Lawton, R.J., V. Messmer, M.S. Pratchett and L.K. Bay. 2011. High gene flow across large geographic scales reduces extinction risk for a highly specialised coral feeding butterflyfish. Mol. Ecol. 20: 3584–3598.

Lecchini, D. and R. Galzin. 2005. Spatial repartition and ontogenetic shifts in habitat use by coral reef fishes (Moorea, French Polynesia). Mar. Biol. 147: 47–58.

Lecchini, D., S. Planes and R. Galzin. 2007. The influence of habitat characteristics and conspecifics on attraction and survival of coral reef fish juveniles. J. Exp. Mar. Biol. Ecol. 341: 85–90.

Liberman, T., A. Genin and Y. Loya. 1995. Effects on growth and reproduction of the coral *Stylophora pistillata* by the mutualistic damselfish *Dascyllus marginatus*. Mar. Biol. 121: 741–746.

Liem, K.F. 1980. Adaptive significance of intra- and interspecific differences in the feeding repertoires of cichlid fishes. Am. Zool. 20: 295–314.

Litsios, G. and N. Salamin. 2014. Hybridisation and diversification in the adaptive radiation of clownfishes. BMC Evol. Biol. 14: 245.

Litsios, G., C.A. Sims, R.O. Wüest, P.B. Pearman, N.E. Zimmermann and N. Salamin. 2012. Mutualism with sea anemones triggered the adaptive radiation of clownfishes. BMC Evol. Biol. 12: 212.

Lobel, P.S. 1980. Herbivory by damselfishes and their role in coral reef community ecology. Bull. Mar. Sci. 30(Supplement 1): 273–289.

McCormick, M.I. and A.S. Hoey. 2004. Larval growth history determines juvenile growth and survival in a tropical marine fish. Oikos 106: 225–242.

McKinney, M.L. 1997. Extinction vulnerability and selectivity: combining ecological and paleontological views. Annu. Rev. Ecol. Evol. Syst. 495–516.

McNally, R.C. 1995. Ecological Versatility and Community Ecology. Cambridge University Press, Cambridge.

Meekan, M.G., A.D.L. Steven and M.J. Fortin. 1995. Spatial patterns in the distribution of damselfishes on a fringing coral reef. Coral Reefs 14: 151–161.

Messmer, V., G.P. Jones, P.L. Munday, S.J. Holbrook, R.J. Schmitt and A.J. Brooks. 2011. Habitat biodiversity as a determinant of fish community structure on coral reefs. Ecology 92(12): 2285–2298.

Munday, P.L. 2004. Habitat loss, resource specialisation, and extinction on coral reefs. Glob. Change Biol. 10: 1642–1647.

Munday, P.L., G.P. Jones and M.J. Caley. 1997. Habitat specialisation and the distribution and abundance of coral-dwelling gobies. Mar. Ecol.-Prog. Ser. 152: 227–239.

Munday, P.L., L. van Herwerden and C.L. Dudgeon. 2004. Evidence for sympatric speciation by host shift in the sea. Curr. Biol. 14: 1498–1504.

Munday, P.L., G.P. Jones, M.S. Pratchett and A. Williams. 2008. Climate change and the future for coral reef fishes. Fish Fish. 9: 261–285.

Munday, P.L., M.I. McCormick and G.E. Nilsson. 2012. Impact of global warming and rising CO_2 levels on coral reef fishes: what hope for the future? J. Exp. Biol. 215: 3865–3873.

Noonan, S.H., G.P. Jones and M.S. Pratchett. 2012. Coral size, health and structural complexity: effects on the ecology of a coral reef damselfish. Mar. Ecol.-Prog. Ser. 456: 127–137.

Öhman, M.C., P.L. Munday, G.P. Jones and M.J. Caley. 1998. Settlement strategies and distribution patterns of coral-reef fishes. J. Exp. Mar. Biol. Ecol. 225: 219–238.

Pandolfi, J.M., S.R. Connolly, D.J. Marshall and A.L. Cohen. 2011. Projecting coral reef futures under global warming and ocean acidification. Science 333: 418–422.

Pratchett, M.S. 2001. Influence of coral symbionts on feeding preferences of crown-of-thorns starfish. Mar. Ecol.-Prog. Ser. 214: 111–119.

Pratchett, M.S. 2005. Dietary overlap among coral-feeding butterflyfishes (Chaetodontidae) at Lizard Island, northern Great Barrier Reef. Mar. Biol. 148: 373–382.

Pratchett, M.S. 2013. Feeding preferences and dietary specialisation among obligate coral-feeding butterflyfishes. pp. 140–179. *In*: M.S. Pratchett, M.L. Berumen and B. Kapoor (eds.). Biology of Butterflyfishes. CRC Press, Boca Raton.

Pratchett, M.S., S.K. Wilson, M.L. Berumen and M.I. McCormick. 2004. Sublethal effects of coral bleaching on an obligate coral feeding butterflyfish. Coral Reefs 23: 352–356.

Pratchett, M.S., P.L. Munday, S.K. Wilson, N.A.J. Graham, J.E. Cinner and D.R. Bellwood. 2008. Effects of climate-induced coral bleaching on coral-reef fishes-ecological and economic consequences. Oceanogr. Mar. Biol.: An Annual Review 46: 251–296.

Pratchett, M.S., L.K. Bay, P.C. Gehrke, J.D. Koehn, K. Osborne, R.L. Pressey, H.P.A. Sweatman and D. Wachenfeld. 2011a. Contribution of climate change to degradation and loss of critical fish habitats in Australian marine and freshwater environments. Mar. Freshw. Res. 62: 1062–1081.

Pratchett, M.S., A.S. Hoey, S.K. Wilson, V. Messmer and N.A.J. Graham. 2011b. Changes in the biodiversity and functioning of reef fish assemblages following coral bleaching and coral loss. Diversity 3: 424–452.

Pratchett, M.S., D.J. Coker, G.P. Jones and P.L. Munday. 2012. Specialisation in habitat use by coral reef damselfishes and their susceptibility to habitat loss. Ecol. Evol. 2: 2168–2180.

Pratchett, M.S., A.S. Hoey, C. Cvitanovic, J.A. Hobbs and C.J. Fulton. 2014. Abundance, diversity, and feeding behaviour of coral reef butterflyfishes at Lord Howe Island. Ecol. Evol. 4: 3612–3625.

Rasher, D.B., A.S. Hoey and M.E. Hay. 2013. Consumer diversity interacts with prey defenses to drive ecosystem function. Ecology 94: 1347–1358.

Robertson, D.R. 1996. Interspecific competition controls abundance and habitat use of territorial Caribbean damselfishes. Ecology 77: 885–899.

Robertson, D.R. and B. Lassig. 1980. Spatial distribution patterns and coexistence of a group of territorial damselfishes from the Great Barrier Reef. Bull. Mar. Sci. 30: 187–203.

Robinson, B.W. and D.S. Wilson. 1998. Optimal foraging, specialisation, and a solution to Liem's paradox. Am. Nat. 151: 223–235.

Ross, S.T. 1986. Resource partitioning in fish assemblages: a review of field studies. Copeia 1982(2): 352–388.

Saenz-Agudelo, P., G.P. Jones, S.R. Thorrold and S. Planes. 2011. Detrimental effects of host anemone bleaching on anemonefish populations. Coral Reefs 30(2): 497–506.

Sale, P.F. 1977. Maintenance of high diversity in coral reef fish communities. Am. Nat. 111(978): 337–359.

Santini, S. and G. Polacco. 2006. Finding Nemo: molecular phylogeny and evolution of the unusual life style of anemonefish. Gene 385: 19–27.

Schmitt, R.J. and S.J. Holbrook. 1999. Settlement and recruitment of three damselfish species: larval delivery and competition for shelter space. Oecologia 118: 76–86.

Shpigel, M. 1982. Niche overlap among two species of coral dwelling fishes of the genus *Dascyllus* (Pomacentridae). Environ. Biol. Fishes 7(1): 65–68.

Steneck, R.S., M.H. Graham, B.J. Bourque, D. Corbett, J.M. Erlandson, J.A. Estes and M.J. Tegner. 2002. Kelp forest ecosystems: biodiversity, stability, resilience and future. Environ. Conserv. 29: 436–459.

Sweatman, H.P.A. 1985. The influence of adults of some coral reef fishes on larval recruitment. Ecol. Monogr. 55: 469–485.

Vázquez, D.P. and D. Simberloff. 2002. Ecological specialisation and susceptibility to disturbance: conjectures and refutations. Am. Nat. 159: 606–623.

Veron, J.E.N. 2000. Corals of the World. Australian Institute of Marine Science, Townsville.

Vitousek, P.M. 1997. Human-domination of Earth's ecosystems. Science 275: 494–499.

Wellington, G.M. 1982. Depth zonation of corals in the Gulf of Panama: control and facilitation by resident reef fishes. Ecol. Monogr. 52: 223–241.

Wellington, G.M. 1992. Habitat selection and juvenile persistence control the distribution of two closely related Caribbean damselfishes. Oecologia 90: 500–508.

Williams, Y.M., S.E. Williams, R.A. Alford, M. Waycott and C.N. Johnson. 2006. Niche breadth and geographical range: ecological compensation for geographical rarity in rainforest frogs. Biol. Lett. 2: 532–535.

Wilson, S.K. and D.R. Bellwood. 1997. Cryptic dietary components of territorial damselfishes (Pomacentridae, Labroidei). Mar. Ecol.-Prog. Ser. 153: 299–310.

Wilson, S.K., N.A. Graham, M.S. Pratchett, G.P. Jones and N.V.C. Polunin. 2006. Multiple disturbances and the global degradation of coral reefs: are reef fishes at risk or resilient? Glob. Change Biol. 12: 2220–2234.

Wilson, S.K., S.C. Burgess, A.J. Cheal, M. Emslie, R. Fisher, I. Miller, N.V.C. Polunin and H.P. Sweatman. 2008. Habitat utilization by coral reef fish, implications for specialists vs. generalists in a changing environment. J. Anim. Ecol. 77: 220–228.

Wilson, S.K., M. Depczynski, R. Fisher, T.H. Holmes, R.A. O'Leary and P. Tinkler. 2010. The importance of coral in the habitat use of juvenile reef fish. PLoS ONE 5: e15185.

Wilson, S.K., N.A.J. Graham and M.S. Pratchett. 2013. Susceptibility of butterflyfish to habitat disturbance: Do 'chaets' ever prosper? pp. 226–245. *In*: M.S. Pratchett, M.L. Berumen and B. Kapoor (eds.). Biology of Butterflyfishes. CRC Press, Boca Raton.

6

Farming Behaviour of Territorial Damselfishes

Hiroki Hata[1],* and *Daniela M. Ceccarelli*[2]

Introduction

Herbivory on coral reefs is an essential factor in preventing competitive dominance of macroalgae over corals (e.g., Bellwood et al. 2004). Numerous herbivore exclusion experiments have shown dramatic shifts from corals to macroalgae (e.g., Hughes et al. 2007), and reefs where herbivores are fishery targets are expected to be at greater risk of general degradation (Mumby 2006). Most grazing fishes on coral reefs are not strictly herbivorous, but also ingest detritus or animal matter (Choat et al. 2002). On coral reefs, we can distinguish between two main groups of herbivores: large roving grazers such as labrids (subfamily Scarinae), acanthurids and siganids, which forage across wide areas, often in schools, and suppress algal growth; and territorial herbivores (almost exclusively damselfishes, commonly referred to as "farmers"), which defend rich algal turfs against roving herbivores and harvest the algae inside the territory (Ceccarelli 2007, Feitosa et al. 2012).

Research on farming damselfishes spans four decades, with much progress being made in our ability to discern different behaviours, farming modes and effects on benthic communities. Early research focused on describing the behavioural phenomenon of farming and territoriality (e.g., Ogden and Lobel 1978), and later efforts characterised the composition of farmed turfs (e.g., Lassuy 1980, Klumpp et al. 1987, Ferreira et al. 1998). Ceccarelli et al.'s (2001) review highlighted the

[1] Graduate School of Science and Engineering, Ehime University, Bunkyo 2-5, Matsuyama, Ehime, Japan.
[2] ARC Centre of Excellence for Coral Reef Studies, James Cook University, Townsville, Australia.
  Email: dmcecca@gmail.com
* Corresponding author: hata@sci.ehime-u.ac.jp

fragmented nature of the research, and subsequent studies have investigated the diversity of farmer territories, highlighting the differences between species and across spatial scales (Ceccarelli 2007, Barneche et al. 2009). Following on from behavioural and experimental studies, the most recent research uses sophisticated methods such as stable isotope analysis to discern dietary details in territorial damselfishes (Frédérich et al. 2009, Hata and Umezawa 2011, Dromard et al. 2013).

In terrestrial ecosystems, farming is known to change the magnitude and flow of energy and materials. For example, humans use around 1.5 billion ha of land, which amounts to 10% of the potentially arable land globally, to produce 6.6 billion tons of primary production per year (2012 estimate, FAOSTAT 2014). Before we acquired agriculture, the hunter-gatherer lifestyle supported only about 4 million people globally, but modern agriculture now feeds 6,000 million people (Tilman et al. 2002). Several authors have suggested that damselfish territories similarly enhance primary production and nitrogen fixation on coral reefs (Russ 1987, Ceccarelli et al. 2005), indirectly (and perhaps unintentionally) supporting varied populations of herbivores. On the other hand, the high densities of farming damselfish have been associated with reef degradation (Sandin et al. 2008). In this chapter, we describe farming behaviour of damselfishes and the known consequences of this behaviour for coral reef ecosystems. We explore the implications for the farmers themselves, in the form of their digestion and behaviour, and the effects on other members of the reef community. We also present the current knowledge about farming effects on benthic communities, especially on algae, corals, and other invertebrates.

Farming Behaviour

The mechanisms of farming by damselfishes are still being uncovered, but it is thought that damselfishes manage their farms in at least three ways. Firstly and most obviously, territorial defence alters the intensity and type of herbivory that occurs inside territories, leading to an algal assemblage which differs from that found outside territory boundaries. Some damselfishes exclude even sea-urchins, picking them up by their spines and depositing them outside the territory boundaries (e.g., *Stegastes nigricans, S. planifrons*, Fig. 1; Williams 1979, Sammarco and Williams 1982, Hata and Kato 2004). Experimental removals of damselfishes can cause a rapid invasion of roving herbivores into the territory, resulting in the consumption of the palatable turf algae and the demise of the algal farm within a few days (e.g., *Stegastes fasciolatus, S. altus*, Mahoney 1981, Kohda 1984, Hourigan 1986). Secondly, territorial damselfishes prepare substratum for their farm, most notably by killing corals (e.g., *Plectroglyphidodon dickii*, Kaufman 1977, Wellington 1982, Jones et al. 2006). Damselfishes also invest in farm maintenance, "cleaning up" debris that land in their farms (e.g., *Pomacentrus alexandrae*, Doropoulos et al. 2013). Finally, some damselfish species selectively remove unpalatable algae ("weeding") in order to promote the growth of their preferred algae (Fig. 1; Irvine 1980, Lassuy 1980, Hata and Kato 2002, Hoey and Bellwood 2010). *Stegastes nigricans*, for example, weed the less digestible corticated or calcified algae from their territories, promoting

Fig. 1. A territorial damselfish, *Stegastes nigricans*, and its algal farm dominated by a filamentous rhodophyta, *Polysiphonia* sp. (A) Weeding behaviour by *S. nigricans* that eliminates unpalatable algae from its territory (B), and defending behaviour against sea-urchins by *S. lividus* that discards sea urchins from its territory (C). Photos B and C courtesy of Hiroyuki Masuhara.

the growth of fine filamentous algae (Hata and Kato 2002). However, it appears that weeding is not necessary to maintain algal farms, and many damselfish species have not acquired this behaviour.

Territorial Damselfish Diets and the Function of Feeding Territories

Farming behaviour has appeared at least twice in the evolutionary history of the damselfish family; within the Stegastinae and the Pomacentrinae (Frédérich et al. 2013). Within the territories, the damselfishes establish and cultivate algal farms on which they feed. The primary components of the algal farms are usually filamentous turf-forming algae (e.g., Ceccarelli et al. 2001 and its citations). Contrary to the original belief that all farming damselfishes are strictly herbivorous, successive studies have found a range of diets (e.g., Moreno-Sanchez et al. 2011), resulting in various explanations for the role played by the algal turfs defended within the territories. The majority of species for which stomach contents have been examined reveal a preference for a narrow range of filamentous algal species (Cleveland and Montgomery 2003, Hata and Umezawa 2011), but some species also appear to ingest and assimilate significant amounts of detritus and/or animal matter (e.g., Wilson and Bellwood 1997, Ceccarelli 2007). This suggests that along with providing food, the turf may also serve as a trap for detritus (Wilson et al. 2003) or as a "ranch" for invertebrates (Ferreira et al. 1998, Hata and Umezawa 2011, Dromard et al. 2013). It is hypothesised that the algal mats function as a refuge from predation and as a rich food source for the invertebrates (Ferreira et al. 1998). In the territories of *Stegastes fuscus*, the densities of invertebrates were 2.6 times higher than outside the territories (Table 1; Ferreira et al. 1998, Dromard et al. 2013).

To the best of our knowledge, the adaptation of territorial damselfishes to herbivory seems limited, as they are unable to break down algal cellulose. Therefore, these damselfishes crop the fine filamentous algae using their incisiform teeth (see Chapter XIV), lyse the cell membranes through porous cell walls in highly acidic stomachs for subsequent digestion of cell contents (Zemke-White and Clements 1999, Zemke-White et al. 2000), and absorb them in the long intestines (Horn 1989, Cleveland and Montgomery 2003). Animal and detrital matter may serve to supplement their diets at different stages of their life cycles.

Farm Composition: Intensive Monoculture and Extensive Mixed-culture

Algal farms maintained by different territorial damselfishes vary in size, algal biomass, and species composition (Table 1). At one end of the spectrum is *Stegastes nigricans* in Okinawa, Japan, which manages small farms (0.27 m^2) largely dominated by only one highly palatable algal species, *Polysiphonia* sp. (Hata and Kato 2002), which has never been found outside the territory of this damselfish species (Hata and Kato 2006, Hata et al. 2010). The damselfish effectively creates the specific conditions for its growth, defending it against competitive algae by weeding, and defending it against roving herbivores by aggressive exclusion (Hata and Kato 2004). At the other end of the spectrum, the sympatric damselfish *Stegastes obreptus* maintains much larger farms (6.3 m^2), comprised of a small amount of filamentous rhodophytes and a diverse assemblage of indigestible macroalgae. *Stegastes obreptus* appears to ingest palatable filamentous rhodophytes selectively from its farm, but does not engage in intensive weeding and defence (see also Hoey and Bellwood 2010). *Stegastes nigricans* therefore applies intensive farming, with increased effort and yield per unit area, and *S. obreptus* applies extensive farming with scattered effort over a large area and a smaller harvest yield per unit area (Fig. 2). In other parts of the world, *S. nigricans* territories range from a similar *Polysiphonia* spp. monoculture (e.g., Lizard Island, Great Barrier Reef; Hoey and Bellwood 2010) to mixed – species algal turfs (e.g., Fiji, Solomons, Tonga; Gobler et al. 2006), suggesting a high degree of behavioural plasticity across broad spatial scales.

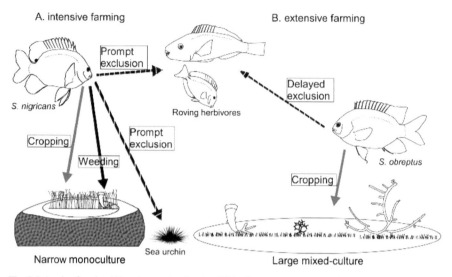

Fig. 2. Intensive farming (A) and extensive farming (B) by *Stegastes nigricans* and *S. obreptus*, respectively. Arrows denote farming behaviours.

Hata and Kato (2004) classified the known territorial species into intensive and extensive farmers, based on their behaviour (e.g., weeding), the composition of their farms (monoculture vs. mixed culture) and mean territory size. This was later expanded by Emslie et al. (2012) in their Great Barrier Reef – wide farmer distribution study (information included in Table 1). They added the category of "indeterminate" farmers (Table 1): species whose farms cannot be readily distinguished from the surrounding environment with the naked eye. Despite the lack of an obvious, discrete algal turf, sample collections have revealed that small but significant farming effects exist in these territories, in the form of greater quantities of palatable filamentous algae (Ceccarelli 2007) and detrimental effects on coral settlement (Casey et al. 2014). This classification is not yet complete, but can help to predict the degree of influence a damselfish community has across a given area from analysing the local damselfish species composition.

Consequences of Farming

Algal biomass

Herbivore pressure on algae inside damselfish territories is expected to be suppressed through territorial defence, and, as a result, the algal turf biomass is expected to increase (Ceccarelli et al. 2001). This is often the case when comparing damselfish territories with the surrounding grazed areas, with algal turf tending to be higher inside territories by varying ratios (between 1.15 and 29 times higher inside territories than outside, Ceccarelli et al. 2001). Caging experiments have shown that, in reality, damselfish territory biomass is somewhere in between the heavily grazed surfaces outside territories and the ungrazed (usually caged) areas (see below). However, there can be high levels of variability between damselfish species, locations and even seasons. For instance, *Stegastes fasciolatus* was recorded to have algal biomass between 1.15 and 3.2 times higher inside its territories than outside (Hixon and Brostoff 1981, 1996, Klumpp et al. 1987), but *Hemiglyphidodon plagiometopon* was recorded as having between 0.2 and nearly 30 times the biomass in its territories than that on the adjacent benthos on the GBR (Wilkinson and Sammarco 1983). The relationship between grazing pressure and algal biomass inside territories also remains unresolved. For example, *Microspathodon dorsalis* and *Stegastes fasciolatus* were found to exert higher grazing pressure within their territories than was recorded from other grazing fishes outside their territories. The effects of this elevated grazing pressure on algal biomass were not uniform: *M. dorsalis* territories had lower biomass than adjacent areas (Montgomery 1980b), but *S. fasciolatus* territories held higher biomass (Russ 1987).

Algal diversity, succession and composition

Early studies used damselfish territories as an example of the intermediate disturbance hypothesis, where cropping by damselfish was seen as intermediate between a complete lack of herbivory and intensive grazing by larger roving herbivores (Hixon and Bristoff 1983). In many cases, damselfish territories contain algal assemblages that are significantly more diverse than the surrounding heavily grazed areas, or caged areas that exclude roving herbivores (De Ruyter van Stevenink 1984, Hinds and Ballantine 1987, Hixon and Brostoff 1996), confirming the validity of the application of the intermediate disturbance hypothesis; territories can even contain species of algae otherwise not found in the vicinity (Hata and Kato 2006). However, this is not universal among farming damselfish, with some species hosting algal communities similar in diversity as the surrounding areas (e.g., indeterminate farmers, Ceccarelli et al. 2001) and other species cultivating monocultures (intensive farmers, Hata and Kato 2004).

Herbivory can play a role in determining the trajectory of algal succession, as late algal colonisers are often species of lower palatability (e.g., *Sargassum* spp.), and these tend to be weeded out by intensive farmers such as *S. nigricans*. In this way, some farmers can alter the successional trajectories of algal assemblages, by halting or slowing succession, either at an early stage dominated by fast-growing palatable filamentous algae (Hata et al. 2002, Ceccarelli 2007), or at an intermediate stage, defined by a species-rich community (Hixon and Brostoff 1996). In contrast, roving herbivores deflect succession to a highly disturbed, species-poor community dominated by herbivore-resistant algae (Hixon and Brostoff 1996). However, where roving herbivores are scarce, the role of roving herbivores can be slightly different. For instance, damselfish such as *Stegastes apicalis* can arrest succession at an early stage dominated by filamentous algae, while roving herbivores decelerate it, but not enough to avoid the growth of large fleshy macroalgae, resulting in a reef assemblage where only damselfish territories are free from the stands of fleshy macroalgae (Ceccarelli et al. 2011).

In general, algal assemblages inside damselfish territories are dominated by filamentous rhodophytes of the genus *Polysiphonia*, other algae of the family Rhodomelaceae and algae of the family Ceramiaceae (Table 1; Ceccarelli et al. 2001, Hata and Kato 2004). *Polysiphonia* spp. are turf-forming algae with both upright and creeping axes, highly productive and palatable (Hay 1981, Sousa et al. 1981, Airoldi et al. 1995, Airoldi and Virgilio 1998, Rindi et al. 1999). Additionally, *Polysiphonia* spp. are also nutrient rich and highly digestible (Montgomery and Gerking 1980, Zemke-White et al. 2000, Hata and Kato 2002). Polysiphonaceous algae are especially found in the territories of intensive farmers, whereas extensive farmers usually host a range of algal species including calcified algae such as *Jania* spp., *Amphiroa* spp. and Peyssonnellidae (Lobel 1980, Hoey and Bellwood 2010, Ceccarelli et al. 2011, Feitosa et al. 2012). Calcified and fleshy algae can be used as a substratum for more palatable epiphytic algae (Ceccarelli et al. 2005). Some damselfishes may exclude certain types of algae from their territories, such as fleshy macroalgae (Ceccarelli et al. 2001) or crustose coralline algae (Doropoulos et al. 2013), perhaps to maximise the space available for palatable or preferred species.

Algal productivity

The farming and feeding activity of territorial damselfishes tends to enhance the productivity of algal assemblages inside their territories (Montgomery 1980b, Klumpp et al. 1987, Russ 1987). Productivity was found to be higher not only per unit area (e.g., rates of carbon production per m^2), but also per unit algal biomass (e.g., rates of carbon production per gram, Montgomery 1980b, Klumpp et al. 1987, Russ 1987). This may be due to the intrinsic productivity of the selected species, whereby weeding and cropping may work as a selective process of gradually removing less productive algal species, resulting in higher productivity of the farm assemblage overall (Montgomery 1980a). Continuous cropping by the territory-holder may also keep the cropped algae at a more productive growth stage. Growth rates of filamentous algae such as *Polysiphonia* are fastest when the thalli are young, slowing down as the filaments grow larger; by keeping them closely cropped, the farmers are encouraging a continuous state of fast growth (Montgomery 1980a,b). Faeces of the resident damselfish may enhance productivity through fertilisation, but this has never been directly measured.

Effect on the broader benthic community

Algal farms actively maintained by damselfish, often for more than 10 years, are more stable than algal assemblages outside the territories, as the constant management by the resident farmer buffers them against environmental changes that can affect benthic communities outside the territories (Ferreira et al. 1998, Letourneur 2000, Hata et al. 2002). Additionally, algal farms of most damselfish species are biomass-rich, and the entangled branches trap sediment and detritus, which in themselves provide habitat and food sources for other organisms (Ferreira et al. 1998, Hata and Nishihira 2002, Hata and Umezawa 2011). As a result, farming damselfish can enhance the species diversity of small invertebrates such as crustaceans, annelids, molluscs, cnidarians, echinoderms, bryozoans, and benthic foraminifera in their farms (Lobel 1980, Robertson and Polunin 1981, Zeller 1988, Ferreira et al. 1998, Hata and Nishihira 2002). On the other hand, farming damselfish can also ingest these benthic animals and detritus as an important nitrogen source (Ferreira et al. 1998, Hata and Umezawa 2011, Dromard et al. 2013, Letourneur et al. 2013).

Farming damselfishes can influence coral zonation and survival, but there is no consensus on what effect farming has on the coral community as a whole. On the one hand, farmers can kill adult coral tissue in order to prepare substrate for their farms, and recruiting and juvenile corals can be inhibited either directly or indirectly by the resident farmer and the algal turf (Wellington 1982, Suefuji and van Woesik 2001, Casey et al. 2014, Gordon et al. 2014). On the other hand, some species defend their territories against roving herbivores and corallivores, which can enhance the survivorship of corals (Sammarco and Carleton 1981, Sammarco et al. 1986, Glynn and Colgan 1988, Gleason 1996, Letourneur et al. 1997). Differential survivorship within and outside territories can result in an overall increase in coral diversity at the scale of an entire reef (Gochfeld 2010).

Despite the general view that they kill coral tissue in the establishment of the farms, territorial damselfish require living corals for their farms. On a coral reef in Japan, farming damselfishes disappeared within a few years after most of the living corals were killed by the seastar *Acanthaster planci* and stopped providing adequate substrata (Sano et al. 1987). Many damselfishes rely on the stability and integrity of the substrata provided by the framework of (mostly) live coral colonies (Risk and Sammarco 1982, Sammarco and Carleton 1981, Sammarco et al. 1987, Stromberg and Kvarnemo 2005). Future research should assess whether damselfish species that farm among live coral colonies kill only a portion of the colony, ensuring that the survival of the remainder continues to provide the structural framework for the farm.

Global, Regional, and Local Variation in Farming

Algal farm structure and farming behaviour of damselfishes vary at global, regional, and local scales. For example, *Stegastes nigricans* is common throughout the Indo-Pacific, and maintains farms dominated by several species of *Polysiphonia* closely related to each other, but the proportion of the territory taken up by *Polysiphonia* varies: 14%–18% in Mauritius, the Red Sea and the Great Barrier Reef, 25%–26% in Kenya and the Maldives, and 87% in Okinawa, Japan (Gobler et al. 2006, Jones et al. 2006, Hata et al. 2010). Territory size is also variable; for example, territories of *S. nigricans* in Papua New Guinea are 13 times larger than those in Réunion Island (Barneche et al. 2009; Table 1). This difference in algal farms is partly caused by the variation of farming behaviour within a species. Indeed, the frequency of weeding (as evidenced by the fish ejecting the algae outside the territory after biting it) varies, and was observed to be 0, 0.5, and 0.7 times per 10 min in the Maldives, Okinawa, and Kenya, respectively (Hata et al. 2010). Barneche et al. (2009) suggest that bite rates are higher and body size is smaller where the sea surface temperature is higher.

Further study on the variation in behaviour, defence efficiency in each area, and the genetic basis of farming and territorial defence are needed in order to explain this geographic variation. Emslie et al. (2012) revealed that the species composition and density of farming damselfishes varies along the latitude and across the shelf of the Great Barrier Reef. Herbivorous fish density, herbivore intensity and farmer density and species composition differ among reefs and reef sites (Bay et al. 2001, Ceccarelli 2007, Vergés et al. 2011, 2012, Doropoulos et al. 2013, Emslie et al. 2012), and algal recruitment, especially that of turf algae, is different between reef sites (Doropoulos et al. 2013). This results in a rich mosaic of areas influenced by different types of territorial and non-territorial damselfishes that could be one of the key drivers of overall coral reef benthic diversity.

Conclusion

On coral reefs, territorial damselfishes defend territories in which they cultivate algal assemblages dominated by the preferred food species. The intensity of farming and consequent structure of the farms varies among damselfish species. Territory size, aggression and feeding behaviour and the species composition of the algal assemblage also vary at global, regional, and local scales. The background environment, such as density and species composition of the competitors and intruders, availability and growth rate of algae, physiological and behavioural differences in the damselfishes themselves, controlled by genetic background and/or environmental factors, are all interacting drivers. Further study is required in order to adequately unravel these patterns.

Farming damselfishes can occupy a large proportion of the space in many coral reef habitats (up to 90%), effectively enhancing the productivity and species diversity of the whole system. Together, the patchwork of non-territory areas and the territories of different damselfish species create a mosaic that offers different microhabitats to both benthic organisms and the fish feeding there. Despite the common association between degraded coral reefs and higher densities of territorial damselfishes, the highly diverse set of influences from the different damselfish species paint a more complex picture. The effects of territorial damselfishes on coral reef communities, positive or negative, are far from settled. Studies on territorial damselfishes are therefore necessary in order to deepen our understanding of the complicated network created by primary producers and consumers in coral reef ecosystems.

Acknowledgements

We appreciate our two editors, Dr. Eric Parmentier and Dr. Bruno Frédérich for assembling this book. This research is partly supported by JSPS KAKENHI Grant Numbers 25840159, 15H02420, and 15H05230 for HH.

Table 1. Territorial damselfishes and their farming strategies.

Species	Standard length (mm) [TL, total length]	Algal farm area (m²)	Weed or not	Species composition in algal farm	Effect on corals	Effect on Invertebrates	Predominant diet	Algal Productivity	Algal Succession	Aggression effects	Substratum	Farming intensity	Experiments	Locality	References
Microspathodon dorsalis	220	0.51	No	*Polysiphonia* sp. monoculture			*Polysiphonia* spp.	Inside/outside territory = 2.5 times, 21 mg O_2 m^{-2} h^{-1} (vs. 8.4 outside)			Granite boulder	Intensive	Caging, oxygen production experiments	Baja California Sur, Mexico	Montgomery 1980a,b
Microspathodon chrysurus	210	6		Mixed culture			Diatoms				Elkhorn coral-rock	Extensive	Removal	San Blas Islands, Caribbean, Panama	Robertson 1984
Plectroglyphidodon lacrymatus			No	Mixed culture of filamentous, corticulated rhodophytes, and phaeophytes							Coral-rock			Manukan Island, Borneo	Hata et al. 2010
Plectroglyphidodon lacrymatus	110 (Maximum TL)			Filamentous monoculture	High cover of live corals						Coral-rock	Intensive	*Sargassum* transplant	Lizard Island, GBR	Hoey and Bellwood 2010
Plectroglyphidodon lacrymatus			No	Mixed culture of filamentous, corticulated rhodophytes, and phaeophytes							Coral-rock			Mombasa, Malindi, Watamu, Kenya	Hata et al. 2010

Table 1. contd....

Table 1. contd.

Species	Standard length (mm) [TL, total length]	Algal farm area (m²)	Weed or not	Species composition in algal farm	Effect on corals	Effect on Invertebrates	Predominant diet	Algal Productivity	Algal Succession	Aggression effects	Substratum	Farming intensity	Experiments	Locality	References
Plectroglyphidodon lacrymatus			No	Mixed culture of filamentous, corticulated, and calcaleous rhodophytes							Coral-rock			North Male Atoll, Maldives	Hata et al. 2010
Plectroglyphidodon lacrymatus			No	Mixed culture of filamentous rohodphytes and phaeophytes							Coral-rock			Okinawa, Japan	Hata et al. 2010
Plectroglyphidodon lacrymatus	100	0.95		Filamentous, corticated rhodophytes and cyanobacteria	Higher coral cover inside territories than outside		Algae (high proportion of diatoms and detritus)				Massive coral-rock	Intensive	Removal	Kimbe Bay, New Britain, PNG	Jones et al. 2006, Ceccarelli 2007
Plectroglyphidodon lacrymatus			No	Mixed culture of filamentous, corticulated rhodophytes, and phaeophytes							Coral-rock			Trang, Thailand	Hata et al. 2010
Plectroglyphidodon lacrymatus		0.43 ± 0.08						1.98–2.00 g C m^{-2} day^{-1}			Coral reef		Settlement plates	Motupore Island, PNG	Polunin 1988

Species									Location	Reference
Plectroglyphidodon lacrymatus		Mixed culture, high occurrence of diatoms						Coral reef	Guam, Yap	Navarro and Lobban 2009
Plectroglyphidodon lacrymatus	0.25–1.5	High filamentous algal cover on coral branches	Increased density of some bioeroding polychaetes, decreased density of others			Branching coral		Coral trans-plants	Zanzibar, Tanzania	Stromberg and Kvarnemo 2005
Plectroglyphidodon dickii	110	Filamentous monoculture	Higher coral cover inside territories than outside	Algae (high proportion of *Polysiphonia* spp., corticated rhodophytes and diatoms)		Ramose coral	Intensive		Kimbe Bay, New Britain, PNG	Jones et al. 2006
Plectroglyphidodon dickii	76–88			Filamentous algae (~30%), coral polyps and sea anemones		Hard and soft corals			Nanwan, Taiwan	Ho et al. 2009
Plectroglyphidodon dickii			Higher coral cover and diversity inside territories during *Acanthaster planci* outbreak		Defend corals from *Acanthaster planci*	Coral reef			American Samoa	Glynn and Colgan 1988

Table 1. contd....

Table 1. contd.

Species	Standard length (mm) [TL, total length]	Algal farm area (m²)	Weed or not	Species composition in algal farm	Effect on corals	Effect on Invertebrates	Predominant diet	Algal Productivity	Algal Succession	Aggression effects	Substratum	Farming intensity	Experiments	Locality	References
Stegastes acapulcoensis					Higher coral cover and diversity inside territories during *Acanthaster planci* outbreak					Defend corals from *Acanthaster planci*	Coral reef			Uva Island, Gulf of Chiriqui, Panama	Glynn and Colgan 1988
Stegastes adustus	150	1		Mixed culture			Diatoms				Ramose coral-rock	Extensive	Removal	San Blas Islands, Caribbean, Panama	Robertson 1984
Stegastes adustus	82–120 (TL)	1.18–2.83		Mixed culture of filamentous rhodophytes and macroalgae			Algae (33%), detritus, invertebrates				Rock	Extensive		Guadeloupe, Caribbean	Dromard et al. 2013
Stegastes apicalis	120	0.12–1.43	No	Mixed culture			Filamentous rhodophytes and cyanobacteria	3.116 and 3.732 g C territory^{-1} day^{-1}			Elkhorn coral-rock	Extensive		GBR, the Gulf of Thailand	Klumpp and Polunin 1989, Kamura and Choonhabandit 1986

Species															Reference
Stegastes apicalis	130	1.5	Yes	Mixed culture dominated by *Polysiphonia*, often growing on *Amphiroa* spp.	Higher coral cover inside territories than outside		Filamentous algae, some invertebrates				Coral plate, massive coral-rock	Intensive		Magnetic Island, GBR	Ceccarelli 2007
Stegastes arcifrons				Filamentous phaeophytes		Remove urchins from territories					Boulders		Algal translocation	Galapagos Islands	Irving and Witman 2009
Stegastes fasciolatus	100	1	No	Mixed culture inside (high algal diversity), crustose corallines outside				1.20 (95% CI 0.23) g C m^{-2} day^{-1} inside, 0.69 (95 % CI 0.22) g C m^{-2} day^{-1}, outside	Decelerated	Decreased grazing intensity inside territories; algal removal greater or equal inside territories	Massive coral-rock	Extensive	Caging, settlement plates	Myrmidon Reef, GBR; Oahu, Hawaii	Russ 1987; Hixon and Brostoff 1996
Stegastes fascio-latus	150	7–17		Mixed culture						Facilitates coexistence with territorial acanthurids	Massive coral-rock	Extensive		Western India	Robertson and Polunin 1981
Stegastes fuscus	76–112 (TL)	1–2		Mixed culture of calcareous algae, filamentous rhodophytes, diatoms, and others			Algae (>70%), detritus and invertebrates (15%)				Rocky shore	Extensive		Tamandre, Brazil	Feitosa et al. 2012

Table 1. contd....

Table 1. contd.

Species	Standard length (mm) [TL, total length]	Algal farm area (m²)	Weed or not	Species composition in algal farm	Effect on corals	Effect on Invertebrates	Predominant diet	Algal Productivity	Algal Succession	Aggression effects	Substratum	Farming intensity	Experiments	Locality	References
Stegastes fuscus	138	1.64		Mixed culture		Overall density higher inside territories.	Filamentous rhodophytes (70%), animal material (30%)	0.86 to 2.34 g C day⁻¹ territory⁻¹; not different from outside	Keep algal community at early succession		Mound of coral-rock	Extensive		Florida, USA	Ferreira et al. 1998
Stegastes fuscus	90–110 (TL)	1.27–3.07					Algae (*Gelidium*)				Rocky reef			Areia Vermelha Marine State Park, Brazil	Osorio et al. 2006
Stegastes leucostictus	120	0.5–3.6		Mixed culture						Change home range and foraging behaviour of wrasses, decrease the settlement of some fish species	Coral rubble, granite boulder	Extensive		St. Croix, US Virgin Islands; Discovery bay, Jamaica	Ebersole 1980, Shulman et al. 1983, Itzkowitz and Slocum 1995, Jones 2005
Stegastes limbatus	150 (Maximum TL)			Mixed culture							Ramose or massive coral rock	Extensive		Mauritius Island	Hata unpublished
Stegastes lividus				Mixed culture of *Polysiphonia* spp., other filamentous, corticated rhodophytes							Ramose coral rock	Intermediate		Kota Kinabalu, Borneo	Hata et al. 2010

Stegastes lividus	110	Yes	Polysiphonia scopulorum and Gelidiopsis intricata, 40%			Coral plate	Intermediate	Caging	Guam	Lassuy 1980
Stegastes lividus			Mixed culture of Polysiphonia spp., other filamentous, corticated rhodophytes			Ramose coral rock	Intermediate		Mauritius Island	Hata et al. 2010
Stegastes lividus			Mixed culture of Polysiphonia spp., other filamentous, corticated rhodophytes			Ramose coral rock	Intermediate		Okinawa, Japan	Hata et al. 2010
Stegastes lividus	100		Filamentous chlorophytes		Reduced diversity of fish fauna where territories are dense	Coral-rock	Intermediate	Caging, settlement plates	Port Sudan, Red Sea	Vine 1974
Stegastes lividus			Mixed culture of Polysiphonia spp., other filamentous, corticated rhodophytes			Ramose coral rock	Intermediate		Trang, Thailand	Hata et al. 2010

Table 1. contd....

Table 1. contd.

Species	Standard length (mm) [TL, total length]	Algal farm area (m²)	Weed or not	Species composition in algal farm	Effect on corals	Effect on Invertebrates	Predominant diet	Algal Productivity	Algal Succession	Aggression effects	Substratum	Farming intensity	Experiments	Locality	References
Stegastes lividus					Lower coral cover but higher coral diversity inside territories than outside; some coral species found only inside territories					Reduced predation on corals by corallivores	Rock and rubble		Coral transplants	Moorea, French Polynesia	Gochfeld 2010
Stegastes lividus				Mixed turf, dominated by *Polysiphonia* and *Ceramium* spp., but higher species richness than outside							Coral rock		Caging, nutrient enrichment	Solomon Islands	Gobler et al. 2006
Stegastes nigricans				Mixed culture of *Polysiphonia* sp., other filamentous rhodophytes, and phaeophytes							Coral rock	Intensive		Dahab, Egypt	Hata et al. 2010

	Filamentous monoculture	High cover of live corals	Detritus, filamentous algae			Ramose Acropora	Intensive	Sargassum transplant		
Stegastes nigricans	150 (Maximum TL)	Yes					Intensive	Sargassum transplant	Lizard Island, GBR	Wilson and Bellwood 1997, Hoey and Bellwood 2010
Stegastes nigricans	Mixed culture of *Polysiphonia* spp., other filamentous, corticated rhodophytes					Coral rock	Intensive		Cairns, GBR	Hata et al. 2010
Stegastes nigricans	Yes / Mixed culture of *Polysiphonia* spp., other filamentous, corticated rhodophytes					Coral rock	Intensive		Malindi, Kenya	Hata et al. 2010
Stegastes nigricans	Mixed culture of *Polysiphonia* spp., other filamentous, corticated rhodophytes					Massive or ramose coral rock	Intensive		North Male Atoll, Maldives	Hata et al. 2010
Stegastes nigricans	Mixed culture of *Polysiphonia* sp., other filamentous, corticated rhodophytes					Coral rock	Intensive		Mauritius Island	Hata et al. 2010

Table 1. contd....

Table 1. contd.

Species	Standard length (mm) [TL, total length]	Algal farm area (m²)	Weed or not	Species composition in algal farm	Effect on corals	Effect on Invertebrates	Predominant diet	Algal Productivity	Algal Succession	Aggression effects	Substratum	Farming intensity	Experiments	Locality	References
Stegastes nigricans	83.6	0.27	Yes	*Polysiphonia* sp. monoculture		Higher densities and diversity of foraminiferans inside territories, especially free-living and sedentary species	Algae and detritus, some animal matter		Deflects succession by selective weeding		Massive coral-rock	Intensive	Caging, settlement plates	Okinawa, Japan	Hata and Nishiura 2002, Hata et al. 2002, Hata and Kato 2002, 2004, Hata and Umezawa 2011
Stegastes nigricans	150 (Maximum TL)	0.25–1.5		Filamentous monoculture	Higher coral cover inside territories than outside		Algae (high proportion of diatoms)				Ramose coral-rock	Intensive		Kimbe Bay, New Britain, PNG	Jones et al. 2006
Stegastes nigricans	114 (TL)	0.1–0.15		Mixed culture			Algae (dominated by *Polysiphonia* spp.)				Ramose coral-rock	Intensive		Réunion	Lison de Loma and Ballesteros 2002, Letourneur 2000
Stegastes nigricans				Mixed culture, high occurrence of diatoms							Coral reef			Guam, Yap	Navarro and Lobban 2009

Species							Reference
Stegastes nigricans	High filamentous algal cover on coral branches	Increased density of some bio-eroding polychaetes, decreased density of others		Branching corals	Coral transplants	Zanzibar, Tanzania	Strömberg and Kvarnemo 2005
Stegastes nigricans	Dense turf of primarily *Polysiphonia* spp.	Coral recruitment higher inside than outside territories; lower coral cover but higher coral diversity inside territories than outside; some coral species found only inside territories	Reduced predation on corals by corallivores; higher aggression towards herbivores, corallivores and egg predators but no overall effects on fish abundance. Sixbar wrasse recruits into damselfish territories	Rock and rubble	Settlement plates, coral transplants	Moorea, French Polynesia	Gleason 1996, Shima 2001, Gochfeld 2010, Johnson et al. 2011
Stegastes nigricans	Changed coral community structure by benefit to some coral species and detriment to others			*Porites* patch reefs	Fish removal, turf removal, caging, coral transplants	Moorea, French Polynesia	White and O'Donnell 2010

Table 1. contd....

Table 1. contd.

Species	Standard length (mm) [TL, total length]	Algal farm area (m²)	Weed or not	Species composition in algal farm	Effect on corals	Effect on Invertebrates	Predominant diet	Algal Productivity	Algal Succession	Aggression effects	Substratum	Farming intensity	Experiments	Locality	References
Stegastes nigricans					Higher coral cover and diversity inside territories during *Acanthaster planci* outbreak					Defend corals from *Acanthaster planci*	Coral reef			American Samoa	Glynn and Colgan 1988
Stegastes nigricans				Mixed turf, dominated by *Polysiphonia* and *Ceramium* spp., but higher species richness than outside							Dead coral rock		Caging, nutrient enrichment	Fiji, Solomons	Gobler et al. 2006
Stegastes obreptus	108.9	6.3	No	Mixed culture of corticated rhodophytes, phaeophytes, and cyanobacteria							Sand with rubble	Extensive		Okinawa, Japan	Hata and Kato 2004
Stegastes obreptus			No	Mixed culture							Ramose or massive coral rock	Extensive		Trang, Thailand	Hata unpublished
Stegastes partitus	60		No	Mixed culture							Ramose coral-rock	Extensive		Florida, USA	De Ruyter van Steveninck 1984

Species														
Stegastes planifrons	96	1	No	Mixed culture (high species diversity)						Ramose coral-rock	Extensive		Florida & Caribbean	Hinds and Ballantine 1987, De Ruyter van Steveninck 1984, Brawley and Adey 1977
Stegastes planifrons							Diatoms			Elkhorn coral-rock	Intensive	Removal	San Blas Islands, Caribbean, Panama	Robertson 1984
Stegastes planifrons	82–120 (TL)	1.08–1.33		Mixed culture of filamentous rhodophytes and macroalgae			Algae (33%), detritus, invertebrates			Massive coral rock	Extensive		Guadeloupe, Caribbean	Dromard et al. 2013
Stegastes planifrons	96	0.6	Yes	Mixed culture	Kills coral tissue to clear substratum		*Polysiphonia*, diatoms	Facilitates coexistence with territorial acanthurid		Loose rocks	Intermediate	Removal, caging, settlement plates	Panama	Robertson et al. 1981, Irvine 1980
Stegastes planifrons								Drives schooling behaviour of *Scarus croicensis*				Removal	San Blas Islands, Caribbean	Robertson et al. 1976
Stegastes rectifraenum	120	2		Mixed culture			Rhodophytes and chlorophytes			Granite boulder	Extensive	Oxygen production experiments	California, USA	Montgomery 1980a

Table 1. contd....

Table 1. contd.

Species	Standard length (mm) [TL, total length]	Algal farm area (m²)	Weed or not	Species composition in algal farm	Effect on corals	Effect on Invertebrates	Predominant diet	Algal Productivity	Algal Succession	Aggression effects	Substratum	Farming intensity	Experiments	Locality	References
Stegastes rectifraenum							Invertebrates (benthic copepods) and algae (*Bryopsis* spp. and *Ectocarpus* spp.)				Rocky reef	Extensive		Baja California Sur, Mexico	Moreno-Sanchez et al. 2011
Stegastes rocasensis	110 ± 11.2 (TL)	0.26–13.86		Mixed culture							Rocky shore	Extensive		Fernando de Noronha Archipelago, Brazil	Souza et al. 2011
Stegastes variabilis	77–92 (TL)	1–2		Mixed culture of calcareous algae, filamentous rhodophytes, diatoms, and others			Algae (>70%), detritus and invertebrates (15%)				Rocky shore	Extensive		Tamandre, Brazil	Feitosa et al. 2012
Dischistodus chrysopoecilus				Mixed culture							Sand gravel, coral rock	Extensive		Kota Kinabalu, Borneo	Hata unpublished
Dischistodus perspicillatus	200 (Maximum TL)		Yes	Cyanobacteria monoculture	Coral tissue dies in competition with algal turf		Detritus, cyanobacteria				Sand	Intensive	*Sargassum* transplant, coral transplant	Heron Island, Lizard Island, GBR	Potts 1977, Wilson and Bellwood 1997, Hoey and Bellwood 2010
Dischistodus perspicillatus				Mixed culture							Coral-rock	Extensive		Trang, Thailand	Hata unpublished

Species	Size		Culture		Effects		Diet	Substrate	Intensity	Manipulation	Location	Reference
Dischistodus prosopotaenia	190 (Maximum TL)		Mixed culture					Coral-rock	Extensive	*Sargassum* transplant	Lizard Island, GBR	Hoey and Bellwood 2010
Dischistodus prosopotaenia			Mixed culture					Sand gravel	Extensive		Okinawa, Japan	Hata unpublished
Dischistodus pseudochrysopoecilus	180 (Maximum TL)		Filamentous monoculture	High cover of live corals				Coral-rock	Intensive	*Sargassum* transplant	Lizard Island, GBR	Hoey and Bellwood 2010
Chrysiptera unimaculata			Mixed culture					Coral rock			Trang, Thailand	Hata unpublished
Hemiglyphidodon plagiometopon		No	Mixed culture of filamentous and/or corticated rhodophytes, phaeophytes					Massive or ramose coral rock			Kota Kinabalu, Borneo	Hata et al. 2010
Hemiglyphidodon plagiometopon	150		Mixed culture, high diversity	Reduced external bioerosion by grazing fishes	Increase in some boring sponges			Coral-rock	Extensive	Caging, settlement plates	Myrmidon Reef, GBR; Oahu, Hawaii	Sammarco 1983, Sammarco et al. 1986, Sammarco et al 1987
Hemiglyphidodon plagiometopon	200 (Maximum TL)	Yes	Mixed culture				Detritus, filamentous algae	Ramose *Acropora*	Intermediate	*Sargassum* transplant	Lizard Island, GBR	Wilson and Bellwood 1997, Hoey and Bellwood 2010

Table 1. contd....

Table 1. contd.

Species	Standard length (mm) [TL, total length]	Algal farm area (m²)	Weed or not	Species composition in algal farm	Effect on corals	Effect on Invertebrates	Predominant diet	Algal Productivity	Algal Succession	Aggression effects	Substratum	Farming intensity	Experiments	Locality	References
Hemiglyphidodon plagiometopon	180	2.15	Yes	Mixed culture							Coral-rock	Intermediate		Orpheus Island, GBR	Ceccarelli 2007
Hemiglyphidodon plagiometopon			No	Mixed culture of *Polysiphonia* spp., filamentous and/or corticated rhodophytes							Massive coral rock			Okinawa, Japan	Hata et al. 2010
Hemiglyphidodon plagiometopon			No	Mixed culture of filamentous and/or corticated rhodophytes							Massive coral rock			Trang, Thailand	Hata et al. 2010
Hemiglyphidodon plagiometopon	150		Yes	*Polysiphonia scopulorum* and *Gelidiopsis intricata*, 76%			Algae (*Polysiphonia* spp.)				Ramose *Acropora* and *Porites*	Intermediate	Caging	Yap	Lassuy 1980
Neoglyphidodon nigroris	110	0.9		Mixed culture							Massive coral-rock	Extensive		Kimbe Bay, New Britain, PNG	Ceccarelli 2007
Pomacentrus adelus	85	1.7	No	Mixed culture (thin turf)							Massive coral-rock	Indeterminate		Orpheus Island, GBR	Ceccarelli 2007

Species												Reference
Pomacentrus adelus	85	1.08	No	Mixed culture (thin turf)			*Hypnea* spp. and *Polysiphonia* spp., some detritus		Massive coral-rock	Indeterminate	Kimbe Bay, New Britain, PNG	Ceccarelli 2007
Pomacentrus aquilus			No	Mixed culture					Coral rock	Indeterminate	Malindi, Mombasa, Kenya	Hata unpublished
Pomacentrus bankanensis	100	1.19	No	Mixed culture (thin turf)			*Hypnea* spp. and *Polysiphonia* spp., some detritus		Massive coral-rock	Indeterminate	Kimbe Bay, New Britain, PNG	Ceccarelli 2007
Pomacentrus burroughi	80	1.4		Mixed culture (thin turf)			*Hypnea* spp. and *Polysiphonia* spp., some detritus		Ramose coral and rubble	Indeterminate	Kimbe Bay, New Britain, PNG	Ceccarelli 2007
Pomacentrus chrysurus	90	2.88		Mixed culture (thin turf)					Coral-rock	Indeterminate	Orpheus Island, GBR	Ceccarelli 2007
Pomacentrus tripunctatus	100	0.23	No	Mixed culture (thin turf)		Algae and detritus			Sand and rubble	Indeterminate	Magnetic Island, GBR	Ceccarelli 2007
Pomacentrus tripunctatus	100	0.3	No	Mixed culture (thin turf)					Sand and rubble	Indeterminate	Orpheus Island, GBR	Ceccarelli 2007
Pomacentrus tripunctatus	100	1.45	No	Mixed culture (thin turf)					Sand and rubble	Indeterminate	Kimbe Bay, New Britain, PNG	Ceccarelli 2007

Table 1. contd....

Table 1. contd.

Species	Standard length (mm) [TL, total length]	Algal farm area (m²)	Weed or not	Species composition in algal farm	Effect on corals	Effect on Invertebrates	Predominant diet	Algal Productivity	Algal Succession	Aggression effects	Substratum	Farming intensity	Experiments	Locality	References
Pomacentrus wardi	100	0.6		Mixed culture				Algae, some detritus			Coral-rock	Intermediate		Magnetic Island, GBR	Ceccarelli 2007
Pomacentrus wardi	100	1.4		Mixed culture							Coral-rock	Intermediate		Orpheus Island, GBR	Ceccarelli 2007

References

Airoldi, L. and M. Virgilio. 1998. Responses of turf-forming algae to spatial variations in the deposition of sediments. Mar. Ecol. Prog. Ser. 165: 271–282.

Airoldi, L., F. Rindi and F. Cinelli. 1995. Structure, seasonal dynamics and reproductive phenology of a filamentous turf assemblages on a sediment influenced, rocky subtidal shore. Bot. Mar. 38: 227–237.

Barneche, D.R., S.R. Floeter, D.M. Ceccarelli, D.M.B. Frensel, D.F. Dinslaken, H.F.S. Mario and C.E.L. Ferreira. 2009. Feeding macroecology of territorial damselfishes (Perciformes: Pomacentridae). Mar. Biol. 156: 289–299.

Bay, L.K., G.P. Jones and M.I. McCormick. 2001. Habitat selection and aggression as determinants of spatial segregation among damselfish on a coral reef. Coral Reefs 20: 289–298.

Bellwood, D.R., T.P. Hughes, C. Folke and M. Nystrom. 2004. Confronting the coral reef crisis. Nature 429: 827–833.

Brawley, S.H. and W.H. Adey. 1977. Territorial behavior of threespot damselfish (*Eupomacentrus planifrons*) increases reef algal biomass and productivity. Environ. Biol. Fishes 2: 45–51.

Casey, J.M., J.H. Choat and S.R. Connolly. 2014. Coupled dynamics of territorial damselfishes and juvenile corals on the reef crest. Coral Reefs 1–11.

Ceccarelli, D.M. 2007. Modification of benthic communities by territorial damselfish: a multi-species comparison. Coral Reefs 26: 853–866.

Ceccarelli, D.M., G.P. Jones and L.J. McCook. 2001. Territorial damselfishes as determinants of the structure of benthic communities on coral reefs. Oceanogr. Mar. Biol. Annu. Rev. 39: 355–389.

Ceccarelli, D.M., G.P. Jones and L.J. McCook. 2005. Effects of territorial damselfish on an algal-dominated coastal coral reef. Coral Reefs 24: 606–620.

Ceccarelli, D.M., G.P. Jones and L.J. McCook. 2011. Interactions between herbivorous fish guilds and their influence on algal succession on a coastal coral reef. J. Exp. Mar. Biol. Ecol. 399: 60–67.

Choat, J.H., K.D. Clements and W.D. Robbins. 2002. The trophic status of herbivorous fishes on coral reefs - I: Dietary analyses. Mar. Biol. 140: 613–623.

Cleveland, A. and W.L. Montgomery. 2003. Gut characteristics and assimilation efficiencies in two species of herbivorous damselfishes (Pomacentridae: *Stegastes dorsopunicans* and *S. planifrons*). Mar. Biol. 142: 35–44.

De Ruyter van Steveninck, E.D. 1984. The composition of algal vegetation in and outside damselfish territories on a Florida reef. Aquat. Bot. 20: 11–19.

Doropoulos, C., G. Hyndes, D. Abecasis and A. Vergés. 2013. Herbivores strongly influence algal recruitment in both coral- and algal-dominated coral reef habitats. Mar. Ecol. Prog. Ser. 486: 153–164.

Dromard, C.R., Y. Bouchon-Navaro, S. Cordonnier, M.-F. Fontaine, M. Verlaque, M. Harmelin-Vivien and C. Bouchon. 2013. Resource use of two damselfishes, *Stegastes planifrons* and *Stegastes adustus*, on Guadeloupean reefs (Lesser Antilles): inference from stomach content and stable isotope analysis. J. Exp. Mar. Biol. Ecol. 440: 116–125.

Ebersole, J.P. 1980. Food density and territory size: an alternative model and a test on the reef fish *Eupomacentrus leucostictus*. Am. Nat. 115: 492–509.

Emslie, M.J., M. Logan, D.M. Ceccarelli, A.J. Cheal, A.S. Hoey, I. Miller and H.P.A. Sweatman. 2012. Regional-scale variation in the distribution and abundance of farming damselfishes on Australia's Great Barrier Reef. Mar. Biol. 159: 1293–1304.

FAOSTAT. 2014. FAOSTAT online database (http://faostat3.fao.org/home).

Feitosa, J.L.L., A.M. Concentino, S.F. Teixeira and B.P. Ferreira. 2012. Food resource use by two territorial damselfish (Pomacentridae: *Stegastes*) on South-Western Atlantic algal-dominated reefs. J. Sea Res. 70: 42–49.

Ferreira, C.E.L., J.E.A. Gonçalves, R. Coutinho and A.C. Peret. 1998. Herbivory by the dusky damselfish *Stegastes fuscus* (Cuvier, 1830) in a tropical rocky shore: effects on the benthic community. J. Exp. Mar. Biol. Ecol. 229: 241–264.

Frédérich, B., G. Fabri, G. Lepoint, P. Vandewalle and E. Parmentier. 2009. Trophic niches of thirteen damselfishes (Pomacentridae) at the Grand Récif of Toliara, Madagascar. Ichthyol. Res. 56: 10–17.

Frédérich, B., L. Sorenson, F. Santini, G.J. Slater and M.E. Alfaro. 2013. Iterative ecological radiation and convergence during the evolutionary history of damselfishes (Pomacentridae). Am. Nat. 181: 94–113.

Gleason, M.G. 1996. Coral recruitment in Moorea, French Polynesia: the importance of patch type and temporal variation. J. Exp. Mar. Biol. Ecol. 207: 79–101.

Glynn, P.W. and M.W. Colgan. 1988. Defense of corals and enhancement of coral diversity by territorial damselfishes. Proc. 6th Int. Coral Reef Symp., Australia 2: 157–163.

Gobler, C.J., D.B. Thibault, T.W. Davis, P.B. Curran, B.J. Peterson and L.B. Liddle. 2006. Algal assemblages associated with *Stegastes* sp. territories on Indo-Pacific coral reefs: characterization of diversity and controls on growth. J. Exp. Mar. Biol. Ecol. 336: 135–145.

Gochfeld, D.J. 2010. Territorial damselfishes facilitate survival of corals by providing an associational defense against predators. Mar. Ecol. Prog. Ser. 398: 137–148.

Gordon, T.A.C., B. Cowburn and R.D. Sluka. 2014. Defended territories of an aggressive damselfish contain lower juvenile coral density than adjacent non-defended areas on Kenyan lagoon patch reefs. Coral Reefs 1–4.

Hata, H. and M. Kato. 2002. Weeding by the herbivorous damselfish *Stegastes nigricans* in nearly monocultural algae farms. Mar. Ecol. Prog. Ser. 237: 227–231.

Hata, H. and M. Nishihira. 2002. Territorial damselfish enhances multi-species co-existence of foraminifera mediated by biotic habitat structuring. J. Exp. Mar. Biol. Ecol. 270: 215–240.

Hata, H. and M. Kato. 2004. Monoculture and mixed-species algal farms on a coral reef are maintained through intensive and extensive management by damselfishes. J. Exp. Mar. Biol. Ecol. 313: 285–296.

Hata, H. and M. Kato. 2006. A novel obligate cultivation mutualism between damselfish and *Polysiphonia* algae. Biol. Lett. 2: 593–596.

Hata, H. and Y. Umezawa. 2011. Food habits of the farmer damselfish *Stegastes nigricans* inferred by stomach content, stable isotope, and fatty acid composition analyses. Ecol. Res. 26: 809–818.

Hata, H., M. Nishihira and S. Kamura. 2002. Effects of habitat-conditioning by the damselfish *Stegastes nigricans* (Lacépède) on the community structure of benthic algae. J. Exp. Mar. Biol. Ecol. 280: 95–116.

Hata, H., K. Watanabe and M. Kato. 2010. Geographic variation in the damselfish-red alga cultivation mutualism in the Indo-West Pacific. BMC Evol. Biol. 10: 185.

Hay, M.E. 1981. The functional morphology of turf-forming seaweeds: persistence in stressful marine habitats. Ecology 62: 739–750.

Hinds, P.A. and D.L. Ballantine. 1987. Effects of the Carribian threespot damselfish, *Stegastes planifrons* (Cuvier), on algal lawn composition. Aquat. Bot. 27: 299–308.

Hixon, M.A. and W.N. Brostoff. 1981. Fish grazing and community structure of Hawaiian reef algae. Proc. 3rd. Int. Coral Reef Symp., Miami. 2: 507–514.

Hixon, M.A. and W.N. Brostoff. 1983. Damselfish as keystone species in reverse: intermediate disturbance and diversity of reef algae. Science 220: 511–513.

Hixon, M.A. and W.N. Brostoff. 1996. Succession and herbivory: effects of differential fish grazing on Hawaiian coral-reef algae. Ecol. Monogr. 66: 67–90.

Ho, C.-T., Y.-C. Fu, C.-L. Sun, S.-j. Kao and R.-Q. Jan. 2009. Plasticity of feeding habits of two *Plectroglyphidodon* damselfishes on coral reefs in Southern Taiwan: evidence from stomach content and stable isotope analyses. Zool. Stud. 48: 649–656.

Hoey, A.S. and D.R. Bellwood. 2010. Damselfish territories as a refuge for macroalgae on coral reefs. Coral Reefs 29: 107–118.

Horn, M.H. 1989. Biology of marine herbivorous fishes. Oceanogr. Mar. Biol. Annu. Rev. 27: 167–272.

Hourigan, T.F. 1986. An experimental removal of a territorial pomacentrid: efffects on the occurrence and behavior of competitors. Environ. Biol. Fishes 15: 161–169.

Hughes, T.P., M.J. Rodrigues, D.R. Bellwood, D. Ceccarelli, O. Hoegh-Guldberg, L. McCook, N. Moltschaniwskyj, M.S. Pratchett, R.S. Steneck and B. Willis. 2007. Phase shifts, herbivory, and the resilience of coral reefs to climate change. Curr. Biol. 17: 360–365.

Irvine, G.V. 1980. Fish as farmers: an experimental study of herbivory in a territorial coral reef damselfish *Eupomacentrus planifrons*. Am. Zool. 20: 822.

Irving, A.D. and J.D. Witman. 2009. Positive effects of damselfish override negative effects of urchins to prevent an algal habitat switch. J. Ecol. 97: 337–347.

Itzkowitz, M. and C.J. Slocum. 1995. Is the amount of algae related to territorial defense and reproductive success in the beaugregory damselfish? Mar. Behav. Physiol. 24: 243–250.

Johnson, M.K., S.J. Holbrook, R.J. Schmitt and A.J. Brooks. 2011. Fish communities on staghorn coral: effects of habitat characteristics and resident farmerfishes. Environ. Biol. Fishes 91: 429–448.

Jones, G.P., L. Santana, L.J. McCook and M.I. McCormick. 2006. Resource use and impact of three herbivorous damselfishes on coral reef communities. Mar. Ecol. Prog. Ser. 328: 215–224.

Jones, K.M.M. 2005. The effect of territorial damselfish (family Pomacentridae) on the space use and behaviour of the coral reef fish, *Halichoeres bivittatus* (Bloch, 1791) (family Labridae). J. Exp. Mar. Biol. Ecol. 324: 99–111.

Kamura, S. and S. Choonhabandit. 1986. Algal communities within territories of the damselfish *Stegastes apicalis* and the effects of grazing by the sea-urchin *Diadema* spp. in the Gulf of Thailand. Galaxea 5: 175–194.

Kaufman, L. 1977. The three spot damselfish: effects on benthic biota of Caribbean coral reefs. Proc. 3rd. Int. Coral Reef Symp., Miami. 1: 559–564.

Klumpp, D.W. and N.V.C. Polunin. 1989. Partitioning among grazers of food resources within damselfish territories on a coral reef. J. Exp. Mar. Biol. Ecol. 125: 145–170.

Klumpp, D.W., D. McKinnon and P. Daniel. 1987. Damselfish territories: zones of high productivity on coral reefs. Mar. Ecol. Prog. Ser. 40: 41–51.

Kohda, M. 1984. Intra- and interspecific territoriality of a temperate damselfish *Eupomacentrus altus* (Teleostei: Pomacentridae). Physiol. Ecol. Japan 21: 35–52.

Lassuy, D.R. 1980. Effects of "farming" behavior by *Eupomacentrus lividus* and *Hemiglyphidodon plagiometopon* on algal community structure. Bull. Mar. Sci. 30: 304–312.

Letourneur, Y. 2000. Spatial and temporal variability in territoriality of a tropical benthic damselfish on a coral reef (Réunion Island). Environ. Biol. Fishes 57: 377–391.

Letourneur, Y., R. Galzin and M. Harmelin-Vivien. 1997. Temporal variations in the diet of the damselfish *Stegastes nigricans* (Lacépède) on a Réunion fringing reef. J. Exp. Mar. Biol. Ecol. 217: 1–18.

Letourneur, Y., T. Lison de Loma, P. Richard, M.L. Harmelin-Vivien, P. Cresson, D. Banaru, M.F. Fontaine, T. Gref and S. Planes. 2013. Identifying carbon sources and trophic position of coral reef fishes using diet and stable isotope (δ^{15}N and δ^{13}C) analyses in two contrasted bays in Moorea, French Polynesia. Coral Reefs 32: 1091–1102.

Lison de Loma, T. and E. Ballesteros. 2002. Microspatial variability inside epilithic algal communities within territories of the damselfish *Stegastes nigricans* at La Réunion (Indian Ocean). Bot. Mar. 45: 316–323.

Lobel, P.S. 1980. Herbivory by damselfishes and their role in coral reef community ecology. Bull. Mar. Sci. 30: 273–289.

Mahoney, B.N. 1981. An examination of interspecific territoriality in the dusky damselfish, *Eupomacentrus dorsopunicans* Poey. Bull. Mar. Sci. 31: 141–146.

Montgomery, W.L. 1980a. Comparative feeding ecology of two herbivorous damselfish (Pomacentridae: Teleostei) from the Gulf of California, Mexico. J. Exp. Mar. Biol. Ecol. 47: 9–24.

Montgomery, W.L. 1980b. The impact of non-selective grazing by the giant blue damselfish, *Microspathodon dorsalis*, on algal communities in the Gulf of California, Mexico. Bull. Mar. Sci. 30: 290–303.

Montgomery, W.L. and S.D. Gerking. 1980. Marine macroalgae as foods for fishes: an evaluation of potential food quality. Environ. Biol. Fishes 5: 143–153.

Moreno-Sánchez, X.G., L.A. Abitia-Cárdenas, O. Escobar-Sánchez and D.S. Palacios-Salgado. 2011. Diet of the Cortez damselfish *Stegastes rectifraenum* (Teleostei: Pomacentridae) from the rocky reef at Los Frailes, Baja California Sur, Mexico. Mar. Biodivers. Rec. 4: 1–5.

Mumby, P.J. 2006. The impact of exploiting grazers (Scaridae) on the dynamics of Caribbean coral reefs. Ecol. Appl. 16: 747–769.

Navarro, J.N. and C.S. Lobban. 2009. Freshwater and marine diatoms from the western Pacific islands of Yap and Guam, with notes on some diatoms in damselfish territories. Diatom Res. 24: 123–157.

Ogden, J.C. and P.S. Lobel. 1978. The role of herbivorous fishes and urchins in coral reef communities. Environ. Biol. Fishes 3: 49–63.

Osorio, R., I.L. Rosa and H. Cabral. 2006. Territorial defence by the Brazilian damsel *Stegastes fuscus* (Teleostei: Pomacentridae). J. Fish Biol. 69: 233–242.

Polunin, N.V.C. 1988. Efficient uptake of algal production by a single resident herbivorous fish on the reef. J. Exp. Mar. Biol. Ecol. 123: 61–76.

Potts, D.C. 1977. Suppression of coral populations by filamentous algae within damselfish territories. J. Exp. Mar. Biol. Ecol. 28: 207–216.

Rindi, F., M.D. Guiry and F. Cinelli. 1999. Morphology and reproduction of the adventive Mediterranean rhodophyte *Polysiphonia setacea*. Hydrobiologia 398/399: 91–100.

Risk, M.J. and P.W. Sammarco. 1982. Bioerosion of corals and the influence of damselfish territoriality: a preliminary study. Oecologia 52: 376–380.

Robertson, D.R. 1984. Cohabitation of competing territorial damselfishes on a Caribbean coral reef. Ecology 65: 1121–1135.

Robertson, D.R. and N.V.C. Polunin. 1981. Coexistence: symbiotic sharing of feeding territories and algal food by some coral reef fishes from the Western Indian Ocean. Mar. Biol. 62: 185–195.

Robertson, D.R., H.P.A. Sweatman, E.A. Fletcher and M.G. Cleland. 1976. Schooling as a mechanism for circumventing the territoriality of competitors. Ecology 57: 1208–1220.

Robertson, D.R., S.G. Hoffman and J.M. Sheldon. 1981. Availability of space for the territorial Caribbean damselfish *Eupomacentrus planifrons*. Ecology 62: 1162–1169.

Russ, G.R. 1987. Is rate of removal of algae by grazers reduced inside territories of tropical damselfishes? J. Exp. Mar. Biol. Ecol. 110: 1–17.

Sammarco, P.W. 1983. Effects of fish grazing and damselfish territoriality on coral reef algae. I. Algal community structure. Mar. Ecol. Prog. Ser. 13: 1–14.

Sammarco, P.W. and J.H. Carleton. 1981. Damselfish territoriality and coral community structure: reduced grazing, coral recruitment, and effects on coral spat. Proc. 4th Int. Coral Reef Symp. 2: 525–535.

Sammarco, P.W. and A.H. Williams. 1982. Damselfish territoriality: influence on *Diadema* distribution and implications for coral community structure. Mar. Ecol. Prog. Ser. 8: 53–59.

Sammarco, P.W., J.H. Carleton and M.J. Risk. 1986. Effects of grazing and damselfish territoriality on bioerosion of dead corals: direct effects. J. Exp. Mar. Biol. Ecol. 98: 1–19.

Sammarco, P.W., M.J. Risk and C. Rose. 1987. Effects of grazing and damselfish territoriality on intertidal bioerosion of dead corals: indirect effects. J. Exp. Mar. Biol. Ecol. 112: 185–199.

Sandin, S.A., J.E. Smith, E.E. DeMartini, E.A. Dinsdale, S.D. Donner, A.M. Friedlander, T. Konotchick, M. Malay, J.E. Maragos, D. Obura, O. Pantos, G. Paulay, M. Richie, F. Rohwer, R.E. Schroeder, S. Walsh, J.B.C. Jackson, N. Knowlton and E. Sala. 2008. Baselines and degradation of coral reefs in the Northern Line Islands. PLoS ONE 3: e1548.

Sano, M., M. Shimizu and Y. Nose. 1987. Long-term effects of destruction of hermatypic corals by *Acanthaster planci* infestation on reef fish communities at Iriomote Island, Japan. Mar. Ecol. Prog. Ser. 37: 191–199.

Shima, J.S. 2001. Recruitment of a coral reef fish: roles of settlement, habitat, and postsettlement losses. Ecology 82: 2190–2199.

Shulman, M.J., J.C. Ogden, J.P. Ebersole, W.N. McFarland, S.L. Miller and N.G. Wolf. 1983. Priority effects in the recruitment of juvenile coral reef fishes. Ecology 64: 1508–1513.

Sousa, W., S. Schroeter and S. Gaines. 1981. Latitudinal variation in intertidal algal community structure: the influence of grazing and vegetative propagation. Oecologia 48: 297–307.

Souza, A., M. Ilarri and I. Rosa. 2011. Habitat use, feeding and territorial behavior of a Brazilian endemic damselfish *Stegastes rocasensis* (Actinopterygii: Pomacentridae). Environ. Biol. Fishes 91: 133–144.

Stromberg, H. and C. Kvarnemo. 2005. Effects of territorial damselfish on cryptic bioeroding organisms on dead *Acropora formosa*. J. Exp. Mar. Biol. Ecol. 327: 91–102.

Suefuji, M. and R. van Woesik. 2001. Coral recovery from the 1998 bleaching event is facilitated in *Stegastes* (Pisces: Pomacentridae) territories, Okinawa, Japan. Coral Reefs 20: 385–386.

Tilman, D., K.G. Cassman, P.A. Matson, R. Naylor and S. Polasky. 2002. Agricultural sustainability and intensive production practices. Nature 418: 671–677.

Vergés, A., M.A. Vanderklift, C. Doropoulos and G.A. Hyndes. 2011. Spatial patterns in herbivory on a coral reef are influenced by structural complexity but not by algal traits. PLoS ONE 6: e17115.

Vergés, A., S. Bennett and D.R. Bellwood. 2012. Diversity among macroalgae-consuming fishes on coral reefs: a transcontinental comparison. PLoS ONE 7: e45543.

Vine, P.J. 1974. Effects of algal grazing and aggressive behaviour of the fishes *Pomacentrus lividus* and *Acanthurus sohal* on coral-reef ecology. Mar. Biol. 24: 131–136.

Wellington, G.M. 1982. Depth zonation of corals in the gulf of Panama: control and facilitation by resident reef fishes. Ecol. Monogr. 52: 223–241.

White, J.-S.S. and J.L. O'Donnell. 2010. Indirect effects of a key ecosystem engineer alter survival and growth of foundation coral species. Ecology 91: 3538–3548.

Wilkinson, C.R. and P.W. Sammarco. 1983. Effects of fish grazing and damselfish territoriality on coral reef algae. II. Nitrogen fixation. Mar. Ecol. Prog. Ser. 13: 15–19.

Williams, A.H. 1979. Interference behavior and ecology of threespot damselfish (*Eupomacentrus planifrons*). Oecologia 38: 223–230.

Wilson, S. and D.R. Bellwood. 1997. Cryptic dietary components of territorial damselfishes (Pomacentridae, Labroidei). Mar. Ecol. Prog. Ser. 153: 299–310.

Wilson, S.K., D.R. Bellwood, J.H. Choat and M.J. Furnas. 2003. Detritus in the epilithic algal matrix and its use by coral reef fishes. Oceanogr. Mar. Biol. Annu. Rev. 41: 379–309.

Zeller, D.C. 1988. Short-term effects of territoriality of a tropical damselfish and experimental exclusion of large fishes on invertebrates in algal turfs. Mar. Ecol. Prog. Ser. 44: 85–93.

Zemke-White, W.L. and K.D. Clements. 1999. Chlorophyte and rhodophyte starches as factors in diet choice by marine herbivorous fish. J. Exp. Mar. Biol. Ecol. 240: 137–149.

Zemke-White, W.L., K.D. Clements and P.J. Harris. 2000. Acid lysis of macroalgae by marine herbivorous fishes: effects of acid pH on cell wall porosity. J. Exp. Mar. Biol. Ecol. 245: 57–68.

Trophic Ecology of Damselfishes

Bruno Frédérich,* Damien Olivier, Laura Gajdzik and
Eric Parmentier

Resources (e.g., food and habitat) partitioning may be viewed as one of the key factors in the diversifying process, promoting the coexistence of closely related and ecologically equivalent species (Colwell and Fuentes 1975). The trophic niche is defined as the place of an organism in the environment in relation to its food (Silvertown 2004). Generally speaking, the trophic ecology of coral reef fishes has been broadly studied. For instance, extensive evidences of trophic adaptations and a great diversity of diets were highlighted in Labridae (Wainwright 1988, Wainwright et al. 2004) and Chaetodontidae (Motta 1988, Pratchett 2007).

Pomacentrids are probably the most conspicuous inhabitants of coral reefs (Allen 1991). Consequently, diverse topics on coral reef ecology related to the diet of some damselfishes have been investigated. For instance, the effects of herbivorous damselfishes on benthic communities are increasingly being studied (Hata and Kato 2002, Gobler et al. 2006, Ceccarelli 2007) (see Chapter VI). Damselfishes have also served as a fish model for the study of planktivory. For example, the effects of biotic factors like rank within social groups (Coates 1980, Forrester 1991), zooplankton dynamics (Noda et al. 1992) and current speeds (Mann and Sancho 2007) on the feeding behavior of planktivorous species were illustrated in damselfishes. In this chapter, we aim to review the diversity of trophic niches present in the family Pomacentridae. After a brief overview of the methods commonly used to investigate the feeding habits of fishes, we describe the various trophic groups in damselfishes. We report

Laboratoire de Morphologie Fonctionnelle et Evolutive, AFFISH - Research Center, University of Liège, Quartier Agora, Allée du six Août 15, Bât. B6C, 4000 Liège (Sart Tilman), Belgium.
* Corresponding author: bruno.frederich@ulg.ac.be

geographic and ontogenetic variation in their diet. Moreover, we also discuss the role of detritus in the diet of some damselfishes, and individual variation/specialization within populations of damselfish species. Finally, we provide some perspectives in the study of trophic ecology in pomacentrids.

Methods Commonly Used for Studying the Trophic Ecology of Fishes

Direct feeding observations can be used to determine feeding strategies (e.g., biting, grazing, sucking) and the type of prey selected by fishes but their diet has been traditionally studied through stomach content analysis (Hyslop 1980). Stomach content data allow the establishment of the range and proportional consumption of different dietary items. Sometimes, the ingested prey can be identified to varied taxonomic-level and even to the species-level. Disadvantages of using stomach content data are that they represent snapshots of the feeding habits and may not be representative of the overall diet. Moreover, some food items may be unidentifiable and, in some cases, stomachs are empty because fish were not caught during their feeding activities. Finally, different dietary items may be digested and assimilated at different rates or not assimilated at all, such that stomach content analysis can give good information on ingestion but misleading data on the relative assimilation of different prey items. Stable isotope analysis has increasingly been used to complement stomach content analysis (e.g., Cocheret de la Morinière et al. 2003, Parmentier and Das 2004). Stable isotope analysis of C, N and S (hereafter noted $\delta^{13}C$, $\delta^{15}N$, $\delta^{34}S$) are based on the statement that the stable isotopic composition of a consumer is the weighted average of the stable isotopic composition of its preys. This weighted average is modified during biochemical reactions occurring in an organism (isotopic discrimination) and leads to isotopic composition change (isotopic fractionation). The stable isotope composition of an animal is therefore linked to the stable isotope composition of the primary food sources at the basis of the food web but is also linked to its trophic level (Peterson and Fry 1987). Because the N isotopic composition is generally more affected by fractionation than the C isotopic composition, it is often used as trophic level proxy (e.g., Post 2002, Navarro et al. 2011). Consequently, stable isotope analysis has emerged as a powerful tool for tracing dietary sources. This method provides an integrated measure of the dietary components over a much longer period of time than does stomach content analysis. Although stable isotope analysis does not provide a detailed picture of dietary preferences, it gives an average estimate of an organism's preferred diet that is much less subject to temporal bias (Pinnegar and Polunin 1999). Stable isotope analyses were recently revealed as a powerful tool for assessing the trophic niche width of species (Bearhop et al. 2004, Jackson et al. 2011). There are also certain limitations in using stable isotope analysis to quantify diet. The use of stable isotopes as source tracers relies on the fact that potential food sources may differ in their isotopic compositions, which is rarely the case and constitute a caveat of this approach. Moreover, the methods rarely provide the diet information at the species-level. Therefore, when the isotopic compositions of prey species overlap, it can be difficult to estimate their relative contribution to consumer diets (Phillips et al. 2005).

Similar to stable isotopes, the fatty acids composition also has been successfully used as biomarkers to identify trophic relationships (e.g., Graeve et al. 2001, Nyssen et al. 2005). Fatty acids are the primary constituents of most lipids. They generally remain intact through the digestion process and can be deposited in the consumer's tissue with minimal modification and in a predictable way (Lee et al. 1971). Certain fatty acids have specific known sources and essential fatty acids, which animals cannot synthesize and must generally ingest with their diets, can thus act as good biomarkers.

At least, when you want to study variations in the trophic ecology among fish species, you can hypothesize differences in their physiology and thus compare absorption efficiencies of nutrients. Accordingly, some authors estimated nutrient-specific absorption efficiencies in some damselfishes (e.g., Montgomery and Gerking 1980, Galetto and Bellwood 1994). Briefly, this technique measures the assimilation of organic nutrients relative to a substance (ash) which experienced slight or no assimilation (Montgomery and Gerking 1980).

Generally speaking, the combination of these approaches in the same study of trophic niche analysis has the advantage of compensating for the inaccuracy/caveats of each method. Table 1 summarizes various studies devoted to the diet of damselfishes and records the used quantitative approaches. Stomach content analysis was by far the

Table 1. Diversity of the studies devoted to the diet of damselfishes since the 70's. N refers to the number of species studied by the author(s).

References	N	Feeding observations	Stomach contents	Stable isotopes	Fatty acids	Ash-marker methods
Emery 1973	13	X				
Gerber and Marshall 1974	6		X			
Fricke 1977	25	X	X			
de Boer 1978	1	X				
Coates 1980	1	X	X			
Montgomery and Gerking 1980	2					X
Tribble and Nishikawa 1982	4		X			
Sano et al. 1984	30		X			
Thresher and Colin 1986	7	X				
Forrester 1991	1	X				
Kuo and Shao 1991	40		X			
Galetto and Bellwood 1994	2					X
Jennings et al. 1995	11	X				
Letourneur et al. 1997	1		X			
Wilson and Bellwood 1997	3	X	X			
Ferreira et al. 1998	1	X	X			
Lison de Loma et al. 2000	1		X			X
Ferreira et al. 2001	5	X	X			
Elliott and Bellwood 2003	15	X	X			

Table 1. contd....

Table 1. contd.

References	N	Feeding observations	Stomach contents	Stable isotopes	Fatty acids	Ash-marker methods
Cleveland and Montgomery 2003	2					X
Feitoza et al. 2003	5	X				
Jones et al. 2006	3		X			
Kent et al. 2006	1		X			
Pinnegar et al. 2007	1		X	X		
Mann and Sancho 2007	1	X	X			
Lin et al. 2007	1		X	X		
Mill et al. 2007	1			X		
Ho et al. 2007	1		X	X		
Mill 2007	8			X		
Carassou et al. 2008	7		X	X		
Vonk et al. 2008	1			X		
Frédérich et al. 2009	13		X	X		
Buckle and Booth 2009	2		X			
Ho et al. 2009	2		X	X		
Meideros et al. 2010	2	X				
Greenwood et al. 2010	1			X		
Moreno-Sanchez et al. 2011	1		X		X	
Hata and Umezawa 2011	1		X	X	X	
Curtis-Quick et al. 2012	2	X		X		
Feitosa et al. 2012	2	X				
Dromard et al. 2013	2		X	X		
Letourneur et al. 2013	1		X	X		
Kramer et al. 2013	3		X			

most commonly used method to study the feeding habits of damselfishes (Table 1). Since 2007, authors started to combine the measurement of stable isotope ratios and the analysis of stomach contents (Table 1). To the best of our knowledge, only one study explored the diet of species combining three different methods: gut contents, stable isotopes and fatty acids composition (Hata and Umezawa 2011).

Trophic Guilds of Damselfishes

Accordingly to their morphology (i.e., relatively small body and small mouth size), pomacentrids mainly feed on small-sized items. However, the taxonomic diversity of preys caught by damselfishes is relatively high: copepods, decapod larvae, apendicularians, *Oikopleura* spp., fish and invertebrate eggs, fish larvae, medusas, sea anemones, isopods, polychaetes, halacarids, mollusks, coral polyps and filamentous algae. The stomach contents of some damselfishes can also include a large proportion of amorphous organic matter (Kuo and Shao 1991, Wilson and Bellwood 1997, Jones et al. 2006). "Herbivores", "planktivores" and "omnivores" are the usual terms used by authors to define the diet of damselfishes (e.g., Allen 1991, Cooper and Westneat 2009). Among these, "omnivores" is certainly too vague and could be misleading. Frédérich et al. (2009) decided to use guilds in order to characterize the trophic habits of the family. A guild is a group of species that are similar in some way that is ecologically relevant (Wilson 1999). The definition of these feeding guilds relies mainly on functional demands, and refers to the different ecosystem compartments where preys are caught (Frédérich et al. 2009): (1) the pelagic feeders that feed mainly on planktonic copepods, (2) the benthic feeders that mainly graze on filamentous algae and (3) an intermediate group including species that forage for their prey in the pelagic and benthic environments in variable proportions (e.g., planktonic and benthic copepods, small vagile invertebrates and filamentous algae). At least, three damselfishes are known to be corallivorous species: *Cheiloprion labiatus, Plectroglyphidodon johnstonianus* and *Plectroglyphidodon dickii* (Allen 1991, Kuo and Shao 1991, Ho et al. 2009), and may be grouped within the benthic feeders. The division among the three trophic guilds is not strict; a continuum exists between exclusive zooplankton feeders and algivorous species (Fig. 1). It can be difficult to precisely assign some species to one of the three categories due to feeding plasticity (see below). Arbitrarily, we suggest that the species having more than 70% of zooplanktonic preys or filamentous algae in their stomachs should be grouped within the pelagic and the benthic feeders, respectively.

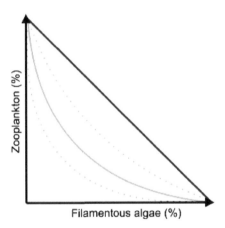

Fig. 1. Schematic representation of the continuum existing between pelagic feeding damselfish, feeding on planktonic copepods and benthic feeding damselfish, grazing on filamentous algae. Linear (dark line) or exponential (grey lines) models may represent the transition between these two trophic specializations. Conceptually, various exponential models (here, only three curves are exemplified) look more appropriate when small benthic invertebrates (zoobenthos) can contribute to the diet of damselfish.

During their evolutionary history, damselfish taxa experienced trophic shifts along such a continuum defined by these two extremes: zooplanktivory and algivory (Fig. 1). *Lepidozygus tapeinosoma*, the lonely representative of Lepidozyginae, exclusively feeds on zooplanktonic preys. The subfamily Stegastinae mainly includes benthic feeders (Frédérich et al. 2009, 2013). The Chrominae mainly contains pelagic feeders when the three trophic guilds are represented in the Abudefdufinae and the Pomacentrinae.

The dietary diversity among sympatric damselfishes is relatively low in comparison with the diverse trophic niches of Labridae (e.g., zooplanktivory, phytoplanktivory, herbivory, durophagy, piscivory, corallivory, gnathiid feeders) (Wainwright and Bellwood 2002). The trophic diversity of Pomacentridae is more similar to that of other conspicuous coral reef fish families such as the Chaetodontidae (e.g., corallivory and species sucking benthic invertebrates) (Ferry-Graham et al. 2001, Pratchett 2007), the Pomacanthidae (e.g., algivory, spongivory, omnivory and zooplanktivory) (Konow and Bellwood 2011) or the Acanthuridae (algivory and zooplanktivory) (Hobson 1975, Purcell and Bellwood 1993).

The trophic diversity of damselfishes may be mainly related to behavior and habitat choice (Waldner and Robertson 1980). In their study of 13 sympatric damselfish species, Frédérich et al. (2009) highlighted that the species feeding on algae are solitary and live close to the reef bottom, in shelters among living or dead coral, or rubble area of shallow water. Some species (e.g., *Pomacentrus baenschi* and *P. sulfureus*) appear to have different shelters in a relatively large area (approximately 3–5 m², B.F. personal observation) but in other species (e.g., *Plectroglyphidodon lacrymatus*), an individual is associated with one shelter (B.F. personal observation). Species feeding mainly on small pelagic preys are generally found in aggregations (schooling species). Exclusive zooplankton feeders, such as *Abudefduf vaigiensis*, *A. sexfasciatus*, *A. sparoides*, *Chromis ternatensis*, *C. chromis*, *C. punctipinnis* and *Azurina hirundo* remain just above the reef, in the water column (Pinnegar and Polunin 2006, Frédérich et al. 2009, Aguilar-Medrano et al. 2011). The adults of *D. trimaculatus* occur in small to large aggregations, and their juveniles are often commensal with large sea anemones or small coral heads (Allen 1991). *Dascyllus aruanus* form social groups living in association with branched corals. *Chromis dimidiata* schools just over the reef, and *P. caeruleus* swims near the substrate and does not usually move away from the reef. Various studies conducted in different geographic areas (e.g., Emery 1973, Frédérich et al. 2009) suggest that diet, habitat and behavior are strongly correlated in damselfishes.

It is generally admitted that phenotypes or any kind of specialization could influence the evolutionary history of a clade (Schoener 2011). Accordingly, we could assume that the trophic specialization of damselfish may have constrained or promoted their diversification during evolution. Recently, Litsios et al. (2012) defined an environmental niche for 169 damselfish species using data such as temperature, oxygen saturation, nitrate, oxygen, phosphate and silicate concentrations, and demonstrated that the trophic ecology of damselfishes influenced their rate of environmental niche evolution. The rates of environmental niche evolution of benthic feeders (i.e., herbivores in Litsios et al. 2012) are slower than those of pelagic feeders. Such differences in the environmental expansion capabilities could be explained by intrinsic physiological tolerances of fish or by the ecological requirement of their algal crops (Litsios et al. 2012). The higher rates of environmental niche evolution in zooplanktivorous species might be related to a current larger geographic distribution than that of mainly algivorous damselfishes.

Detritus as Food for Grazing Damselfishes?

The «weeding» or «gardening» behavior of many territorial damselfishes suggests that they selectively modify the algal assemblages within their territories (Chapter VI). According to this behavior, it is often assumed that these farming species are herbivores feeding on filamentous algae. However, some studies showed that amorphic detritus might account for over 85% of the organic matter ingested by territorial damselfishes (Wilson et al. 2003). Here, detritus is used in a broad sense, i.e., to describe the organic material with no visible structure (Wilson and Bellwood 1997). Consequently, detritus may include faeces, other amorphous material but also live material such as microscopic algae, fungi and/or bacteria (Wilson and Bellwood 1997). Detritus could be the primary source of nutrients for these farming fishes and these species could also be classified as detritivores (Wilson et al. 2003). However, as explained in our first paragraph dealing with the methods commonly used to study the diet of fishes, one of the major limitations of gut content analysis is that it can only identify what the fish ingests, not whether the item is digested and ultimately assimilated. By using biomarkers like fatty acids, Wilson et al. (2001) provided direct evidence that ingested detritus is assimilated by the coral reef blenny *Salarias patzneri*. However, the same kind of proof was never provided for damselfishes until recently. Hata and Umezawa (2011) provided one of the most comprehensive studies of the diet of the farming *Stegastes nigricans*. By combining stomach contents, stable isotopes and fatty acid composition, they demonstrated that *S. nigricans* ingests red filamentous algae (*Polysiphonia* sp.), small benthic invertebrates and detritus, and they showed that all items contribute to the production of the fish biological material. This study supports the hypothesis that detritus contributes to the diet of such farming damselfishes but also demonstrates that these species assimilate filamentous algae and benthic animals providing essential fatty acids and a rich nitrogen source, respectively. These findings support our functional segregation of trophic guilds in Pomacentridae. Indeed the benthic feeders include the farming species ingesting variable proportion of elements coming from the benthic compartment: filamentous algae, detritus and benthic invertebrates.

Ontogenetic Diet Shift

During their pelagic larval phase, damselfishes mainly feed on calanoid and oithonid copepods (Sampey et al. 2007). At the adult stage, they are divided into the three previously described trophic guilds: pelagic feeders, benthic feeders and the "intermediate" group. Consequently, we expect that species belonging to the last two guilds observed diet shift(s) during their ontogeny. Many examples were reported and some are listed hereafter. On a fringing coral reef at Iriomote Island (Japan), the juveniles of *Dischistodus prosopotaenia* (9–13 mm SL) fed mainly on harpacticoid copepods whereas larger fish (32–68 mm SL) predominantly fed on filamentous algae (Nakamura et al. 2003). Ontogenetic diet shifts, characterized by an increase in herbivory, were observed in two temperate damselfish species *Parma microlepis* and *P. unifasciata* (Buckle and Booth 2009) but also in the tropical species *Stegastes nigricans* (Letourneur et al. 1997). The ontogenetic shifts in the diet of these four species are likely related to a transitional phase between a carnivorous diet during larval life and an algivorous diet as adults.

The Humbug damselfish, *Dascyllus aruanus* is generally considered to be a mainly zooplanktivorous species (Coates 1980, Forrester 1991). However ontogenetic diet shift was recently observed in this species at the Toliara Reef in Madagascar (Frédérich et al. 2010): smaller fish fed on benthic prey such as isopods and copepods, and the larger fish foraged in the water column on planktonic copepods. It is also expected that most of the pelagic feeders select larger-sized planktonic prey during their growth. In the same region of Toliara, Frédérich et al. (2009) observed a variation in the $\delta^{13}C$ compositions of *Amphiprion akallopisos* which is size-related. It could reflect ontogenetic changes in food preference which were not revealed by stomach content analysis, or it could be related to ontogenetic variability in diet-tissue fractionation and/or physiology (Bearhop et al. 2004).

Until now, some ontogenetic diet shifts were reported but we have no precise information about the dynamic of diet change: does it last some days, weeks or months? Future ecomorphological studies should also explore the relationship between the morphological changes and the dynamic of diet shift in various species.

Diet Plasticity and Individual Specialization

Trophic switching has occurred many times during the evolutionary history of Pomacentridae (Cooper and Westneat 2009, Frédérich et al. 2013). For evolutionary ecologists, it would be really interesting to understand the factors driving such diet shifts but also to appreciate their consequences on ecosystem functioning. Within the following paragraphs, we illustrate that the damselfishes show feeding plasticity. We discuss about opportunism and diet variation among and within populations of damselfish species.

Opportunism

Most of the ichthyologists who are specialized on the biology of damselfishes will state that these fishes are really opportunistic, especially about their food. This opportunism is clearly obvious during the mass coral spawning events. In addition to planktivorous species picking up permanently in the water column, benthic feeders (e.g., *Stegastes apicalis*, *Plectroglyphidodon lacrymatus*) and species of the "intermediate" group (e.g., *Abudefduf bengalensis*) switch to a pelagic feeding mode and feed almost exclusively on coral propagules during mass coral spawning (Westneat and Resing 1988, Pratchett et al. 2001).

The discordance among results from stomach contents, stable isotopes analysis, and fatty acid compositions or the study of diet using a large number of specimens enable the demonstration of some feeding opportunisms. For example, Frédérich et al. (2009) observed that one specimen of the mainly pelagic feeders *Abudefduf vaigiensis* and *Chromis ternatensis* had their stomach totally filled with demersal fish eggs. This observation provides evidence that these planktivorous species may shift to benthic foraging for exploiting nutrient-rich food.

Large-scale and local-scale geographic variation in diet

Numerous damselfishes have a wide geographic distribution: some extending from the East coast of Africa to French Polynesia (e.g., *Dascyllus aruanus, D. trimaculatus, Abudefduf sexfasciatus...*). The comparisons of results from various studies revealed diet variation among populations from different regions. For example, *Abudefduf vaigiensis* was identified as an algivorous species by Kuo and Shao (1991) in Taiwan, whereas the combined results of stomach contents and stable isotopes allowed Frédérich et al. (2009) to group it with pelagic feeders at the Great Barrier Reef of Toliara (Madagascar). Diet variation between populations of a species living at two different locations on the same coral reef was also reported. Fatty acid compositions support an increase in reef-derived benthic nutrition for *Abudefduf sexfasciatus* and *Stegastes fasciolatus* while comparing populations from the reef slope to the reef flat of the fringing reef at Ningaloo Reef, Western Australia (Wyatt et al. 2012). By combining stomach contents and stable isotopes analysis, Ho et al. (2009) illustrated that the benthic feeder *Plectroglyphidodon dickii* usually consumes filamentous algae and coral polyps on zones dominated by *Acropora* branching corals in Southern Taiwan. However, on a zone dominated by the sea anemone *Mesactinia genesis*, it is able to switch to feed on sea anemones.

Variation in diet over time

Relatively few studies looked at the temporal variation in the diet of damselfishes. At the Réunion Island (Indian Ocean), Letourneur et al. (1997) showed that the proportion of algal diet in the benthic feeder *Stegastes nigricans* increases during the summer peak. This feeding pattern coincides with seasonal fluctuations in biomass of algae, which increases strongly during summer on some Réunion reef flats.

Using the method described in Frédérich et al. (2009), we collected adult specimens of *Pomacentrus trilineatus* on the reef flat of the Great Barrier Reef of Toliara in November 2005 (n = 9), October 2006 (n = 25), June 2010 (n = 22) and June 2011 (n = 14). The entire digestive tracts of all the individuals were removed and conserved in 70% alcohol. We analyzed the stomach contents following the methodology described in Frédérich et al. (2012) and we calculated the percentage of benthic algae, zooplankton and small benthic invertebrates within stomach contents. Percentages were arcsine-square root transformed for statistical analysis. Qualitatively, *Pomacentrus trilineatus* feeds always on a large proportion of filamentous algae and complements its diet with planktonic and small benthic animals. However, *P. trilineatus* showed significant temporal variation in its diet (Kruskall-Wallis test: 16.59; *P-value* = 0.0009). The populations collected in 2005 (mean = 77%) and 2011 (mean = 82%) consumed significantly less filamentous algae than the population of 2006 (mean = 98%; Fig. 2). Instead of *S. nigricans*, the variation in the diet of *P. trilineatus* is not related to seasonality (October-November vs. June). Our results strengthen the idea that a fish diet has a multifactorial explanation. Other factors like inter-specific competition or population density may explain the diet variation observed in *P. trilineatus*.

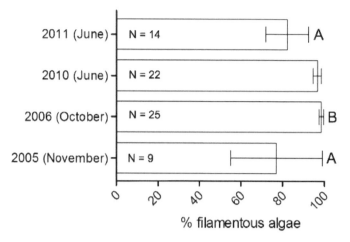

Fig. 2. Temporal variation in the percentage of filamentous algae present in the stomach contents of *Pomacentrus trilineatus*. The populations collected in 2005 and 2011 (A) consumed significantly less filamentous algae than the population of 2006 (B; post-hoc tests: *P-value* < 0.05). N refers to the number of studied specimens.

Individual specialization

Ecologists have long realized that individuals within populations can differ substantially in their resource use. In many species, co-occurring individuals actively select different prey from their shared environment (reviewed in Araújo et al. 2011). As exemplified earlier, trophic niche variation can be attributed to ontogenetic niche shifts. However, resource-use variation can be observed even among individuals of a given age. This individual specialization—in which individuals use a small subset of the population's resource base—has been shown to be a widespread phenomenon in many vertebrate and invertebrate taxa (Bolnick et al. 2003). The ecology of individuals is clearly understudied in damselfishes. Some trophic specialization was clearly highlighted in *Dascyllus aruanus* at the Toliara Reef (Frédérich et al. 2010). This highly site-attached fish living in permanent social groups associated with branched corals showed a density-dependent specialization in prey size selectivity (Frédérich et al. 2010). In *D. aruanus*, the density of the social group drives specialization on prey size. The level of intra-group competition or social interactions within groups, or both, could be linked to group density, leading to increased division of the resources among all individuals within the colony (Svanback and Bolnick 2007).

   This short overview provides evidences that diet variation among populations living in different locations or within populations exists in damselfish species. However, the available data enable us to make limited comparisons. The question arises whether or not the ability to switch guilds is rare, common, driven by prevailing environmental conditions, food availability or inter-specific interactions? Further studies are really required in order to quantify the amplitude of such kind of diet variation, diet specialization and to understand the factors underlying these variations.

Intermediate Trophic Group of Damselfishes: What Kind of Generalism is?

Among the three trophic guilds of damselfishes, we defined the "intermediate group" as including species that forage for their prey in the pelagic and the benthic environments in variable proportions. The diet of these generalist species (also called omnivorous species by Allen 1991, Cooper and Westneat 2009) is generally composed of planktonic and benthic copepods, small vagile invertebrates (e.g., annelids, tiny molluscs, etc.) and filamentous algae. Two types of generalization are commonly recognized (Fig. 3; Van Valen 1965, Bearhop et al. 2004): Type A generalization where the population is composed of generalist individuals all taking a wide range of food types or Type B generalization with individuals each specializing in a different but narrow range of food types. Discriminating between the types of generalization requires laborious sampling of individuals over extended time periods followed by integration of the information, which is often difficult to achieve (Bearhop et al. 2004). Nowadays, some methods relying on the measurement of stable isotopes and statistical approaches are currently being developed to assess these questions (see details in Bearhop et al. 2004, Jackson et al. 2011). However, to the best of our knowledge, no study has questioned whether these two kinds of generalist damselfish species may be encountered in coral reefs. In their discussion, Frédérich et al. (2009) compared the isotopic compositions and the significance of two generalist species: *Pomacentrus baenschi* and *P. trilineatus*. *Pomacentrus baenschi* may be a type A generalist, and the very low variations in $\delta^{13}C$ in its muscle tissue may suggest that the prey components in its diet are very consistent over time and among individuals. Conversely, a larger variation of $\delta^{13}C$ values in *P. trilineatus* could be related to diet-switching over time or perpetual intra-population variability in prey choice (i.e., type B generalization).

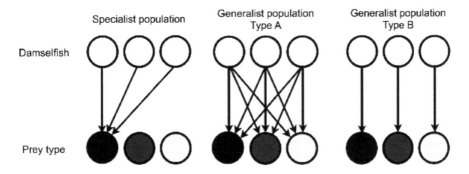

Fig. 3. Schematic representation of predator-prey systems, exemplifying specialist, type A generalist and type B generalist damselfish. For each example, three individuals of one population (white circles) and three types of prey (black, dark grey and light grey) are represented (Modified from Bearhop et al. 2004).

Conclusion and Perspectives

To conclude, most of the studies combining various methods suggest that three trophic guilds are present in Pomacentridae: pelagic feeders catching mainly planktonic animal preys, benthic feeders including algivorous, corallivorous (at least 3 species) and detritivorous species and an intermediate group made of generalist species. Most of the damselfishes appear as highly opportunistic and show feeding plasticity. However, although we have a good amount of general knowledge about their trophic diversity, we still need to answer relevant questions regarding the trophic ecology of Pomacentridae. What are the main factors driving their feeding plasticity: environmental conditions, food availability, biotic interactions or internal factors? What is the prevalence of type A/B generalists in the intermediate group? Is the plasticity of each trophic guild of damselfishes predictable? Does the diet of a damselfish have an impact on its evolvability, i.e., the ability of a population to evolve through natural selection? In the context of global changes, identifying how factors influence the niche breadth of damselfishes and other reef fishes is essential if we wish to understand the effects of declining reef quality on ecosystem function.

Acknowledgements

We thank P. Vandewalle and G. Lepoint for helping us to collect *Pomacentrus trilineatus* from Toliara (Madagascar) during our field works in 2005, 2006, 2010 and 2011. The authors also thank V.P. Andrinirina, who partially contributed to the analysis of the diet of *Pomacentrus trilineatus*. We also gratefully acknowledge two reviewers for their insightful comments and helpful criticism of the original manuscript. This research was supported by FRS-FNRS grants (FRFC no. 2.4.583.05). BF is a Postdoctoral Researcher at the F.R.S.-FNRS (Belgium). DO and LG are Research Fellows at the FRIA and F.R.S.-FNRS (Belgium), respectively.

References

Aguilar-Medrano, R., B. Frédérich, E. De Luna and E.F. Balart. 2011. Patterns of morphological evolution of the cephalic region in damselfishes (Perciformes: Pomacentridae) of the Eastern Pacific. Biol. J. Linnean Soc. 102(3): 593–613.

Allen, G.R. 1991. Damselfishes of the World. Mergus, Melle, Germany.

Araújo, M.S., D.I. Bolnick and C.A. Layman. 2011. The ecological causes of individual specialisation. Ecol. Lett. 14(9): 948–958.

Bearhop, S., C.E. Adams, S. Waldron, R.A. Fuller and H. Macleod. 2004. Determining trophic niche width: a novel approach using stable isotope analysis. J. Anim. Ecol. 73(5): 1007–1012.

Bolnick, D.I., R. Svanback, J.A. Fordyce, L.H. Yang, C.D. Hulsey and M.L. Forister. 2003. The ecology of individuals: incidence and implications of individual specialization. Am. Nat. 161(1): 1–28.

Buckle, E.C. and D.J. Booth. 2009. Ontogeny of space use and diet of two temperate damselfish species, *Parma microlepis* and *Parma unifasciata*. Mar. Biol. 156(7): 1497–1505.

Ceccarelli, D.M. 2007. Modification of benthic communities by territorial damselfish: a multi-species comparison. Coral Reefs 26(4): 853–866.

Coates, D. 1980. Prey-size intake in humbug damselfish, *Dascyllus aruanus* (Pisces, Pomacentridae) living within social groups. J. Anim. Ecol. 49(1): 335–340.

Cocheret de la Morinière, E., B.J.A. Pollux, I. Nagelkerken, M.A. Hemminga, A.H.L. Huiskes and G. Van der Velde. 2003. Ontogenetic dietary changes of coral reef fishes in the mangrove-seagrass-reef continuum: Stable isotopes and gut-content analysis. Mar. Ecol. Prog. Ser. 246: 279–289.

Colwell, R.K. and E.R. Fuentes. 1975. Experimental studies of the niche. Annu. Rev. Ecol. Evol. Syst. 6: 281–310.

Cooper, W.J. and M.W. Westneat. 2009. Form and function of damselfish skulls: rapid and repeated evolution into a limited number of trophic niches. BMC Evol. Biol. 9: 24.

Emery, A.R. 1973. Comparative ecology and functional osteology of fourteen species of damselfish (Pisces: Pomacentridae) at Alligator Reef, Florida Keys. Bull. Mar. Sci. 23: 649–770.

Ferry-Graham, L.A., P.C. Wainwright, C.D. Hulsey and D.R. Bellwood. 2001. Evolution and mechanics of long jaws in butterflyfishes (family Chaetodontidae). J. Morph. 248(2): 120–143.

Forrester, G.E. 1991. Social rank, individual size and group composition as determinants of food-consumption by Humbug Damselfish, *Dascyllus aruanus*. Anim. Behav. 42: 701–711.

Frédérich, B., G. Fabri, G. Lepoint, P. Vandewalle and E. Parmentier. 2009. Trophic niches of thirteen damselfishes (Pomacentridae) at the Grand Récif of Toliara, Madagascar. Ichthyol. Res. 56(1): 10–17.

Frédérich, B., O. Lehanse, P. Vandewalle and G. Lepoint. 2010. Trophic niche width, shift, and specialization of *Dascyllus aruanus* in Toliara lagoon, Madagascar. Copeia (2): 218–226.

Frédérich, B., O. Colleye, G. Lepoint and D. Lecchini. 2012. Mismatch between shape changes and ecological shifts during the post-settlement growth of the surgeonfish, *Acanthurus triostegus*. Front. Zool. 9: 8.

Frédérich, B., L. Sorenson, F. Santini, G.J. Slater and M.E. Alfaro. 2013. Iterative ecological radiation and convergence during the evolutionary history of damselfishes (Pomacentridae). Am. Nat. 181(1): 94–113.

Galetto, M.J. and D.R. Bellwood. 1994. Digestion of algae by *Stegastes nigricans* and *Amphiprion akindynos* (Pisces, Pomacentridae), with an evaluation of methods used in digestibility studies. J. Fish Biol. 44(3): 415–428.

Gobler, C.J., D.B. Thibault, T.W. Davis, P.B. Curran, B.J. Peterson and L.B. Liddle. 2006. Algal assemblages associated with *Stegastes* sp. territories on Indo-Pacific coral reefs: characterization of diversity and controls on growth. J. Exp. Mar. Biol. Ecol. 336(1): 135–145.

Graeve, M., P. Dauby and Y. Scailteur. 2001. Combined lipid, fatty acid and digestive tract content analyses: a penetrating approach to estimate feeding modes of Antarctic amphipods. Polar Biol. 24(11): 853–862.

Hata, H. and M. Kato. 2002. Weeding by the herbivorous damselfish *Stegastes nigricans* in nearly monocultural algae farms. Mar. Ecol. Prog. Ser. 237: 227–231.

Hata, H. and Y. Umezawa. 2011. Food habits of the farmer damselfish *Stegastes nigricans* inferred by stomach content, stable isotope, and fatty acid composition analyses. Ecol. Res. 26(4): 809–818.

Ho, C.T., Y.C. Fu, C.L. Sun, S.J. Kao and R.Q. Jan. 2009. Plasticity of feeding habits of two *Plectroglyphidodon* damselfishes on coral reefs in southern Taiwan: evidence from stomach content and stable isotope analyses. Zool. Stud. 48(5): 649–656.

Hobson, E.S. 1975. Feeding patterns among tropical reef fishes. Am. Sci. 63(4): 382–392.

Hyslop, E.J. 1980. Stomach contents analysis—a review of methods and their application. J. Fish Biol. 17: 411–429.

Jackson, A.L., R. Inger, A.C. Parnell and S. Bearhop. 2011. Comparing isotopic niche widths among and within communities: SIBER—Stable Isotope Bayesian Ellipses in R. J. Anim. Ecol. 80(3): 595–602.

Jones, G.P., L. Santana, L.J. McCook and M.I. McCormick. 2006. Resource use and impact of three herbivorous damselfishes on coral reef communities. Mar. Ecol. Prog. Ser. 328: 215–224.

Konow, N. and D.R. Bellwood. 2011. Evolution of high trophic diversity based on limited functional disparity in the feeding apparatus of marine angelfishes (f. Pomacanthidae). PLoS ONE 6(9): e24113.

Kuo, S.-R. and K.-T. Shao. 1991. Feeding habits of damselfishes (Pomacentridae) from the southern part of Taiwan. J. Fish. Soc. Taiwan 18(3): 165–176.

Lee, R.F., J.C. Nevenzel and G.A. Paffenhöffer. 1971. Importance of wax esters and other lipids in the marine food chain: phytoplankton and copepods. Mar. Biol. 9: 99–108.

Letourneur, Y., R. Galzin and M. Harmelin-Vivien. 1997. Temporal variations in the diet of the damselfish *Stegastes nigricans* (Lacepede) on a Reunion fringing reef. J. Exp. Mar. Biol. Ecol. 217(1): 1–18.

Litsios, G., L. Pellissier, F. Forest, C. Lexer, P.B. Pearman, N.E. Zimmermann and N. Salamin. 2012. Trophic specialization influences the rate of environmental niche evolution in damselfishes (Pomacentridae). Proc. R. Soc. B-Biol. Sci. 279(1743): 3662–3669.

Mann, D.A. and G. Sancho. 2007. Feeding ecology of the domino damselfish, *Dascyllus albisella*. Copeia (3): 566–576.

Montgomery, W.L. and S.D. Gerking. 1980. Marine macroalgae as foods for fishes: an evaluation of potential food quality. Environ. Biol. Fishes 5(2): 143–153.

Motta, P.J. 1988. Functional morphology of the feeding apparatus of ten species of Pacific butterflyfishes (Perciformes, Chaetodontidae): an ecomorphological approach. Environ. Biol. Fishes 22(1): 39–67.

Nakamura, Y., M. Horinouchi, T. Nakai and M. Sano. 2003. Food habits of fishes in a seagrass bed on a fringing coral reef at Iriomote Island, southern Japan. Ichthyol. Res. 50(1): 15–22.

Navarro, J., M. Coll, M. Louzao, I. Palomera, A. Delgado and M.G. Forero. 2011. Comparison of ecosystem modelling and isotopic approach as ecological tools to investigate food webs in the NW Mediterranean Sea. J. Exp. Mar. Biol. Ecol. 401(1-2): 97–104.

Noda, M., K. Kawabata, K. Gushima and S. Kakuda. 1992. Importance of zooplankton patches in foraging ecology of the planktivorous reef fish *Chromis chrysurus* (Pomacentridae) at Kuchinoerabu Island, Japan. Mar. Ecol. Prog. Ser. 87(3): 251–263.

Nyssen, F., T. Brey, P. Dauby and M. Graeve. 2005. Trophic position of Antarctic amphipods-enhanced analysis by a 2-dimensional biomarker assay. Mar. Ecol. Prog. Ser. 300: 135–145.

Parmentier, E. and K. Das. 2004. Commensal vs. parasitic relationship between Carapini fish and their hosts: some further insight through delta C-13 and delta N-15 measurements. J. Exp. Mar. Biol. Ecol. 310(1): 47–58.

Peterson, B.J. and B. Fry. 1987. Stable isotopes in ecosystem studies. Annu. Rev. Ecol. Evol. Syst. 18: 293–320.

Phillips, D.L., S.D. Newsome and J.W. Gregg. 2005. Combining sources in stable isotope mixing models: alternative methods. Oecologia 144(4): 520–527.

Pinnegar, J.K. and N.V.C. Polunin. 1999. Differential fractionation of delta C-13 and delta N-15 among fish tissues: implications for the study of trophic interactions. Funct. Ecol. 13(2): 225–231.

Pinnegar, J.K. and N.V.C. Polunin. 2006. Planktivorous damselfish support significant nitrogen and phosphorus fluxes to Mediterranean reefs. Mar. Biol. 148(5): 1089–1099.

Post, D.M. 2002. Using stable isotopes to estimate trophic position: models, methods, and assumptions. Ecology 83(3): 703–718.

Pratchett, M.S. 2007. Dietary selection by coral-feeding butterflyfishes (Chaetodontidae) on the Great Barrier Reef, Australia. Raffles Bull. Zool.: 171–176.

Pratchett, M.S., N. Gust, G. Goby and S.O. Klanten. 2001. Consumption of coral propagules represents a significant trophic link between corals and reef fish. Coral Reefs 20(1): 13–17.

Purcell, S.W. and D.R. Bellwood. 1993. A functional analysis of food procurement in two surgeonfish species, *Acanthurus nigrofuscus* and *Ctenochaetus striatus* (Acanthuridae). Environ. Biol. Fishes 37(2): 139–159.

Sampey, A., A.D. McKinnon, M.G. Meekan and M.I. McCormick. 2007. Glimpse into guts: overview of the feeding of larvae of tropical shorefishes. Mar. Ecol. Prog. Ser. 339: 243–257.

Schoener, T.W. 2011. The newest synthesis: understanding the interplay of evolutionary and ecological dynamics. Science 331(6016): 426–429.

Silvertown, J. 2004. Plant coexistence and the niche. Trends Ecol. Evol. 19(11): 605–611.

Svanback, R. and D.I. Bolnick. 2007. Intraspecific competition drives increased resource use diversity within a natural population. Proc. R. Soc. B-Biol. Sci. 274(1611): 839–844.

Van Valen, L. 1965. Morphological variation and width of ecological niche. Am. Nat. 99: 377–390.

Wainwright, P.C. 1988. Morphology and ecology – functional basis of feeding constraints in Carribean labrid fishes. Ecology 69(3): 635–645.

Wainwright, P.C. and D.R. Bellwood. 2002. Ecomorphology of feeding in coral reef fishes. pp. 33–56. *In*: P.F. Sale (ed.). Coral Reef Fishes: Dynamics and Diversity in a Complex Ecosystem. Academic Press, San Diego.

Wainwright, P.C., D.R. Bellwood, M.W. Westneat, J.R. Grubich and A.S. Hoey. 2004. A functional morphospace for the skull of labrid fishes: patterns of diversity in a complex biomechanical system. Biol. J. Linnean Soc. 82(1): 1–25.

Waldner, R.E. and D.R. Robertson. 1980. Patterns of habitat partitioning by eight species of territorial caribbean damselfishes (Pisces: Pomacentridae). Bull. Mar. Sci. 30: 171–186.

Westneat, M.W. and J.M. Resing. 1988. Predation on coral spawn by planktivorous fish. Coral Reefs 7(2): 89–92.

Wilson, J.B. 1999. Guilds, functional types and ecological groups. Oikos 86(3): 507–522.

Wilson, S. and D.R. Bellwood. 1997. Cryptic dietary components of territorial damselfishes (Pomacentridae, Labroidei). Mar. Ecol. Prog. Ser. 153: 299–310.

Wilson, S.K., D.R. Bellwood, J.H. Choat and M.J. Furnas. 2003. Detritus in the epilithic algal matrix and its use by coral reef fishes. Oceanogr. Mar. Biol. 41: 279–309.

Wilson, S.K., K. Burns and S. Codi. 2001. Sources of dietary lipids in the coral reef blenny *Salarias patzneri*. Mar. Ecol. Prog. Ser. 222: 291–296.

Wyatt, A.S.J., A.M. Waite and S. Humphries. 2012. Stable isotope analysis reveals community-level variation in fish trophodynamics across a fringing coral reef. Coral Reefs 31(4): 1029–1044.

Ontogeny and Early Life Stages of Damselfishes

Kathryn Kavanagh[1],* and *Bruno Frédérich*[2]

The early life stages of damselfish have figured prominently in the ecological studies of coral reef fish communities, most likely due to their diversity and abundance on many reef systems and also due to their substrate brooding, which allows access to the embryonic stages and hatchlings. Despite their importance in advancing ecological understanding, specific detailed studies of morphological, behavioral, and physiological development in this family are sparse. Although there are approximately 360 species of coral reef damselfishes, only about 15% have had their early life stages described in the literature (Murphy et al. 2007) and even fewer have had their embryonic development described (Leis 2015).

In the present chapter, we review the growth and development from fertilization to the post-settlement stages in Pomacentridae, illustrating some of the known variations among taxa (see Chapter III for a detailed description of the bipartite life cycle). Indeed, this family contains taxa that show unusually large variations in the early life history stage durations and behavioral variations as compared to other marine fish taxa (Kavanagh and Alford 2003). After a brief description of embryogenesis, we describe the development of various morphological traits during the pelagic stage and then, we discuss ontogenetic shape variation (allometric growth) observed during post-settlement and recruitment. However, this review clearly demonstrates the lack of studies devoted to the study of larval development in damselfishes and these works are required in order to understand the diversity and the evolution of the early life stages in this family.

[1] Biology Department, University of Massachusetts, Dartmouth, USA.
[2] Laboratoire de Morphologie Fonctionnelle et Evolutive, AFFISH - Research Center, University of Liège, Quartier Agora, Allée du six Août 15, Bât. B6C, 4000 Liège (Sart Tilman), Belgium.
* Corresponding author: kkavanagh@umassd.edu

Early Embryogenesis and Hatching Variation

Damselfish eggs are oval-shaped, with greater elongation in species with longer egg stages and larger yolk reserves. The eggs have adhesive tendrils at one end that attach individual eggs to the substratum and single or multiple oil droplets that make them buoyant (Fig. 1A–D). Most species lay eggs in distinct patches on flat surfaces like dead coral, and others including many *Chromis* lay them individually amongst algal fronds on the reef. As the great majority of pomacentrids are found in tropical oceans, the typical developmental temperatures are high – 26°C to 28°C—and the developmental metabolic rates are consequently fast. Gastrulation and somitogenesis occur during the first 24 hours. By Day 2, the heart and eyes begin to differentiate and the otoliths appear. By Day 3 of development, the jaw and gut are forming and the incipient organs are initiated (Murphy et al. 2007). Cooler temperatures will slow down these rates.

Pomacentrid eggs take 2–16 days to hatch at these tropical temperatures, depending on the species (Kavanagh and Alford 2003). Amphiprionini typically take 7–8 days to hatch in tropical waters, while many tropical *Chromis* species take

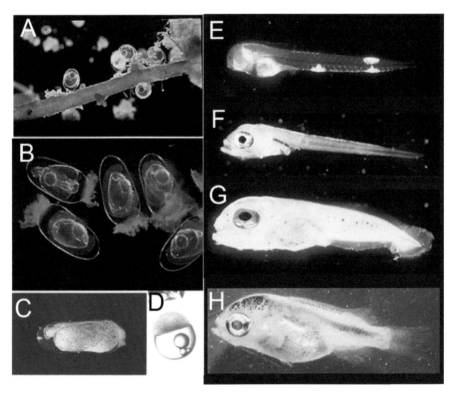

Fig. 1. Damselfish eggs and early embryogenesis. (A) Eggs of *Chromis weberi* on an algal stem. (B) Eggs and embryos of *Chrysiptera taupou* at somitogenesis stage. (C) Early embryo of *Acanthochromis polyacanthus*. (D) Egg of *Chromis dispilis* at gastrulation stage (from Kingsford 1985). (E–H) Hatchlings of 4 species with varying egg stage durations of 2d, 4d, 8d, and 16d, respectively (top to bottom: *Chromis atripectoralis, Pomacentrus amboinensis, Premnas biaculeatus, Acanthochromis polyacanthus*, from Kavanagh and Alford 2003).

only 2 days. The brooding damselfish, *Acanthochromis polyacanthus*, takes 16 days. All other genera are intermediate between these extremes, with many genera taking 4–5 days to hatch at 26–28°C. Because of this large variation, the size and developmental stage at hatching also vary. Hatchlings vary from approximately 2 to 5.5 mm notochord length. Early hatchers (which take 2 days to hatch) such as most *Chromis* hatch very undeveloped, without functional mouth, eyes, or muscle fibers, and they will often hang vertically, head down, in the water for a day or so before beginning to swim horizontally. These early hatchers begin to move and feed within 24 hr post hatching. Genera of tropical pomacentrids (see exceptions below) that hatch about 4–5 days after fertilization, have more developed muscles and sensory systems but still lack fins (Fig. 1). The anemonefishes (*Premnas* and *Amphiprion* spp.) hatch about 7–8 days after fertilization, at approximately the stage of notochord flexion and early fin formation. This relative delay in hatching in this tribe may be associated with their particular symbiosis with sea anemones. Experimental studies of anemonefishes suggest the need for hatchlings to be developed to a sufficient stage in order to imprint on the host anemone species at hatching (Arvedlund et al. 1999). Another pomacentrid species, *Acanthochromis polyacanthus*, hatches well-developed and large at 5.5 mm SL, with a nearly complete fin ray count, after a long incubation period of 15–16 days (Fig. 1H). As in the case of the anemonefishes, this prolonged embryonic stage is associated with a divergent life history. In this case, *A. polyacanthus* is unusual in having no pelagic larval stage and instead directly settling on the coral reef after hatching (Robertson 1973, Kavanagh 2000; Chapter III). *Altrichthys* spp. reportedly also guard their young on the reef, but no developmental studies have been published on this genus (Allen 1999). Molecular phylogenetic studies indicate that larval-brooding behaviors in damselfishes evolved once; both *Acanthochromis* and *Altrichthys* share a common ancestor (Bernardi 2011).

Larval Development (Pelagic Stage)

The larval stage begins at hatching, thus, the large differences in hatching times among damselfishes indicate significant differences in the developmental stage at which the larval period begins (Kavanagh and Alford 2003). Therefore age-specific comparisons of larval development may actually include the later embryonic stages as well. In general, pomacentrids do not have a dramatic metamorphosis but are fairly direct developers. Individual variations in egg 'quality' and larval size, etc. within a species have been recorded in several studies, correlated with maternal size or other environmental features (Kerrigan 1997).

With the exception of *Acanthochromis polyacanthus*, which does not have a pelagic larval stage, the larval stage durations vary among species from approximately 14–30 days, and the total time to settlement varies from 21 to 32 days (Chapter III). Longer egg stage durations are associated with shorter larval stage durations and shorter overall duration to settlement (Fig. 2; Kavanagh and Alford 2003). Bay et al. (2006) measured intraspecific variation in the pelagic larval duration of 10 species of damselfishes, with a range of 5.5 days.

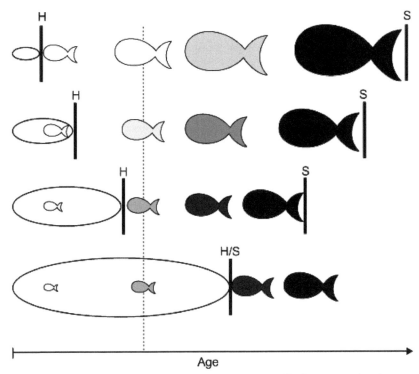

Fig. 2. An illustration of the predicted relationships among early life-history stage durations, growth rates, and developmental rates in the Pomacentridae. The x-axis is age (days after fertilization). Different patterns of fertilization to settlement are shown moving from left to right in each line. The embryonic period is represented by the oval eggs to the left of the first black bar (H = hatching), and settlement (S) is indicated by the black bar furthest to the right. The degree of shading in the fish outline indicates increasing competence, and thus the developmental rate is indicated by the degree of shading at a particular age. The size of the fish outline indicates relative size-at-age (growth rate). In the first line, growth rate is higher, developmental rate is slower, egg stage duration is shorter and pelagic larval stage duration is longer than the other patterns (e.g., fish settle later and at a larger size). As total duration to settlement decreases (line 2 and 3), developmental rate increases, growth rate decreases, and egg stage duration increases. The pattern shown by the bottom line is unique to the benthic brooder *Acanthochromis polyacanthus*: after an extended embryonic period, the young hatch and stay settled on the reef. They are smaller and less developed than other pomacentrids at settlement.

In describing the morphological development during the pelagic larval stages, we focus here on *Pomacentrus amboinensis* from the Great Barrier Reef because of the availability of a complete description of embryonic and larval stages (Kavanagh and Alford 2003, Murphy et al. 2007). Consequently, the data listed below refer to this species, and we also provide available morphological information from other species (see also Leis and Trnski 1989, Kavanagh et al. 2000 and cited references therein).

General growth trajectory and developmental identification

The typical larval development of pomacentrids is direct with few specializations for pelagic life. Most pomacentrids have a tightly coiled and compact triangular gut, inconspicuous gas bladder and weak head spination: small supracleithral, preopercular and interopercular spines and a small opercular spine (Kavanagh et al. 2000). Pomacentrids also generally lack elongate fin spines (Fig. 2; Kavanagh et al. 2000).

Preflexion larvae are moderately elongated with a short gut. Around the time of notochord flexion (approximately 7 mm), the body depth increases and the fins develop. For example, in *P. amboinensis*, body depth is 16% at hatching, reaches 23% during flexion and is 45% by settlement (Murphy et al. 2007).

In all studied pomacentrids (except *Acanthochromis polyacanthus*), there are 26 myomeres, initially 6 preanal + 20 postanal, becoming 11 precaudal + 15 caudal vertebrae in postflexion larvae due to the posterior extension of the gut. *Acanthochromis polyacanthus* has a uniquely increased and variable number of myomeres and vertebrae, 14–15 precaudal + 15–16 caudal (Kavanagh et al. 2000).

Proportions: In *P. amboinensis*, the preanal length is 33%–38% in preflexion larvae and 47%–58% in postflexion larvae as the gut extends posteriorly (Murphy et al. 2007). The gut is coiled and compact, with a gap between the anus and anal fin that becomes smaller but does not completely close before settlement: 8.6% at the end of flexion, decreasing to 3.2% at settlement. The head is initially small (18–27%), but becomes moderate in size by flexion (26%–30%) and is large (32%–39%) in postflexion larvae. The snout is short (less than eye diameter) throughout development. It is blunt at hatching, but quickly becomes concave, and remains so until the end of flexion, after which it is more rounded and convex, but becomes shorter relative to head length. The eye is round and large, constituting 31%–48% of the head length (HL) (Kavanagh et al. 2000).

Growth rates: In *P. amboinensis*, preflexion larvae are reported to grow at an average rate of about 0.3 mm d^{-1}, and postflexion larvae grew nearly 67% faster—about 0.5 mm d^{-1}. In addition, there was considerable variation in growth rate among individuals. Five days after hatching (DAH), the range of standard length (SL) was 3.2–4.8 mm, and the range of size at age varied by 1 to 3 mm until settlement.

Gagliano et al. (2007) back-calculated growth rates over the early life stages of *Pomacentrus amboinensis* on the Great Barrier Reef. They found evidence of differential selection for slower or faster growth rates during different stages. At all stages, growth rates are affected by temperature regimes (Takahashi et al. 2012) and parasites may slow down the growth of larval damselfish (Sun et al. 2012).

External features and integument

i. Color and pigmentation

Damselfishes begin to develop individual pigmented cells during early embryogenesis. Larvae are lightly to moderately pigmented and much of the pigment is somewhat variable among individuals. Young larvae have species-specific patterns of melanophores and xanthophores, which increase in number, become joined together, migrate, and develop iridescent layers. Pigment can be used to differentiate closely related species (e.g., Potthoff et al. 1987, John and Reinhold 2008).

The damselfish larvae are largely transparent, but with a silvery gut. Typical pigment of early damselfish larvae includes melanophores along the ventral midline of the tail, melanophores on the ventral surface of the yolk sac, and over the gut, on the forebrain and midbrain (Kavanagh et al. 2000). Melanic pigment on the head and body increases during development. During the settlement phase and the juvenile stage, species exhibit various color patterns which may be different from adult stage. For example, many species are yellowish with blue lines during the settlement stage (e.g., *Pomacentrus amboinensis*, *Chrysiptera leucopoma*, *Chrysiptera unimaculata*).

ii. Mucus and scales

The mucus layer has UV protection that increases as the fish grows larger (Zamzow and Siebeck 2006). Scales begin to form at about 7 mm ventral to the base of the pectoral fin and in one or two rows along and above the lateral midline. A full complement of scales is present at 9–10 mm in *P. amboinensis* (Murphy et al. 2007).

Musculo-skeletal development

i. Cranial skeleton

The mouth is quite oblique in preflexion larvae, becoming less oblique in postflexion larvae. Once the mouth is open, the posterior tip of the maxilla extends past a vertical through the anterior edge of the eye, but not to the pupil. During settlement, the posterior tip of the maxilla reaches a vertical through the anterior margin of the eye. At settlement, all components of the osseous cranium are present (Fig. 3, Frédérich et al. 2006). However the development of cranial skeleton varies among species. For example, the degree of cranial ossification is less in *Chromis viridis* than in *Dascyllus flavicaudus* (Frédérich et al. 2006). In *P. amboinensis*, small conical teeth are present on both the upper and lower jaw by 9.6 mm, but are not readily visible due to the lips (Murphy et al. 2007). The premaxilla and the mandible of *C. viridis* do not bear teeth, at settlement stage (10 mm SL). On the other hand, the teeth are almost tricuspid in *Chrysiptera glauca* (15 mm SL) and *Plectroglyphidodon lacrymatus* (15 mm SL) at the same development stage (Frédérich et al. 2006).

ii. Postcranial skeleton and fin development

The notochord, medial fin fold and the pectoral fin buds are early stages of the postcranial body parts. A very weak caudal-fin anlage is first present at about 2.8 mm total length (TL) in *P. amboinensis* (Murphy et al. 2007). Anterior vertebrae, cleithrum, and otoliths are the first to ossify (Potthoff et al. 1987). The timing of notochord flexion in damselfish larvae varies among species. Alshuth et al. (1998) report that *Abudefduf saxatilis* notochord flexion occurs between 5.3 and 9.2 mm TL (10–17d), and Murphy et al. (2007) report that in *P. amboinensis* notochord flexion begins at about 4 mm TL (7d at 26–28°C) and is complete at about 5 mm TL. The first rays are present at the start of flexion, and the principal rays are formed by completion of flexion. The caudal fin has 9+7 principal caudal rays. Anlagen of the dorsal and anal fins form during notochord flexion; incipient rays form early during the postflexion stage. In the dorsal and anal fins, soft rays develop from anterior to posterior and begin to form prior to the spines.

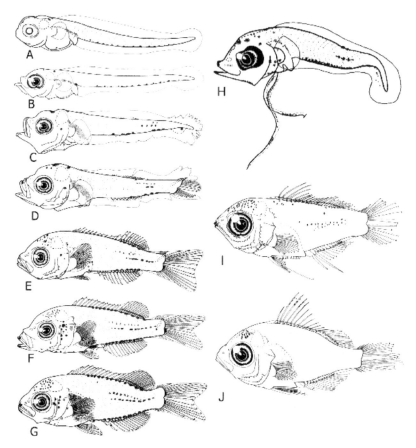

Fig. 3. Left: Larval development of reared *Pomacentrus amboinensis*. Age is in days after hatch (DAH): (A) 2.50 mm standard length (SL) yolk-sac preflexion-stage larva, 0 DAH, AMS I.38139-034. Note incompletely pigmented eye and apparently non-functional mouth. The yolk sac occupies most of the gut region, (B) 2.80 mm LS preflexion-stage larva, 1 DAH, AMS I.38139-035. No separate yolk is visible, (C) 3.65 mm LS preflexion-stage larva, 6 DAH, AMS I.38139-023, (D) 4.57 mm LS flexion stage larva, 7 DAH, AMS I.38139-026, (E) 5.98 mm LS postflexion-stage larva, 11 DAH, AMS I.38139-028, (F) 7.30 mm LS postflexion-stage larva, 13 DAH, AMS I.38139-030. Some scales are present, but are not illustrated and (G) 8.80 mm LS postflexion-stage larva, 15 DAH, AMS I.38139-031. Scales are present over most of the body, but are not illustrated. (Murphy et al. 2007). Right: Illustrations of the chromine pomacentrid genus *Chromis*. (H) 2.03 mm *C. vanderbilti*? (Fowler) from an epibenthic plankton tow in the Great Barrier Reef Lagoon off Lizard Island. (I). 7.5 mm *C. vanderbilti*? from a plankton tow in the East China Sea. (J) unidentified 8.2 mm *Chromis* species from a midwater trawl in the Bismarck Sea.

The first pectoral rays form early during the postflexion stage, and are completed at postflexion. Pelvic-fin buds form during flexion and complete postflexion, at about 8 mm TL in *P. amboinensis* (Kavanagh et al. 2000, Murphy et al. 2007).

iii. Swimming

Damselfish larvae in late pelagic stages are capable swimmers. In swimming tests, damselfishes were able to swim continuously up to 231.7 hours, equivalent to

112.6 km at the test speed of 13.5 cm/s. Compared with other taxa, this capability is moderate (Stobutzki and Bellwood 1997). In another study, the swimming capability during different developmental stages of two damselfishes *Pomacentrus amboinensis* and *Amphiprion melanopus* were measured (Fisher et al. 2000). They found that critical swimming speed steadily increased as the fish aged, while sustained swimming speed increased sharply at a particular stage of ontogenetic development, approximately when the body depth increases and the caudal fin flexion is completed.

The documentation of the morphological development of *P. amboinensis* larvae provided here will facilitate an ecomorphological discussion. For example, Fisher et al. (2000) studied the development of swimming ability in the larvae of *P. amboinensis*, and they portrayed a marked increase in swimming speed at 7–10 mm SL. Morphologically, there was no major developmental event, with the unlikely exception of the completion of the pectoral-fin rays and the completion of scalation, that might be correlated to this increase in speed performance. Additionally, between 7 and 9 mm SL, the relative body length increased by one-third (from 30% to 40% SL) in *P. amboinensis*, resulting in a large increase in muscle mass and probably contributing to speed performance as well.

In addition to the physical capability for swimming, navigational abilities of damselfish larvae are also documented. Presettlement pelagic damselfish larvae seem to be able to use a solar compass and swim directionally (Leis et al. 1996, Leis and Carson-Ewart 2003).

Sensory system development

i. Visual system

Retinal development

Generally speaking, the eye placode differentiates from the cranial tissue at 2 days of embryonic development. The inner nuclear layer of the retina differentiates by 4 days. The pigment layer, cone layer, and all other retinal layers begin to differentiate over the next week. In *P. amboinensis*, rods appear in the retina by 12 DAH (Fig. 4). By 17 DAH, the pigment layer can move, allowing light adaptation of the retina. Following settlement, at 23 DAH, rod and cone layer thicknesses are roughly equal. Kavanagh and Alford (2003) showed some variations in the retinal development among species. For example, *Premnas biaculeatus* showed the earliest development in comparison with *P. amboinensis*, *C. viridis* and *A. polyacanthus*. On the other hand, retinal differentiation started latest in *A. polyacanthus* but developed very rapidly such that it was fully developed several days earlier than *P. bioculeatus*, *P. amboinensis* and *C. viridis*.

Sensitivity

Cones develop before rods in damselfishes and therefore sensitivity to color develops earlier than sensitivity to low light. Job and Bellwood (2000) measured light sensitivity in larval pomacentrids by observing the lowest light irradiance at which

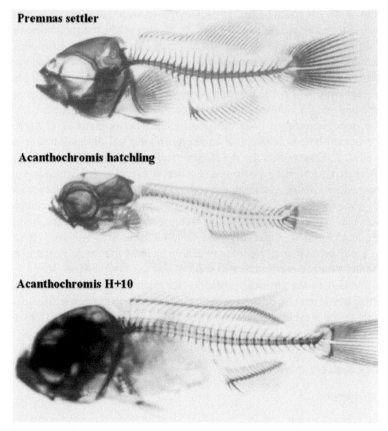

Premnas settler

Acanthochromis hatchling

Acanthochromis H+10

Fig. 4. Skeletal development at settlement in *Premnas biaculeatus* and at hatch plus 10 days in *Acanthochromis polyacanthus*. Fish were cleared and stained with Alcian Blue and Alizarin Red to stain the cartilage and ossified bone respectively.

the larvae would feed. In comparison with apogonids, pomacentrids have relatively lower sensitivity. Light sensitivity increases with age and size, coïncident with the development of rods.

Larvae of two damselfishes (*Pomacentrus amboinensis* and *Premnas biaculeatus*) have been shown to detect and use UV light to feed, at wavelengths that would allow them to use UV light as far down as 90–130 m in oceanic waters (Job and Bellwood 2000, 2007). The sensitivity does not change abruptly at settlement, suggesting a continuous development of UV sensitive cones in the retina throughout the larval stage.

ii. Olfactory system

The rate of development of the olfactory system appears to be similar among damselfish species, with the exception of anemonefishes, which develop relatively faster (Fig. 5, Kavanagh and Alford 2003). In general, the olfactory system does not achieve its final morphological form until well past settlement. The first observable stage of olfactory

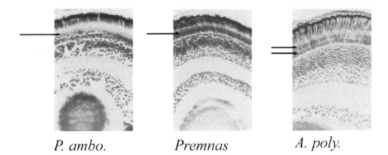

P. ambo.   Premnas   A. poly.

Fig. 5. Vertical histological sections through the eyes of 7-day-old larvae or embryos of three species of pomacentrids demonstrating variation in the developmental rate of retinae. The layers of photoreceptor cell nuclei increase (arrows), as do the external and internal nuclear layers (located below arrows). Left—*Pomacentrus amboinensis* has a thin layer of cone cell nuclei, while *Premnas biaculeatus* (middle) has a thicker layer of cone cell nuclei, and *Acanthochromis polyacanthus* (right) has a double layer of cone cell nuclei (From Kavanagh and Alford 2003).

development occurs at 4 days after fertilization (DAF), where a round patch of cilia is present where the olfactory nares will form. At 6–8 DAF, the nasal pit forms and deepens. The pit elongates and by 16 DAF, divides by "pinching in" of tissue along the middle. By 27 DAF, the nares are cleanly divided. The dorsal-most nostril of the pair then enlarges. In *P. amboinensis*, the nasal pit begins to roof over at about 6.0 mm, and two nostrils are present at 7.0 mm. Two nostrils are still present in the largest settlement-stage larvae (13.0 mm). The adult condition of one nostril on each side develops only following settlement.

In *Premnas biaculeatus* and *Amphiprion melanopus*, olfactory development follows a similar but accelerated ontogeny in comparison with *P. amboinensis*. This faster development of the olfactory system in anemonefishes is correlated with unique behaviors that require olfactory perception. Arvedlund et al. (1999) showed that the anemonefish *Amphiprion melanopus* will imprint on the host sea anemone at hatching and, as later settlers, can use olfactory cues to detect the same anemone species.

iii. Auditory system

Otoliths are among the first mineralized organs to appear in larval pomacentrids. The otoliths and otic sacs are important for balance and orientation, as well as for hearing. There are three sets of otoliths, with the sagitta and lapillus developing well before the asteriscus. Green and McCormick (2001) report that in *Amphiprion melanopus*, the otic capsule and two pairs of otoliths (sagitta and lapillus) form by 3 days after fertilization, and the third pair of otoliths, the asteriscus, forms by 5 days after fertilization.

Leis and Carson-Ewart (2003) showed experimentally that some reef fish larvae including pomacentrids are attracted to reef sounds, suggesting that they can use this sense to orient themselves towards the settlement habitat.

iv. Internal organ development

Details on internal organ development of *Amphiprion* spp. are provided from Yasir and Qin (2007) and Green and McCormick (2001). At Day 3 after fertilization, a single chambered heart begins to beat and circulate colorless blood lacking erythrocytes and hemoglobin. By Day 4, erythrocytes develop in the blood plasma, coloring the heart and blood orange, and the heart has two visible chambers. By Day 5, the gills and opercula are first visible and blood flows to the gills. Pectoral fins and a gut tube appear, and the third set of otoliths—the asteriscus—is visible. By Day 6, the developing jaw, orbital bones and gill rakers appear and blood flows to the branchial region, and the gut loops. By Day 7 four rows of gill rakers appear, and the gills ventilate regularly. By Day 8 eyeshine reflects from the tapetum layer of the eye and the hindgut forms. At 1 day after hatch, *Amphiprion melanopus* gut is differentiated into foregut, midgut, and hindgut. Histological sections show rudimentary liver, pancreas, and kidney. Further differentiation of organs occurs over the next few days (Green and McCormick 2001).

Post-settlement Stages

First week after the reef colonization—Metamorphosis

The great majority of damselfishes (except *Acanthochromis* and *Altrichthys* spp.) have a bipartite life cycle that involves ontogenetic change in morphology, physiology and behavior as their pelagic larval stages colonize benthic habitats (Leis and McCormick 2002). This ecological transition is a crucial phase and the term «metamorphosis» is commonly used by ecologists to encompass the changes in structure and function that occur during settlement (McCormick et al. 2002).

In 2002, McCormick and colleagues explored the changes in pigmentation and external morphology that occur immediately after settlement in different tropical reef fish families, including the Pomacentridae. The most striking change occurring during the settlement phase of pomacentrids is certainly a variation in their color pattern. At settlement, most of the pomacentrids have a silver or transparent body (e.g., *Stegastes*, *Plectroglyphidodon*, *Pomacentrus*, *Chrysiptera* spp.; McCormick et al. 2002) whereas *Chromis viridis* has a bronze body and clear fins. The rate of changes in pigmentation during settlement greatly varies within Pomacentridae. As in chaetodontids, holocentrids, monocanthids and mullids, *Pomacentrus* and *Chrysiptera* spp. show rapid changes (<12 h), while *Plectroglyphidodon lacrymatus* and *Stegastes nigricans* are slower in attaining benthic coloration, taking 3 and 7 days, respectively (McCormick et al. 2002). On the other hand, *Chromis viridis* shows no color change after 5 days post-settlement (McCormick et al. 2002).

Similar to butterflyfishes, the damselfishes show a high level of variability in morphological changes during the first days of post-settlement although the main changes involve changes in the head shape and reduction of spination (McCormick et al. 2002). Especially in the pomacentrids, it appears that species with a shorter larval duration undergo a greater metamorphosis than those that spend more time in the pelagic environment (McCormick et al. 2002).

Beyond the study of external morphology, very few studies have compared functional capabilities between pre- and post-settlement stages in coral reef fishes. Indeed, it is likely that sensory systems are modified during the period of metamorphosis allowing adaptation to the new benthic environment (Lecchini et al. 2005). In the few studies that have compared sensory sensitivity immediately before and after settlement, it appears that some senses adjust during metamorphosis, while others do not. The pre-settlement and post-settlement stages of *Pomcentrus bankanensis* have similar auditory and olfactory senses (Wright et al. 2005). The visual sensitivity of larvae increased during the larval phase (Job and Bellwood 2000) and UV vision seems to be improved gradually during the growth of pomacentrids, as revealed in *Dascyllus albisella* (Losey et al. 2000).

Ontogenetic shape changes after settlement—Allometry

In Pomacentridae, the larvae are zooplanktivorous and feed mainly on copepods (Sampey et al. 2007) while juveniles and adults may have varied diets (Frédérich et al. 2009, Chapter VII). Many studies have illustrated such an ontogenetic diet shift in various damselfishes (e.g., Nakamura et al. 2003, Buckle and Booth 2009, Frédérich et al. 2010). If oral morphology appears to determine the fundamental trophic niche, it is expected that dietary shifts would be associated with morphological modifications of the feeding apparatus (Wainwright 1991, Liem 1993; Chapter IX). Ontogenetic changes in diet could be related to changes in the feeding behavior or feeding mode (Eggold and Motta 1992, Frédérich et al. 2008). Quantitative analysis of ontogenetic shape changes showed that damselfishes observed a high level of allometric[1] variation during post-settlement growth and observed shifts in their feeding mode (Frédérich et al. 2006, 2008, Frédérich and Vandewalle 2011). An ecomorphological interpretation of the shape changes of the head skeleton suggests an improvement of a suction feeding mode in the great majority of studied damselfishes (Frédérich et al. 2008, Frédérich and Vandewalle 2011). Among other traits, the adult damselfishes show higher opercles and suspensoria, and a longer ascending process of the premaxilla than larvae, optimizing suction feeding. On the other hand, algivorous species, such as *Stegastes nigricans* and *Plectroglyphidodon lacrymatus*, show ontogenetic shape changes improving robustness and strength needed for grazing activities (e.g., proportionally shorter mandibles with higher coronoid and primordial processes in adults than in larvae; Frédérich et al. 2006, Frédérich and Vandewalle 2011). Among zooplanktivorous species, *Chromis viridis* and *Chromis atripectoralis* show some striking differences in allometric patterns along with other pelagic feeders (Frédérich and Vandewalle 2011). The small amount of shape changes observed in these species during the post-settlement ontogeny suggest functional similarity with larvae, which are considered to be ram-suction feeders (Liem 1993). Moreover, an exploration of the morphological diversity at the family level revealed recently that *C. viridis* and others evolved through highly specialized zooplanktivory (Frédérich et al. 2014).

---

[1] Allometry is the pattern of covariation among several morphological traits or between measures of size and shape (Klingenberg 1998).

In 2011, Frédérich and Vandewalle clearly demonstrated that the morphological disparity within the Pomacentridae increases during the post-settlement ontogeny. The higher disparity level at the adult stages results from variation in the allometric patterns among species (Frédérich and Vandewalle 2011). Various internal (e.g., functional and developmental) and/or external (ecological) factors may explain such a pattern but it appears that allometric trajectories are species-specific (Frédérich and Vandewalle 2011).

Many ecologists agree that the transition phase between the pelagic and the benthic environment is crucial and may have many consequences at different ecological levels and scales for coral reef fish populations. Therefore, it is surprising how little research has been focused on the physiological and morphological changes in damselfishes during post-settlement ontogeny. The main research on post-settlement ontogenetic morphological changes in damselfishes has focused on the head region, pigmentation and the overall body. Given the diversity of species and ecological niches occupied by damselfishes (Chapter V), it would very interesting to continue the exploration of the ontogenetic variation of various functional traits (i.e., physiology and morphology) around the time of metamorphosis.

Remaining Challenges

Studies of the early life stages of coral reef fishes, including pomacentrids, lag behind those of commercially important fishes and terrestrial vertebrates. Larval reef fish studies have tended to focus primarily on identifications for the purpose of counting individuals in plankton collections (and thus focusing on easily observable traits of fixed larvae under a microscope) or on correlating the timing of functional capabilities related to finding adult settlement habitat (and thus ignoring most details of organ development and physiology). Difficulty in rearing and observing the early stages *in situ* also has led to a dearth of comparative information on preflexion larvae in the wild.

The pomacentrid family is a significant component of the reef fish community and has been an important model system for reef fish researchers because of the relatively easy accessibility of the embryos and older larvae due to the substrate brooding of eggs and settlement behaviors. Additional embryonic and larval studies are needed in order to understand more fully the evolution, ecology, and developmental physiology of both typical and atypical members of this important and diverse family.

Selective forces act at all developmental stages of organisms. The study of embryonic, larval, and post-settlement ontogeny, including the critical period during the first few days following settlement, helps us to understand the factors selecting phenotypes. Future studies should focus on comparative analyses of traits and trends within a phylogenetic context, as well as the intra-populational and intra-species variation of morpho-functional disparity during ontogeny in order to have a better understanding of «when» and «how» selection operates in coral reef fishes including the damselfishes.

Acknowledgements

This research was supported by FRS-FNRS grants (FRFC no. 2.4.583.05) and the University of Massachusetts, Dartmouth. BF is a Postdoctoral Researcher at the F.R.S.-FNRS (Belgium). We thank E. Winiarz and T. Paré for assistance.

References

Allen, G.R. 1999. *Altrichthys*, a new genus of damselfish (Pomacentridae) from Philippine seas with description of a new species. Revue Française d'Aquariologie 26: 23–28.

Alshuth, S.R., J.W. Tucker, Jr. and J. Hateley. 1998. Egg and larval development of laboratory-reared sergeant major, *Abudefduf saxatilis* (Pisces, Pomacentridae). Bull. Mar. Sci. 62(1): 121–133.

Arvedlund, M., M.I. McCormick, D.G. Fautin and M. Bildsoe. 1999. Host recognition and possible imprinting in the anemonefish *Amphiprion melanopus* (Pisces: Pomacentridae). Mar. Ecol. Prog. Ser. 188: 207–218.

Bay, L.K., K. Buechler, M. Gagliano and M.J. Caley. 2006. Intraspecific variation in the pelagic larval duration of tropical reef fishes. J. Fish Biol. 68: 1206–1214.

Bernardi, G. 2011. Monophyletic origin of brood care in damselfishes. Mol. Phylogenet. Evol. 59: 245–248.

Buckle, E.C. and D.J. Booth. 2009. Ontogeny of space use and diet of two temperate damselfish species, *Parma microlepis* and *Parma unifasciata*. Mar. Biol. 156(7): 1497–1505.

Eggold, B.T. and P.J. Motta. 1992. Ontogenetic dietary shifts and morphological correlates in striped mullet, *Mugil cephalus*. Environ. Biol. Fishes 34(2): 139–158.

Fisher, R., D.R. Bellwood and S.D. Job. 2000. Development of swimming abilities in reef fish larvae. Mar. Ecol. Prog. Ser. 202: 163–173.

Frédérich, B. and P. Vandewalle. 2011. Bipartite life cycle of coral reef fishes promotes increasing shape disparity of the head skeleton during ontogeny: an example from damselfishes (Pomacentridae). BMC Evol. Biol. 11: 82.

Frédérich, B., E. Parmentier and P. Vandewalle. 2006. A preliminary study of development of the buccal apparatus in Pomacentridae (Teleostei, Perciformes). Anim. Biol. 56(3): 351–372.

Frédérich, B., D. Adriaens and P. Vandewalle. 2008. Ontogenetic shape changes in Pomacentridae (Teleostei, Perciformes) and their relationships with feeding strategies: a geometric morphometric approach. Biol. J. Linnean Soc. 95(1): 92–105.

Frédérich, B., G. Fabri, G. Lepoint, P. Vandewalle and E. Parmentier. 2009. Trophic niches of thirteen damselfishes (Pomacentridae) at the Grand Récif of Toliara, Madagascar. Ichthyol. Res. 56(1): 10–17.

Frédérich, B., O. Lehanse, P. Vandewalle and G. Lepoint. 2010. Trophic niche width, shift, and specialization of *Dascyllus aruanus* in Toliara Lagoon, Madagascar. Copeia (2): 218–226.

Frédérich, B., D. Olivier, G. Litsios, M.E. Alfaro and E. Parmentier. 2014. Trait decoupling promotes evolutionary diversification of the trophic and acoustic system of damselfishes. Proc. R. Soc. B-Biol. Sci. 281(1789).

Gagliano, M., M.I. McCormick and M.G. Meekan. 2007. Temperature-induced shifts in selective pressure at a critical developmental transition. Oecologia 152: 219–225.

Green, B.S. and M.I. McCormick. 2001. Ontogeny of the digestive and feeding systems in the anemonefish *Amphiprion melanopus*. Environ. Biol. Fishes 61(1): 73–83.

Job, S.D. and D.R. Bellwood. 2000. Light sensitivity in larval fishes: Implications for vertical zonation in the pelagic zone. Limnol. Oceanogr. 45(2): 362–371.

Job, S.D. and D.R. Bellwood. 2007. Ultraviolet photosensitivity and feeding in larval and juvenile coral reef fishes. Mar. Biol. 151: 495–503.

John, H.-C. and H. Reinhold. 2008. Larval development of the Cape Verdean damselfish *Chromis lubbocki*, with a note on Cape Verdean *Chromis cyanea* (Teleostei, Pomacentridae). Cybium 32(3): 217–224.

Kavanagh, K.D. 2000. Larval brooding in the marine damselfish *Acanthochromis polyacanthus* (Pomacentridae) is correlated with highly divergent morphology, ontogeny and life-history traits. Bull. Mar. Sci. 66(2): 321–337.

Kavanagh, K.D. and R.A. Alford. 2003. Sensory and skeletal development and growth in relation to the duration of the embyonoic and larval stages in damselfishes (Pomacentridae). Biol. J. Linnean Soc. 80: 187–206.

Kavanagh, K.D., J.M. Leis and D.S. Rennis. 2000. Pomacentridae. pp. 526–535. *In*: J.M. Leis and B.M. Carson-Ewart (eds.). The Larvae of Indo-Pacific Coastal Fishes—An Identification Guide to Marine Fish Larvae. Fauna Malesiana Handbooks. Brill, Leiden.

Kerrigan, B.A. 1997. Variability in larval development of the tropical reef fish *Pomacentrus amboinensis* (Pomacentridae): The parental legacy. Mar. Biol. 127(3): 395–402.

Klingenberg, C.P. 1998. Heterochrony and allometry: the analysis of evolutionary change in ontogeny. Biol. Rev. 73(1): 79–123.

Lecchini, D., J. Shima, B. Banaigs and R. Galzin. 2005. Larval sensory abilities and mechanisms of habitat selection of a coral reef fish during settlement. Oecologia 143: 326–334.

Leis, J.M. 2015. Taxonomy and systematics of larval Indo-Pacific fishes: a review of progress since 1981. Ichthyol. Res. In Press.

Leis, J.M. and T. Trnski. 1989. The Larvae of Indo-Pacific Shorefishes. New South Wales University Press, Sydney.

Leis, J.M. and M.I. McCormick. 2002. The biology, behaviour and ecology of the pelagic, larval stage of coral reef fishes. pp. 171–199. *In*: P.F. Sale (ed.). Coral Reef Fishes: Dynamics and Diversity in a Complex Ecosystem. Academic Press, San Diego.

Leis, J.M. and B.M. Carson-Ewart. 2003. Orientation of pelagic larvae of coral-reef fishes in the ocean. Mar. Ecol. Prog. Ser. 252: 239–253.

Leis, J.M., H.P.A. Sweatman and S.E. Reader. 1996. What the pelagic stages of coral reef fishes are doing out in blue water: daytime field observations of larval behavioural capabilities. Mar. Freshw. Res. 47(2): 401–411.

Liem, K.F. 1993. Ecomorphology of the teleostean skull. pp. 422–452. *In*: J. Hanken and B.K. Hall (eds.). The Skull: Functional and Evolutionary Mechanisms. The University of Chicago Press, Chicago.

Losey, G.S., P.A. Nelson and J.P. Zamzow. 2000. Ontogeny of spectral transmission in the eye of the tropical damselfish, *Dascyllus albisella* (Pomacentridae), and possible effects on UV vision. Environ. Biol. Fishes 59(1): 21–28.

McCormick, M.I., L. Mackey and V. Dufour. 2002. Comparative study of metamorphosis in tropical reef fishes. Mar. Biol. 141: 841–853.

Murphy, B.F., J.M. Leis and K.D. Kavanagh. 2007. Larval development of the Ambon damselfish *Pomacentrus amboinensis*, with a summary of pomacentrid development. J. Fish Biol. 71: 569–584.

Nakamura, Y., M. Horinouchi, T. Nakai and M. Sano. 2003. Food habits of fishes in a seagrass bed on a fringing coral reef at Iriomote Island, southern Japan. Ichthyol. Res. 50(1): 15–22.

Potthoff, T., S. Kelley, V. Saksena, M. Moe and F. Young. 1987. Description of larval and juvenile yellowtail damselfish, *Microspathodon chrysurus*, Pomacentridae, and their osteological development. Bull. Mar. Sci. 40(2): 330–375.

Robertson, D.R. 1973. Field observations on the reproductive behaviour of a pomacentrid fish, *Acanthochromis polyacanthus*. Z. Tierpsychol. 32: 319–324.

Sampey, A., A.D. McKinnon, M.G. Meekan and M.I. McCormick. 2007. Glimpse into guts: overview of the feeding of larvae of tropical shorefishes. Mar. Ecol. Prog. Ser. 339: 243–257.

Stobutzki, I.C. and D.R. Bellwood. 1997. Sustained swimming abilities of the late pelagic stages of coral reef fishes. Mar. Ecol. Prog. Ser. 149: 35–41.

Sun, D., S.P. Blomberg, T.H. Cribb, M.I. McCormick and A.S. Grutter. 2012. The effects of parasites on the early life stages of a damselfish. Coral Reefs 31: 1065–1075.

Takahashi, M., M.I. McCormick, P.L. Munday and G.P. Jones. 2012. Influence of seasonal and latitudinal temperature variation on early life-history traits of a coral reef fish. Mar. Freshw. Res. 63(10): 856–864.

Wainwright, P.C. 1991. Ecomorphology—experimental functional-anatomy for ecological problems. Am. Zool. 31(4): 680–693.

Wright, K.J., D.M. Higgs, A.J. Belanger and J.M. Leis. 2005. Auditory and olfactory abilities of pre-settlement larvae and post-settlement juveniles of a coral reef damselfish (Pisces: Pomacentridae). Mar. Biol. 147(6): 1425–1434.

Yasir, I. and J.G. Qin. 2007. Embryology and early ontogeny of an anemonefish *Amphiprion ocellaris*. J. Mar. Biol. Assoc. UK 87: 1025–1033.

Zamzow, J.P. and U.E. Siebeck. 2006. Ultraviolet absorbance of the mucus of a tropical damselfish: effects of ontogeny, captivity and disease. J. Fish Biol. 69: 1583–1594.

Ecomorphology and Iterative Ecological Radiation of Damselfishes

Bruno Frédérich,[1,]* W. James Cooper[2] and
Rosalía Aguilar-Medrano[3]

Ecological morphology is a comparative discipline focused on the connections between morphological and ecological diversity (Williams 1972, Motta and Kotrschal 1992, Wainwright and Reilly 1994, Norton et al. 1995). The ecomorphological approach examines the optimization of functional morphology to specific ecological characters at multiple levels: among individuals within a species, among species and higher taxa, and among guilds and communities (Barel 1983, Wainwright 1988, Kotrschal 1989, Motta et al. 1995). Such examinations of the causative connections between form, function and ecology represent an important and a relevant framework for addressing adaptation (Wainwright 1994, Motta et al. 1995).

In this chapter, we describe the ecomorphological diversity of the Pomacentridae and discuss the adaptive significance of specific aspects of their functional morphology of feeding and swimming. We also report morphological variation among populations of various species within the context of local adaptation and/or plasticity. At the family level, we discuss the reticulated pattern of evolutionary diversification in damselfish ecomorphology as revealed by the use of modern phylogenetic comparative methods. The chapter concludes with some perspectives regarding the study of ecomorphology in the Pomacentridae.

[1] Laboratoire de Morphologie Fonctionnelle et Evolutive, AFFISH - Research Center, University of Liège, Quartier Agora, Allée du six Août 15, Bât. B6C, 4000 Liège (Sart Tilman), Belgium.
[2] School of Biological Sciences, Washington State University, Pullman, WA, 99164.
[3] Instituto de Ecología Aplicada, Universidad Autónoma de Tamaulipas, 356 División del Golfo, Col. Libertad, Ciudad Victoria, Tamaulipas, México, 87029.
* Corresponding author: bruno.frederich@ulg.ac.be

Damselfishes are one of the best-studied marine fish groups in terms of their ecology; there have been over 100 scientific publications associated with pomacentrid ecology in the past 20 years alone. To our knowledge, specific interest in their ecomorphology began with Emery (1973), who was the first to publish a comparative osteological study of damselfishes. He illustrated the skeletal anatomy of 13 species from the Caribbean and discussed their morphological diversity within an ecological context. Coughlin and Strickler (1990) published the first explicitly functional study of damselfish feeding nearly two decades later and Gluckman and Vandewalle (1998) added a comparison of the morphology of the suspensorium and the oral jaws of four damselfishes to this body of work. The past decade has witnessed significant acceleration in pomacentrid ecomorphological research and almost all of this work has focused on the functional morphology of damselfish feeding (Frédérich et al. 2006, Frédérich et al. 2008a, 2008b, Cooper 2009, Cooper and Westneat 2009, Aguilar-Medrano et al. 2011, Frédérich et al. 2012, 2013) and swimming (Fulton 2007, Fulton et al. 2013, Aguilar-Medrano et al. 2013, Binning et al. 2014). Recent years have seen a series of taxonomically extensive molecular phylogenetic studies (Quenouille et al. 2004, Cooper et al. 2009, Frédérich et al. 2013) which significantly advanced the examinations of damselfish ecomorphological evolution by allowing the use of phylogenetic comparative methods to study changes in morphology across the entire lineage (Cooper and Westneat 2009, Holzman et al. 2012, Frédérich et al. 2013, 2014).

Morphological Variation among Species in Relation to Diet

As illustrated in Chapter VII, damselfishes can be grouped into three primary trophic groups: algivores, zooplanktivores, and omnivores that feed on various proportions of filamentous algae, zooplankton and small benthic invertebrates respectively. Recent works have shown that there is a clear relationship between morphological and ecological diversity in damselfishes (Frédérich et al. 2008b, Cooper and Westneat 2009, Aguilar-Medrano et al. 2011, Frédérich et al. 2013). Their relatively restricted trophic diversity is associated with what seems to be low morphological variation (Wainwright and Bellwood 2002).

Feeding can be divided into two different tasks: food acquisition and food processing. Ecomorphological changes associated with food acquisition are generally localized on oral jaws, teeth and associated musculature, ligaments and supporting skeletal elements, while alterations in pharyngeal jaws and teeth are strongly associated with evolutionary changes in the mechanical processing of food before it is passed to the gut for chemical digestion (Liem 1993, Wainwright and Bellwood 2002). Evolutionary changes in the digestive tract (e.g., intestinal length, the number of pyloric caeca, etc.) are also associated with important dietary shifts (Buddington and Diamond 1986, Elliott and Bellwood 2003, Konow and Bellwood 2011). Hereafter, we will summarize the body of knowledge related to the variation of the functional morphology of food acquisition in damselfishes. The great majority of these studies have employed landmark-based geometric morphometric analyses (e.g., Adams et al. 2004, 2013, Zelditch et al. 2004). We also discuss new ecomorphological data linked to food processing in damselfishes.

Food acquisition

Studies of damselfish head morphology (Cooper and Westneat 2009, Aguilar-Medrano et al. 2011) and its associated skeletal units (Emery 1973, Gluckmann and Vandewalle 1998, Frédérich and Vandewalle 2011, Frédérich et al. 2006, 2008b, 2013) have shown that there is a strong degree of morphological overlap between omnivores and algivores, while zooplanktivores can be easily segregated from the other two trophic guilds.

Damselfishes that feed primarily from the benthos (most particularly those that feed extensively on algae) have the eyes in an upper position than the snout (Fig. 1; Cooper and Westneat 2009, Aguilar-Medrano et al. 2011). There is a strong trend among multiple lineages of herbivorous marine fishes to evolve more dorsally positioned eyes and it is likely that this markedly increases their ability to detect predators when feeding in a head-down position (Bellwood et al. 2014). Primarily benthic-feeding damselfishes have incisiform teeth distributed in one (e.g., *Stegastes* spp., *Plectroglyphidodon* spp.) or two distinct rows (e.g., *Pomacentrus* spp., *Chrysiptera* spp.; Fig. 2) on the mandible and the premaxillary bone. Some omnivores that consume smaller amounts of algae may have incisiform or caniniform teeth arranged in one, two or three lines (Fig. 2). Benthic feeders also tend to have relatively robust oral jaws (Fig. 3) with a proportionally shorter mandible that has a large primordial process on the articular bone (Frédérich et al. 2008b; bone anatomy after Barel et al. 1976), a longer dentary symphysis between the left and the right mandibles solidifying the lower jaw (Frédérich et al. 2008b), broader suspensoria that support large *pars rictalis* muscles (Gluckmann and Vandewalle 1998, Frédérich et al. 2008b, Cooper

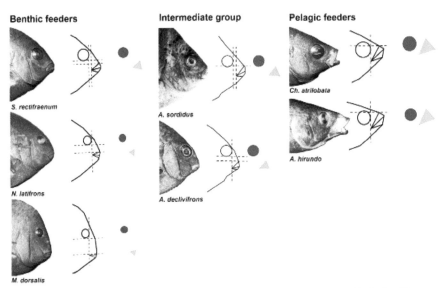

Fig. 1. Illustration of the differences in the distance between eyes and snout among benthic feeders (*Stegastes rectifraenum, Nexilosus latifrons, Microspathodon dorsalis*), pelagic feeders (*Chromis atrilobata, Azurina hirundo*) and the intermediate group (*Abudefduf sordidus, A. declivifrons*). All these species are grouped according to the study of Aguilar-Medrano et al. (2011).

Benthic feeders

Intermediate group

Pelagic feeders

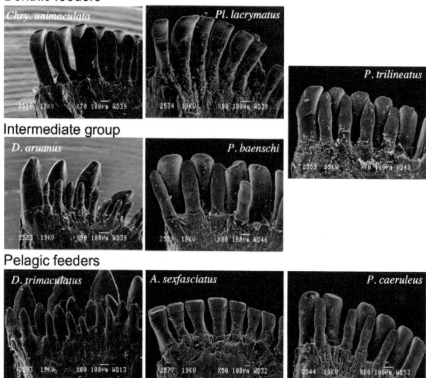

Fig. 2. Scanning electron micrographs of the former parts of the mandible (benthic feeders: *Chrysiptera unimaculata, Plectroglyphidodon lacrymatus*; pelagic feeders: *Dascyllus trimaculatus, Abudefduf sexfasciatus* and *Pomacentrus caeruleus*; "intermediate group": *Dascyllus aruanus, Pomacentrus baenschi*). All these species are grouped according to the study of Frédérich et al. (2009), but *Pomacentrus trilineatus* should be considered as a benthic feeder (see Chapter VII).

and Westneat 2009; the *pars rictalis*, *sensu* Datovo and Vari 2013, is synonymous with the A2 division of Winterbottom 1973), a broader maxillary process of the palatine that is deflected ventrally, and a neurocranium with a ventrally directed vomer that provides points of support for the upper jaws during grazing (Frédérich et al. 2008b). Finite element analyses of fish neurocrania support the idea that the ventrally directed vomerine processes can significantly improve the ability of benthic-feeding fishes to safely employ larger bite forces (Cooper et al. 2011).

Benthic-feeding damselfishes have *pars malaris* muscles (i.e., the A1 division of the *adductor mandibulae*) that insert relatively far from the maxillary-palatine joint (the upper rotation point of the maxilla) and this provides a longer biomechanical lever that improves the transfer of bite forces to the jaws (Cooper and Westneat 2009). The forces generated by the large *pars rictalis* muscles of these fishes are further augmented by the combined effects of having high coronoid processes on the dentary, high primordial process on the articular bone, and relatively short mandible lengths such that this muscle utilizes a relatively high mechanical advantage during biting

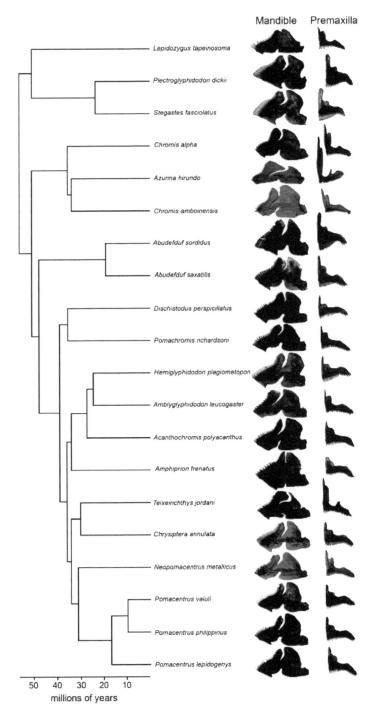

Fig. 3. Illustration of shape diversity for the mandible and the premaxilla across a phylogenetic tree of Pomacentridae (time-tree modified from Frédérich et al. 2013).

(high mechanical advantages enhance the transfer of force in mechanical systems; Cooper and Westneat 2009). All of these traits represent adaptations to optimize bite force production (Liem 1993). Predictions of bite mechanics derived from functional analyses of head morphology indicate that algivorous and omnivorous damselfishes are not distinguished by any biomechanical parameters (e.g., kinematic transmissions coefficient and mechanical advantages associated with oral jaw opening and closing; Cooper and Westneat 2009).

Zooplanktivorous damselfishes tend to have large eyes that are at the same height as the mouth (Fig. 1). These are presumably adaptations for perceiving zooplankton and targeting the feeding strike. The external morphology of zooplanktivores is characterized by long jaws and a lengthened and angular cephalic profile (Fig. 1; Cooper and Westneat 2009, Aguilar-Medrano et al. 2011). The bites of these fishes are characterized by a high degree of upper jaw protrusion. Indeed, zooplanktivores can protrude their upper oral jaws (i.e., premaxillary bones) to a greater extent with a higher efficiency of movement than other damselfishes (Cooper and Westneat 2009). This is enabled by longer ascending processes on the premaxilla (Fig. 3; Emery 1973, Frédérich et al. 2008b, 2014) and higher kinematic transmission coefficients (KT) of the upper jaw protrusion (Cooper and Westneat 2009). Enhanced jaw protrusion is used to decrease the predator-prey distance when feeding on elusive zooplankton such as copepods (Barel 1983, Coughlin and Strickler 1990). These fishes also have *pars malaris* muscles that insert very close to the maxillary-palatine joint and this significantly increases their ability to close their mouths quickly (Cooper and Westneat 2009), which is useful for catching highly elusive prey (Wainwright and Bellwood 2002, Sonnefeld et al. 2014). Fast bite speeds are likewise supported by short primordial processes on the articular and relatively long mandibles that result in a low mechanical advantage of biting associated with the *pars rictalis* (low mechanical advantages enhance the transfer of motion in mechanical systems; Cooper and Westneat 2009). In addition to these traits, some species, especially *Chromis* and *Dascyllus* spp., show well-developed suspensoria and opercles (Frédérich et al. 2008b, Frédérich and Vandewalle 2011) that probably aid in optimizing suction capabilities (Liem 1993, Frédérich et al. 2008a). Oral dentition is not a determinant of zooplanktivory in damselfishes. Indeed, pelagic feeders may have either caniniform (e.g., *Chromis* or *Dascyllus* spp.) or incisiform teeth (e.g., *Pomacentrus* and *Abudefduf* spp.; Fig. 2).

Food processing

In this section, we present new results illustrating morphological differences in the branchial basket (gill rakers and pharyngeal jaws) and digestive tract among damselfish trophic groups. Contrary to the patterns seen in the functional morphology of biting, omnivores and zooplanktiovres share some characters while algivores are easily distinguished.

a. Pharyngeal jaws and gill rakers

Data acquisition—We photographed the lower pharyngeal jaw (LPJ) in 6 species of damselfishes (3 species per trophic group; Table 1) and quantified its morphology

Table 1. Species studied for the diversity of structures involved in prey processing. SL, standard length; n, number of specimens; PJA, pharyngeal jaw apparatus.

Trophic groups	Species	SL (mm)	Intestine length (n)	PJA (n)
Pelagic feeders	*Abudefduf sexfasciatus*	83–99	X (11)	
	Abudefduf sparoides	51–89	X (13)	X (5)
	Abudefduf vaigiensis	93–127	X (9)	
	Chromis dimidiata	33–56	X(12)	X (5)
	Chromis ternatensis	63–82	X (5)	
	Dascyllus trimaculatus	37–79	X (13)	
	Pomacentrus caeruleus	44–67	X (11)	
	Pomacentrus coelestis	53–61		X (5)
Intermediate group	*Dascyllus aruanus*	28–57	X (11)	X (6)
	Pomacentrus baenschi	47–77	X (4)	X (5)
	Pomacentrus trilineatus	47–68	X (10)	X (6)
Benthic feeders	*Amphiprion akallopisos*	34–80	X (10)	
	Chrysiptera unimaculata	48–81	X (15)	X (3)
	Plectroglyphidodon lacrymatus	44–76	X (10)	X (5)
	Stegastes lividus	95–121		X (5)

using the protocol of Hellig et al. (2010). Six homologous landmarks (LMs) were recorded on each LPJ (Fig. 4). From this configuration of LMs, three angles (alpha, beta, gamma) and five distances (LPW, PDW, LPL, PDL*max*, PDL*min*) were measured using the software VistaMetrix (Fig. 4). The function *path* from VistaMetrix was used to calculate the surface of the dentigerous area. Photographs were also taken of one specimen per species with a scanning electron microscope (Jeol, JSM840A) using

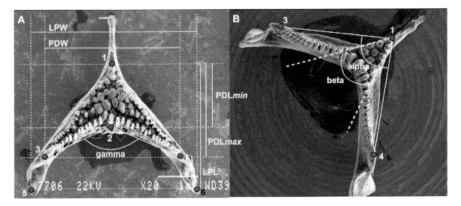

Fig. 4. Dorsal view of the lower pharyngeal jaws in (A) *Abudefduf sparoides* and (B) *Plectroglyphidodon lacrymatus* illustrating the position of landmarks, angles (alpha, beta, gamma) and distances. LPW, width of LPJ; PDW, width of the dentigerous area; LPL, length of LPJ; PDL*max* and PDL*min*, maximum and minimum length of the dentigerous area.

an acceleration voltage of 22 kV. The dentition of every species was then described using the traits detailed in Table 2. In addition to the morphological study of the LPJ, we counted the number of gill rakers supported by the first branchial arch. We also measured the length of the longest gill raker on the first arch and the mean distance separating the three first gill rakers supported by ceratobranchial-1.

Table 2. Description of the traits describing the dentition of the lower pharyngeal jaw (LPJ).

Trait	Description	Code
Lateral view	Form of a line joining the top of the teeth in lateral view	0 : no break
		1 : «step» formed by the last line of teeth
Tooth shape	Shape of the top of the tooth from the anterior region of the LPJ	0 : sharp
		1 : rounded
		2 : flattened
Teeth of the posterior part	Shape of the teeth from the most posterior line of LPJ	0 : straight
		1 : curved
Tilt of the teeth	Tilt of the teeth from the anterior part of the LPJ	0 : not tilted (i.e., right)
		1 : backward tilted

Statistical analyses—All the continuous traits were size corrected by head length and we examined the correlations among them in order to reduce the dataset. We found the following pairs of variables to be highly and significantly correlated: LPL vs. PDL*max* ($r = 0.87$), PDL*min* vs. PDL*max* ($r = 0.62$), LPW vs. PDW ($r = 0.78$). Thus, we only used PDL*max* and PDW in the following statistical analysis because both variables characterize the most effective dentigerous part of the LPJ. Similarly, we did not use the angle alpha due to its high and significant correlation with PDL*max* ($r = 0.77$), PDW ($r = -0.58$), beta ($r = -0.62$) and gamma ($r = 0.67$). We performed a PCA to summarize our dataset and explored the distribution of each trophic group in the morphospace. Finally, we tested the significance of morphological difference among trophic groups by ANOVAs performed on the PC1 and PC2 scores.

Results—The PCA was performed on 12 variables: the two first principal components explain 55.5% of the morphological variance (PC1 = 33.9%, eigenvalue = 4.074; PC2 = 21.6%, eigenvalue = 2.596; Fig. 5). The morphology of the pharyngeal jaws and gill rakers differ among trophic groups (ANOVA on PC1: $F_{2,41} = 56.27$, *P-value* < 0.001; ANOVA on PC2: $F_{2,41} = 1.18$, *P-value* = 0.32). Benthic feeders are significantly segregated from pelagic feeders and the intermediate group along PC1 (Tukey test: *P-value* < 0.001). They mainly differ by the dentition on the LPJ, the length of gill rakers and the beta angle. Indeed, the beta angle is smaller and the gill rakers are shorter in herbivorous damselfishes than in pelagic feeders (Fig. 5). In the benthic feeders (e.g., *Stegastes lividus* and *Chrysiptera unimaculata*; Fig. 6), the teeth have a cylindric shape, truncated and caudally tilted except the teeth from the last row, which are higher and their cusp is frontward curvated. In the pelagic feeders, the teeth show molariform or conical shape.

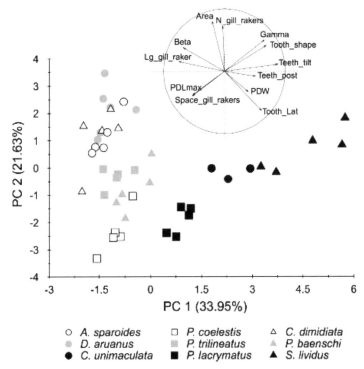

Fig. 5. Scatter plot of principal components 1 and 2 illustrating the morphological variation of the pharyngeal jaw apparatus in nine damselfish species (see Table 1). The variables loading on each axis are also illustrated (See text, Table 2 and Fig. 4 for information about abbreviations). For an illustrative purpose, the species are grouped by trophic guilds: pelagic feeders (white), intermediate (grey) and benthic feeders (black).

b. Length of the intestine

Methods—Longer intestines are generally expected in herbivorous fishes (Al-Hussaini 1947) although Galetto and Bellwood (1994) showed that an acidic stomach could be a more usual physiological adaption of herbivory in Pomacentridae. In order to determine if this general pattern holds true for the damselfishes, we compared intestinal length in the 13 species studied by Frédérich et al. (2009; Table 1). Intestinal length was measured to the nearest mm from the pyloric outlet to the rectum and the relative gut-index (RGI) was calculated as follows:

RGI = Length of intestine/Standard length

Results—We found significant differences among trophic groups (ANOVA: $F_{2,120} = 37.82$, *P-value* < 0.001; Fig. 7). Tukey tests revealed that benthic feeders have a significantly longer intestine length than pelagic feeders and species belonging to the intermediate group (*P-value* < 0.001), but no difference exists between zooplanktivores and omnivores (*P-value* = 0.054). Among the herbivorous species, *Plectroglyphidodon lacrymatus* and *Chrysiptera unimaculata* have the longest intestine (Table 3). The zooplanktivorous *Pomacentrus caeruleus* has the shortest RGI.

Benthic feeders

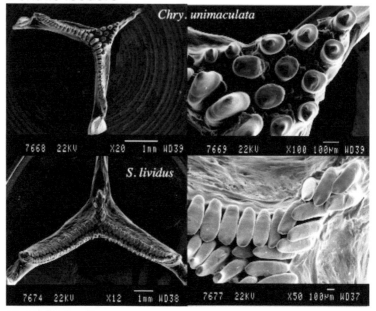

Pelagic feeders

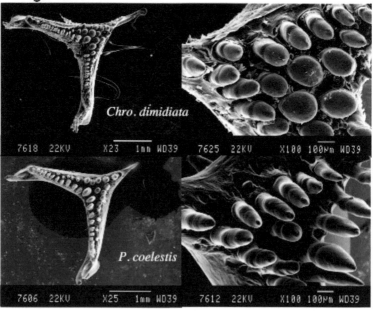

Fig. 6. Scanning electron micrographs of the lower pharyngeal jaw in benthic feeders (*Chrysiptera unimaculata* and *Stegastes lividus*) and pelagic feeders (*Chromis dimidiata* and *Pomacentrus coelestis*).

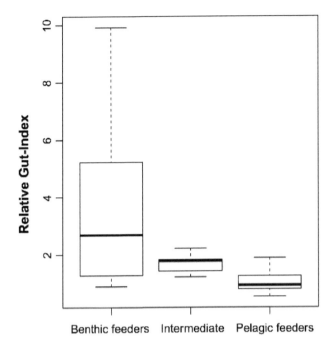

Fig. 7. Box plots of the relative gut-index of every trophic group. Mean, maximum, minimum and percentile values were calculated from species in Table 1.

Table 3. Mean values of the relative intestine length in the 13 studied damselfish species.

Trophic groups	Species	Intestine length/SL (Mean ± SD)
Pelagic feeders	*Abudefduf sexfasciatus*	0.77 ± 0.11
	Abudefduf sparoides	0.82 ± 0.03
	Abudefduf vaigiensis	1.03 ± 0.13
	Chromis dimidiata	1.09 ± 0.13
	Chromis ternatensis	1.53 ± 0.18
	Dascyllus trimaculatus	1.51 ± 0.38
	Pomacentrus caeruleus	0.65 ± 0.05
Intermediate group	*Dascyllus aruanus*	1.86 ± 0.36
	Pomacentrus baenschi	1.35 ± 0.09
	Pomacentrus trilineatus	1.80 ± 0.48
Benthic feeders	*Amphiprion akallopisos*	1.10 ± 0.13
	Chrysiptera unimaculata	2.68 ± 0.36
	Plectroglyphidodon lacrymatus	6.54 ± 1.66

Shape Variation of the Overall Body and the Pectoral Fins

A large number of reef associated fishes, including many damselfishes, employ labriform propulsion, i.e., synchronous flapping of both pectoral fins to power "underwater flying" (Thorsen and Westneat 2005). Labriform propulsion has been shown to be highly efficient for steady "cruising" in the water column over a wide range of swimming speeds, but it does not confer the same degree of maneuverability as the combination of pectoral fin "rowing" and swimming using the caudal fin (Walker and Westneat 2000). Accordingly, Fulton (2007) showed that most damselfishes and especially the zooplanktivorous ones (*Chromis, Dascyllus, Neopomacentrus, Amblyglyphidodon, Acanthochromis* spp.) use their pectoral fins extensively (median-paired fin, MPF) while swimming. Only the herbivorous damselfishes (those most closely associated with the benthos) used a combination of both pectoral fins (MPF) and body-caudal fin (BCF) for swimming (*Dischistodus, Neoglyphidodon, Plectroglyphidodon* and *Stegastes* spp.; Fulton 2007).

Thorsen and Westneat (2005) showed that at least one damselfish species (*Abudefduf saxatilis*) possess a derived morphology for the *abductor profundus* muscle, which originates on the pectoral girdle and inserts on the pectoral fin. This morphology suggests that the biomechanics of pomacentrid swimming using the pectoral fins may be somewhat distinct from that of the other reef fishes. To the best of our knowledge, no quantitative analyses have compared the myology and the osteology associated with pectoral girdle among various damselfish species but the descriptions of Emery (1973) indicate that zooplanktivorous damselfish have more robust pectoral girdles (e.g., larger cleithra) than herbivorous species.

In their study of damselfishes from the Eastern Pacific, Aguilar-Medrano et al. (2013) described the habitat segregation of damselfishes in relation to their position in the water column and the substrate as follows: (A) species that swim over benthic rocks and/or are exposed to the wave action; (B) species that live close to rock walls or boulders, but not directly over the benthic rocks; (C) species that live close to the reef but which are rarely observed near the surface; and (D) species that live in the water column from mid-water to close to the surface. The results of morphometric analyses indicated a strong relationship between habitat use, body morphology and pectoral fin shape (Aguilar-Medrano 2013, Aguilar-Medrano et al. 2013).

Group A: Species living over reef substratum and/or exposed to the wave action

Here we can group territorial species belonging to the genera *Stegastes, Nexilosus, Plectroglyphidodon* and algivorous members of the genus *Abudefduf* (e.g., *A. concolor, A. declivifrons, A. taurus*) that carry out most of their daily activities over a specific substratum. *Stegastes, Nexilosus* and *Plectroglyphidodon* species are territorial and highly site-attached, while *Abudefduf* species are not territorial and may live in small groups. All of the species in these genera live in close association with topographically complex substrates and utilize a high degree of maneuverability as they move through these heterogeneous environments. These species have short pectoral fins with a large insertion and with the most vertical angle of fin attachment (AFA) in the family (Aguilar-Medrano et al. 2013; Fig. 8). Their body form is intermediate between the elongated bodies of *Azurina, Lepidozygus, Neopomacentrus* and some *Chromis* spp. (group D) and the disc-shaped bodies of *Dascyllus* and many *Chromis* (group C) (Aguilar-Medrano 2013; Fig. 9).

Algivorous *Abudefduf* species are included within this group since they share pectoral fin shapes with territorial, herbivorous species. However, they also show similar values of AFA to species from group D, i.e., a more horizontal AFA (see below) (Aguilar-Medrano et al. 2013). This morphology combining maneuverability and speed efficiency allows these *Abudefduf* species to live in a surge pool environment.

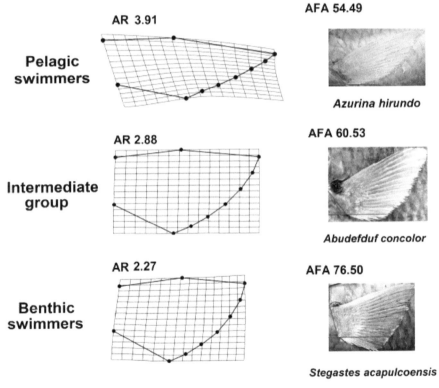

Fig. 8. Pectoral fin shape, angle of fin attachment (AFA) and aspect radio (AR) in pelagic swimmers (*Azurina hirundo*), benthic swimmers (*Stegastes acapulcoensis*) and the intermediate group (*Abudefduf concolor*). The formula of the AR is leading edge^2 area^{-1} (for more details see Aguilar-Medrano et al. 2013).

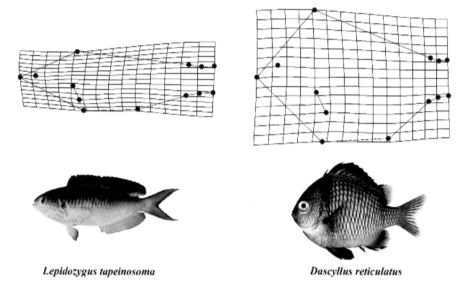

Lepidozygus tapeinosoma   *Dascyllus reticulatus*

Fig. 9. Example of the extreme body shape in damselfishes: the elongated hydrodynamic body shape represented by *Lepidozygus tapeinosoma* and the disc-shaped body represented by *Dascyllus reticulatus*.

Group B: Species living close to rock walls or boulders (but not over the reef substratum)

In this group we include *Microspathodon, Hypsypops* and *Parma* species. Some of these species are territorial but usually not to the degree seen among the species in group A. These species are among the biggest in the family (up to 30 cm SL); their body shape is rounded, high and wide. Their pectoral fins are similar to those from group A, but are somewhat more spanwise. Within group B, there is a high degree of variation in AFA. The AFA of *Hypsypops* is highly similar to that of the species in group A (i.e., AFA quite vertical) and confers a relatively high degree of maneuverability (Fig. 8). This pattern suggests a higher degree of territoriality in *Hypsypops* than in *Microspathodon* species while the morphology of *Microspathodon* suggests faster swimming capabilities relative to *Hypsypops* (Aguilar-Medrano et al. 2013).

Group C: Species living close to the reef and swimming around the reef, but rarely observed near the surface

Species of the genus *Chromis, Dascyllus* and *Chrysiptera* belong to this functional group, though the polyphyletic nature of *Chrysiptera* (Chapter II; Quenouille et al. 2004, Cooper et al. 2009) makes generalization about the species currently assigned to this genus problematic. These are generally schooling species (though multiple species of *Chrysiptera* are largely solitary) and tend to live close to the reef in small groups (15–30 individuals). They are commonly observed picking zooplankton from the water column. These species employ rapid bursts of swimming to hide from predators within the interstices of rocky and coral reefs. The body morphology of this group is highly variable and ranges from the compact disc-shaped bodies of *Dascyllus* spp. to more elongated body shapes (Aguilar-Medrano 2013, Frédérich et al. 2014). The pectoral fins of this group tend to be long and slender with an AFA that is a general intermediate between groups A and D (Aguilar-Medrano et al. 2013; Fig. 8).

Group D: Species living in the water column from midwater to close to the surface

This group includes schooling species within the genera *Chromis, Abudefduf, Lepidozygus, Neopomacentrus* and *Azurina*. These species may congregate in large groups (frequently several hundred individuals in the case of *Chromis dispilus*) and actively pursue zooplankton. Except for *Abudefduf* species, these fishes have highly elongated bodies such as *Lepidozygus tapeinosoma, Azurina* spp., *C. punctipinnis, C. cyanea* and *C. multilineata* (Fig. 9; Aguilar-Medrano 2013, Frédérich et al. 2014). They have slender pectoral fins, with the most horizontal AFA in the family (Aguilar-Medrano et al. 2013). This pectoral fin morphology permits sustained swimming in the water column using labriform propulsion for long periods of time.

Overall, an ecomorphological study of body form and pectoral fins allows the discrimination of four functional groups related to habitat partitioning and swimming mode.

Phenotypic Variation among Populations

Within the previous sections of the present chapter, we discussed interspecific ecomorphological variation, which is the product of evolution via natural selection. Intraspecific variation is the source of evolutionary potential and population differentiation is essential in the speciation process (e.g., Turelli et al. 2001). The phenotypic variation among damselfish populations has been examined recently. Some of this variation concerns color patterns, and multiple examples can be found in the literature. For instance, *Chrysiptera rex* shows striking color variation among enclosed seas in the Western Pacific (Drew et al. 2010), while *Chrysiptera leucopoma* presents color polymorphism in French Polynesia (Allen 1991, Frédérich et al. 2010). Moreover, while the numbers of studies are still limited, recent works illustrated ecomorphological variation among populations. Indeed, the same kind of morphological variation described among species can be observed among populations. Concerning the overall body morphology, Molina et al. (2006) highlighted body shape variation between continental and island populations of *Abudefduf saxatilis*. Fulton and colleagues clearly showed pectoral fin shape variation in *Acanthochromis polyacanthus* among populations exposed to different water flows on the Great Barrier Reef (Binning et al. 2014, Fulton et al. 2013). Concerning trophic morphology, Frédérich et al. (2012) illustrated shape-variation in oral jaws between populations of *Pomacentrus coelestis* from Taiwan and Japan, suggesting different diets. For example, the more robust mandibles observed in Japanese populations may indicate higher levels of grazing and/or biting activities than those from Taiwan.

While it is generally assumed that reef fish populations are highly connected via the larval phase capable of sustaining gene flow (Chapter III), ecomorphological studies of intraspecific variation clearly show that adult populations may be heterogeneous and display local adaptation. Intraspecific responses to environmental variation can arise via developmental plasticity, local selection upon successive generations, or a combination of both mechanisms (West-Eberhard 2003, Schluter 2000). Phenotypic plasticity broadly defines the adjustment of phenotypic values depending on the environment, without genetic changes. Originally, it describes the different phenotypes produced by the same genotype as a function of the environment (Schlichting and Pigliucci 1996). On the other hand, we should refer to local adaptation when heterogeneity in traits of a population is adaptive and genetically based (Kawecki and Ebert 2004). In the future, it will be challenging to test whether such intraspecific morphological variations observed in damselfishes may be caused by phenotypic plasticity and/or local adaptation. For example, studies of genetic variation in morphogenetic genes (Albertson et al. 2005) or genome-scale sequence comparison (Loh et al. 2008) among populations may provide valuable information about the genetic signal underlying morphological variation. In addition, future studies should to the extent possible combine morphological, genetic, ecological and environmental data in order to test the identity of the main forces driving the differentiation of damselfish populations.

Iterative Ecomorphological Evolution

With 394 species (Eschmeyer 2015; see Chapter II), the Pomacentridae are an example of a highly successful adaptive radiation. Phylogenetic comparative analyses have demonstrated high rates of evolutionary change in their trophic morphology and biomechanics (Cooper and Westneat 2009, Frédérich et al. 2013). What is unusual about this radiation is that instead of invading an ever-increasing number of new niches, it has progressed by rapidly and repeatedly converging on similar ecomorphological states. Cooper and Westneat (2009) refer to this pattern as "reticulate adaptive radiation". Frédérich et al. (2013) who confirmed this evolutionary pattern with additional species and using more recent phylogenetic comparative methods, preferred the term "iterative evolution". Both refer to the repetitive occurrence of similar morphologies, ecologies or behaviors during the evolutionary progression of a lineage.

This pattern of repeated convergence was described for overall skull shape (Cooper and Westneat 2009, Aguilar-Medrano et al. 2011), bite mechanics (Cooper and Westneat 2009), oral jaws (Fig. 3; Frédérich et al. 2013), additional anatomical characters (Chapter II), farming behavior (Chapter VI; Frédérich et al. 2013) and trophic ecology (Chapter VII; Cooper and Westneat 2009, Frédérich et al. 2013). Future studies could make useful contributions by determining whether similar patterns have arisen during the evolution of other functionally relevant body parts. The morphological evolution of damselfish body shape, fins and pharyngeal jaws offers promising targets for further studies of damselfish adaptive divergence.

There are numerous examples of convergent evolution from large-scale phylogenetic studies in vertebrates. For instance, in fishes, the capacity of breathing air and moving on land using the pectoral fins is observed in Gobiidae (e.g., *Boleophthalmus boddarti*), Blenniidae (e.g., *Blenniella* genus), Cotidae (e.g., *Clinocottus analis*),

Clariidae (e.g., *Clarias batrachus*) and Anabantidae (e.g., *Anabas cobojius*). Taken as a whole, the adaptive radiation of cichlids in each African Rift Lake constitutes an example of repeated radiation (Kocher 2004, Genner and Turner 2005, Salzburger et al. 2005, Young et al. 2009, Cooper et al. 2010) when convergences can also occur within a single adaptive radiation (Muschick et al. 2012). Replicate divergences of benthic and limnetic forms of sticklebacks in Holarctic postglacial lakes are additional examples of repeated radiations (Cooper et al. 2010). What is uncommon in Pomacentridae is that damselfish lineages evolved within a small number of trophic niches (herbivores, zooplanktivores and "omnivores"; see Chapter VII) while simultaneously evolving at high rates (Cooper and Westneat 2009, Frédérich et al. 2013, Lobato et al. 2014). In the future, it will be highly valuable to understand which factors sustain these rapid and repeated shifts among ecomorphs in Pomacentridae.

Potential factors explaining the diversification of damselfishes

Many factors may explain an iterative evolutionary pattern. Repeated patterns of evolutionary change in phenotypic traits are commonly regarded as evidence of adaptation under common selection pressures in similar environments (Losos 2011). This kind of iterative radiation has been demonstrated for organisms that colonize new islands or lakes (Losos and Ricklefs 2009), but damselfishes illustrate the first case of repeated adaptive radiations occurring in oceans. In coral reef ecosystems, repeated radiations might be expected during events of regionalization or geographic expansion allowing the colonization of new reefs. It is well known that some reef regionalizations were induced by the appearance of physical barriers such as the mid-Atlantic barrier (60–80 Ma), the Red Sea land bridge (terminal Tethys event, 12–18 Ma), the Isthmus of Panama (3–3.5 Ma) or by restricted surface water exchanges between the Indian and Pacific basins during the late Miocene (Kennett et al. 1985, Floeter et al. 2008). The appearance of volcanic islands also provided unoccupied areas for coral reef fishes. All of these events could have influenced the radiation of damselfishes, creating novel ecological opportunity for convergent radiations. These assumptions about geographical factors shaping replicated radiation in the marine realm are strengthened by the recent work of Litsios et al. (2014) on clownfishes. Litsios and collaborators (2014) demonstrated two independent geographic replicates of the radiation of the clownfishes showing the whole range of mutualistic interactions with sea anemones: the primary one occurring in the Central Indo-Pacific region and a second one in the West Indian Ocean (Litsios et al. 2014).

Several factors may lead to convergence and repeated patterns of adaptive radiation (Losos 2011). Factors such as competition could be viewed as an external constraint. Many fish families (Labridae, Pomacentridae, Acanthuridae, Holocentridae, Apogonidae, Gobiidae and Serranidae) were present together during the rise of the recent coral reef ecosystems (Bellwood 1996) leading to a high level of competition that may have constrained the radiation of damselfishes. Developmental processes, pleiotropic effects, morphological integration or peculiar morphological characters can all provide sources of internal constraint leading to a limited phenotypic repertoire in evolutionary radiations (Losos 2011). Trophic diversity in damselfishes is limited while other fishes with similar, highly derived pharyngeal jaws that permit efficient food processing (which should theoretically permit higher rates of oral jaw evolution) such as cichlids or labrids show some of the highest trophic diversity known among all ray-finned fish lineages (Barel 1983, Liem 1993, Wainwright and Bellwood 2002, Wainwright et al. 2004). Pomacetrids possess a ligament joining the mandible and hyoid bar (Stiassny 1981). Olivier et al. (2014) demonstrated that this cerato-mandibular ligament is involved in the feeding and sound production mechanisms of damselfishes (see Chapter XIV). This ligament may have constrained the morphological diversification of Pomacentridae, especially their trophic morphology (Frédérich et al. 2014).

Conclusion

Most of the ecomorphological studies of damselfishes have focused on the cranial region and the oral jaws. These studies have highlighted strong levels of congruence between morphology and diet. Here we have expanded upon this work to examine both pharyngeal jaws and the digestive tract morphology, and we report a similar pattern of congruence between morphology and ecology. Although the first studies illustrated a relationship between habitat and pectoral fin morphology, further investigations are especially needed for body shape variation. Finally, even though we have a respectable amount of general knowledge about the phenotypic diversity in the damselfishes, questions remain regarding the ecomorphology of Pomacentridae. What is the adaptive significance of color diversity in damselfishes? To what extent are the locomotor and feeding apparatus integrated in damselfishes? What is the genetic basis of intraspecific shape variation? What are the main factors driving the morphological differentiation of populations?

Acknowledgements

The authors thank G. Fabry and L. Gilles, who contributed to the analysis of the morphology of the digestive tract and the branchial basket, respectively. We also gratefully acknowledge two reviewers for their insightful comments and helpful criticism of the original manuscript. This research was supported by FRS-FNRS grants (FRFC no. 2.4.583.05). BF is a Postdoctoral Researcher at the F.R.S.-FNRS (Belgium) and RAM is a Postdoctoral Researcher of UC-MEXUS CONACYT (Mexico-USA).

References

Adams, D.C., F.J. Rohlf and D.E. Slice. 2004. Geometric morphometrics: ten years of progress following the 'revolution'. Ital. J. Zool. 71(1): 5–16.
Adams, D.C., F.J. Rohlf and D.E. Slice. 2013. A field comes of age: geometric morphometrics in the 21st century. Hystrix 24(1): 7–14.
Aguilar-Medrano, R. 2013. Body shape evolution of *Chromis* and *Azurina* species (Percifomes, Pomacentridae) of the eastern Pacific. Anim. Biol. 63(2): 217–232.
Aguilar-Medrano, R., B. Frédérich, E. De Luna and E.F. Balart. 2011. Patterns of morphological evolution of the cephalic region in damselfishes (Perciformes: Pomacentridae) of the Eastern Pacific. Biol. J. Linnean Soc. 102(3): 593–613.

Aguilar-Medrano, R., B. Frédérich, E.F. Balart and E. De Luna. 2013. Diversification of the pectoral fin shape in damselfishes (Perciformes, Pomacentridae) of the Eastern Pacific. Zoomorphology 132: 197–213.

Al-Hussaini, A.H. 1947. The feeding habits and the morphology of the alimentary tract of some teleosts living in the neighbourhood of the marine biological station, Ghardaga, Red Sea. Publications of the Marine Biology Station, Ghardaga, Red Sea 5: 1–61.

Albertson, R.C., J.T. Streelman, T.D. Kocher and P.C. Yelick. 2005. Integration and evolution of the cichlid mandible: the molecular basis of alternate feeding strategies. Proc. Natl. Acad. Sci. USA. 102(45): 16287–16292.

Allen, G.R. 1991. Damselfishes of the World. Mergus, Melle, Germany.

Barel, C.D.N. 1983. Towards a constructional morphology of cichlid fishes (Teleostei, Perciformes). Neth. J. Zool. 33(4): 357–424.

Barel, C.D.N., F. Witte and J.P. Van Oijen. 1976. The shape of the skeletal elements in the head of a generalized haplochromis species: *Haplochromis elegans*. Neth. J. Zool. 26: 163–265.

Bellwood, D.R. 1996. The Eocene fishes of Monte Bolca: The earliest coral reef fish assemblage. Coral Reefs 15(1): 11–19.

Bellwood, D.R., C.H.R. Goatley, S.J. Brandl and O. Bellwood. 2014. Fifty million years of herbivory on coral reefs: Fossils, fish and functional innovations. Proc. R. Soc. B-Biol. Sci. 281(1781).

Binning, S.A., D.G. Roche and C.J. Fulton. 2014. Localised intraspecific variation in the swimming phenotype of a coral reef fish across different wave exposures. Oecologia 174(3): 623–630.

Buddington, R.K. and J.M. Diamond. 1986. Aristotle revisited: the function of pyloric caeca in fish. Proc. Natl. Acad. Sci. USA 83(20): 8012–8014.

Cooper, W.J. 2009. The biogeography of damselfish skull evolution: a major radiation throughout the Indo-West Pacific produces no unique skull shapes. 11th International Coral Reef Symposium, 7–11 July, Ft Lauderdale, Florida.

Cooper, W.J. and M.W. Westneat. 2009. Form and function of damselfish skulls: rapid and repeated evolution into a limited number of trophic niches. BMC Evol. Biol. 9: 24.

Cooper, W.J., L.L. Smith and M.W. Westneat. 2009. Exploring the radiation of a diverse reef fish family: phylogenetics of the damselfishes (Pomacentridae), with new classifications based on molecular analyses of all genera. Mol. Phylogenet. Evol. 52(1): 1–16.

Cooper, W.J., K. Parsons, A. McIntyre, B. Kern, A. McGee-Moore and R.C. Albertson. 2010. Bentho-pelagic divergence of cichlid feeding architecture was prodigious and consistent during multiple adaptive radiations within African Rift-Lakes. PLoS ONE 5(3): A38–A50.

Cooper, W.J., J. Wernle, K. Mann and R.C. Albertson. 2011. Functional and genetic integration in the skulls of lake Malawi cichlids. Evol. Biol. 38(3): 316–334.

Coughlin, D.J. and J.R. Strickler. 1990. Zooplankton capture by a coral-reef fish—an adaptive response to evasive prey. Environ. Biol. Fishes 29(1): 35–42.

Datovo, A. and R.P. Vari. 2013. The jaw adductor muscle complex in teleostean fishes: Evolution, homologies and revised nomenclature (osteichthyes: actinopterygii). PLoS ONE 8 (4): e60846.

Drew, J.A., G.R. Allen and M.V. Erdmann. 2010. Congruence between mitochondrial genes and color morphs in a coral reef fish: population variability in the Indo-Pacific damselfish *Chrysiptera rex* (Snyder, 1909). Coral Reefs 29(2): 439–444.

Elliott, J.P. and D.R. Bellwood. 2003. Alimentary tract morphology and diet in three coral reef fish families. J. Fish Biol. 63(6): 1598–1609.

Emery, A.R. 1973. Comparative ecology and functional osteology of fourteen species of damselfish (Pisces: Pomacentridae) at Alligator Reef, Florida Keys. Bull. Mar. Sci. 23: 649–770.

Eschmeyer, W.N. 2015. Catalog of Fishes, http://research.calacademy.org/research/ichthyology/catalog/fishcatmain.asp. Electronic version accessed March, 2015.

Floeter, S.R., L.A. Rocha, D.R. Robertson, J.C. Joyeux, W.F. Smith-Vaniz, P. Wirtz, A.J. Edwards, J.P. Barreiros, C.E.L. Ferreira, J.L. Gasparini, A. Brito, J.M. Falcon, B.W. Bowen and G. Bernardi. 2008. Atlantic reef fish biogeography and evolution. J. Biogeogr. 35(1): 22–47.

Frédérich, B. and P. Vandewalle. 2011. Bipartite life cycle of coral reef fishes promotes increasing shape disparity of the head skeleton during ontogeny: an example from damselfishes (Pomacentridae). BMC Evol. Biol. 11: 82.

Frédérich, B., E. Parmentier and P. Vandewalle. 2006. A preliminary study of development of the buccal apparatus in Pomacentridae (Teleostei, Perciformes). Anim. Biol. 56(3): 351–372.

Frédérich, B., D. Adriaens and P. Vandewalle. 2008a. Ontogenetic shape changes in Pomacentridae (Teleostei, Perciformes) and their relationships with feeding strategies: a geometric morphometric approach. Biol. J. Linnean Soc. 95(1): 92–105.

Frédérich, B., A. Pilet, E. Parmentier and P. Vandewalle. 2008b. Comparative trophic morphology in eight species of damselfishes (Pomacentridae). J. Morphol. 269(2): 175–188.

Frédérich, B., G. Fabri, G. Lepoint, P. Vandewalle and E. Parmentier. 2009. Trophic niches of thirteen damselfishes (Pomacentridae) at the Grand Récif of Toliara, Madagascar. Ichthyol. Res. 56(1): 10–17.

Frédérich, B., S.C. Mills, M. Denoël, E. Parmentier, C. Brie, R. Santos, V.P. Waqalevu and D. Lecchini. 2010. Colour differentiation in a coral reef fish throughout ontogeny: habitat background and flexibility. Aquat. Biol. 9(3): 271–277.

Frédérich, B., S.Y.V. Liu and C.F. Dai. 2012. Morphological and genetic divergences in a coral reef damselfish, *Pomacentrus coelestis*. Evol. Biol. 39(3): 359–370.

Frédérich, B., L. Sorenson, F. Santini, G.J. Slater and M.E. Alfaro. 2013. Iterative ecological radiation and convergence during the evolutionary history of damselfishes (Pomacentridae). Am. Nat. 181(1): 94–113.

Frédérich, B., D. Olivier, G. Litsios, M.E. Alfaro and E. Parmentier. 2014. Trait decoupling promotes evolutionary diversification of the trophic and acoustic system of damselfishes. Proc. R. Soc. B-Biol. Sci. 281(1789).

Fulton, C.J. 2007. Swimming speed performance in coral reef fishes: Field validations reveal distinct functional groups. Coral Reefs 26(2): 217–228.

Fulton, C.J., S.A. Binning, P.C. Wainwright and D.R. Bellwood. 2013. Wave-induced abiotic stress shapes phenotypic diversity in a coral reef fish across a geographical cline. Coral Reefs 32(3): 685–689.

Galetto, M.J. and D.R. Bellwood. 1994. Digestion of algae by *Stegastes nigricans* and *Amphiprion akindynos* (Pisces, Pomacentridae), with an evaluation of methods used in digestibility Studies. J. Fish Biol. 44(3): 415–428.

Genner, M.J. and G.F. Turner. 2005. The mbuna cichlids of Lake Malawi: a model for rapid speciation and adaptive radiation. Fish Fish. 6(1): 1–34.

Gluckmann, I. and P. Vandewalle. 1998. Morphofunctional analysis of the feeding apparatus in four Pomacentridae species: *Dascyllus aruanus*, *Chromis retrofasciata*, *Chrysiptera biocellata* and *C-unimaculata*. Ital. J. Zool. 65: 421–424.

Hellig, C.J., M. Kerschbaumer, K.M. Sefc and S. Koblmüller. 2010. Allometric shape change of the lower pharyngeal jaw correlates with a dietary shift to piscivory in a cichlid fish. Naturwissenschaften 97(7): 663–672.

Holzman, R., D.C. Collar, S.A. Price, C.D. Hulsey, R.C. Thomson and P.C. Wainwright. 2012. Biomechanical trade-offs bias rates of evolution in the feeding apparatus of fishes. Proc. R. Soc. B-Biol. Sci. 279(1732): 1287–1292.

Kawecki, T.J. and D. Ebert. 2004. Conceptual issues in local adaptation. Ecol. Lett. 7(12): 1225–1241.

Kennett, J.P., G. Keller and M.S. Srinavasan. 1985. Miocene planktonic foraminiferal biogeography and paleoceanographic development of the Indo-Pacific region. Geological Society of America Memoirs 163: 197–236.

Kocher, T.D. 2004. Adaptive evolution and explosive speciation: The cichlid fish model. Nat. Rev. Genet. 5(4): 288–298.

Konow, N. and D.R. Bellwood. 2011. Evolution of high trophic diversity based on limited functional disparity in the feeding apparatus of marine angelfishes (f. pomacanthidae). PLoS ONE 6(9): e24113.

Kotrschal, K. 1989. Trophic ecomorphology in Eastern Pacific blennioid fishes – character transformation of oral jaws and associated change of their biological roles. Environ. Biol. Fishes 24(3): 199–218.

Liem, K.F. 1993. Ecomorphology of the teleostean skull. pp. 422–452. *In:* J. Hanken and B.K. Hall (eds.). The Skull: Functional and Evolutionary Mechanisms. The University of Chicago Press, Chicago.

Litsios, G., P.B. Pearman, D. Lanterbecq, N. Tolou and N. Salamin. 2014. The radiation of the clownfishes has two geographical replicates. J. Biogeogr. 41: 2140–2149.

Lobato, F.L., D.R. Barneche, A.C. Siqueira, A.M.R. Liedke, A. Lindner, M.R. Pie, D.R. Bellwood and S.R. Floeter. 2014. Diet and diversification in the evolution of coral reef fishes. PLoS ONE 9(7).

Loh, Y.-H.E., L.S. Katz, M.C. Mims, T.D. Kocher, S.V. Yi and J.T. Streelman. 2008. Comparative analysis reveals signatures of differentiation amid genomic polymorphism in Lake Malawi cichlids. Genome Biol. 9(7).

Losos, J.B. 2011. Convergence, adaptation, and constraint. Evolution 65(7): 1827–1840.

Losos, J.B. and R.E. Ricklefs. 2009. Adaptation and diversification on islands. Nature 457(7231): 830–836.

Molina, W.F., O.A. Shibatta and P.M. Galetti-Jr. 2006. Multivariate morphological analysis in continental and island populations of *Abudefduf saxatilis* (Linnaeus) (Pomacentridae, Perciformes) of Western Atlantic. Panam JAS 1(2): 49–56.

Motta, P.J. and K.M. Kotrschal. 1992. Correlative, experimental, and comparative evolutionary approaches in ecomorphology. Neth. J. Zool. 42(2-3): 400–415.

Motta, P.J., S.F. Norton and J.J. Luczkovich. 1995. Perspectives on the ecomorphology of bony fishes. Environ. Biol. Fishes 44(1-3): 11–20.

Muschick, M., A. Indermaur and W. Salzburger. 2012. Convergent evolution within an adaptive radiation of cichlid fishes. Curr. Biol. 22(24): 2362–2368.

Norton, S.F., J.J. Luczkovich and P.J. Motta. 1995. The role of ecomorphological studies in the comparative biology of fishes. Environ. Biol. Fishes 44(1-3): 287–304.

Olivier, D., B. Frédérich, M. Spanopoulos-Zarco, E.F. Balart and E. Parmentier. 2014. The cerato-mandibular ligament: a key functional trait for grazing in damselfishes (Pomacentridae). Front. Zool. 11: 63.

Quenouille, B., E. Bermingham and S. Planes. 2004. Molecular systematics of the damselfishes (Teleostei: Pomacentridae): Bayesian phylogenetic analyses of mitochondrial and nuclear DNA sequences. Mol. Phylogenet. Evol. 31(1): 66–88.

Salzburger, W., T. Mack, E. Verheyen and A. Meyer. 2005. Out of Tanganyika: genesis, explosive speciation, key-innovations and phylogeography of the haplochromine cichlid fishes. BMC Evol. Biol. 5: 17.

Schlichting, C.D. and M. Pigliucci. 1996. Phenotypic Evolution: A Reaction Norm Perspective. Sinauer Associates, Sunderland, MA.

Schluter, D. 2000. The Ecology of Adaptive Radiation. Oxford University Press, Oxford.

Sonnefeld, M.J., R.G. Turingan and T.J. Sloan. 2014. Functional morphological drivers of feeding mode in marine teleost fishes. Adv. Zool. Bot. 2(1): 6–14.

Stiassny, M.L.J. 1981. The phyletic status of the family Cichlidae (Pisces, Perciformes): a comparative anatomical investigation. Neth. J. Zool. 31(2): 275–314.

Thorsen, D.H. and M.W. Westneat. 2005. Diversity of pectoral fin structure and function in fishes with labriform propulsion. J. Morphol. 263(2): 133–150.

Turelli, M., N.H. Barton and J.A. Coyne. 2001. Theory and speciation. Trends Ecol. Evol. 16(7): 330–343.

Wainwright, P.C. 1988. Morphology and ecology – functional basis of feeding constraints in Caribbean labrid fishes. Ecology 69(3): 635–645.

Wainwright, P.C. 1994. Functional morphology as a tool in ecological research. pp. 42–59. *In*: P.C. Wainwright and S.M. Reilly (eds.). Ecological Morphology: Integrative Organismal Biology. The University of Chicago Press, Chicago.

Wainwright, P.C. and S. Reilly. 1994. Ecological Morphology: Integrative Organismal Biology. University of Chicago Press, Chicago.

Wainwright, P.C. and D.R. Bellwood. 2002. Ecomorphology of feeding in coral reef fishes. pp. 33–56. *In*: P.F. Sale (ed.). Coral Reef Fishes: Dynamics and Diversity in a Complex Ecosystem. Academic Press, San Diego.

Wainwright, P.C., D.R. Bellwood, M.W. Westneat, J.R. Grubich and A.S. Hoey. 2004. A functional morphospace for the skull of labrid fishes: patterns of diversity in a complex biomechanical system. Biol. J. Linnean Soc. 82(1): 1–25.

Walker, J.A. and M.W. Westneat. 2000. Mechanical performance of aquatic rowing and flying. Proc. R. Soc. B-Biol. Sci. 267(1455): 1875–1881.

West-Eberhard, M.J. 2003. Developmental Plasticity and Evolution. Oxford University Press, New York.

Williams, E.E. 1972. The origin of faunas. evolution of lizard congeners in a complex island fauna: a trial analysis. Evol. Biol. 6: 47–89.

Winterbottom, R. 1973. A descriptive synonymy of the striated muscles of the teleostei. Proc. Acad. Nat. Sci. Phila. 125: 225–317.

Young, K.A., J. Snoeks and O. Seehausen. 2009. Morphological diversity and the roles of contingency, chance and determinism in African cichlid radiations. PLoS ONE 4(3): e4740.

Zelditch, M.L., D.L. Swiderski, H.D. Sheets and W.L. Fink. 2004. Geometric Morphometrics for Biologists: A Primer. Elsevier Academic Press, San Diego.

Sound Production in Damselfishes

Eric Parmentier,[1,*] *David Lecchini*[2,3] *and David A. Mann*[4]

Introduction

The damselfishes (Pomacentridae) are one of the most thoroughly investigated and best understood family of acoustic reef fishes (Lobel et al. 2010). Scuba divers can easily hear damselfish sounds without the aid of a hydrophone, particularly when a male is aggressively defending his territory. These features of pomacentrid acoustic behavior make them easily accessible to bioacoustic study (Lobel et al. 2010). In this chapter, we review the different terms that have been used to describe the damselfish calls, summarize the behavior associated with sound production in different genera, and provide some additional information on species in which the sounds have not yet been described. In the next section, we focus on different acoustic characteristics and how they are important in damselfish communication and synthesize data on sound producing mechanisms. Lastly, we postulate about the origin of the sound production mechanism and its use in Pomacentridae.

Damselfishes are a well-known vocal species from the coral reefs. Some species are not only able to make sounds; they can also emit different kinds of sounds that are produced in various behavioral contexts (Mann and Lobel 1998, Parmentier et al. 2010). Different terms (threatening, shaking, click, grunt, etc.) were given to differentiate these

[1] Laboratoire de Morphologie Fonctionnelle et Evolutive, AFFISH - Research Center, University of Liège, Quartier Agora, Allée du six Août 15, Bât. B6C, 4000 Liège (Sart Tilman), Belgium.

[2] USR 3278 CNRS-EPHE-UPVD, CRIOBE, BP1013 Papetoai, 98729 Moorea, French Polynesia.

[3] Laboratoire d'Excellence 'CORAIL'.

[4] Loggerhead Instruments, 6576 Palmer Park Circle, Sarasota, FL 34238, USA.

* Corresponding author: e.parmentier@ulg.ac.be

various sounds but "pops" and "chirps" were the most commonly used (Schneider 1964, Allen 1972). This terminology for sound is however inconsistent since these call nouns were not always supported by empirical descriptions of the physical characteristics of the sounds. Moreover, both terminologies were associated with different kinds of behavior (Amorim 2006), such as courtship, fights, chases or threat displays (Mann and Lobel 1998, Parmentier et al. 2010). The best-characterized sound in damselfishes is the "chirp", produced by the male of several species (e.g., *Abudefduf* spp., *Dascyllus* spp., *Stegastes* spp.) primarily during a stereotyped courtship swimming display called the "signal jump", "dip" or "gamboling". The courtship dip consists of a male rising in the water column and then rapidly swimming downwards near, or to the prospective spawning area, at the same time as making a pulsed sound (Myrberg 1972, Spanier 1979, Lobel and Mann 1995, Mann and Lobel 1998, Lobel and Kerr 1999) and adopting in some species a unique species-specific courtship coloration (Myrberg et al. 1978). "Pops" usually referred to single (less frequently, double) pulsed sounds, while chirps would correspond to multiple pulsed sounds. However, multiple-pulsed "pops" were described in *Stegastes partitus* (Myrberg 1972), *Plectroglyphidodon lacrymatus*, *Dascyllus aruanus* (Parmentier et al. 2006), *Amphiprion akallopisos* (Parmentier et al. 2005) and *A. frenatus* (Colleye and Parmentier 2012). The inconsistent use of the terms is probably due to human perception of the sound. Isolated pulses sound like pop, but a combination of these same pops in multiple-pulse sound like chirps. For example, the sounds made simultaneously to dips (= dip sounds) of *Dascyllus flavicaudus* sound like cooing pigeons but pulses that composed these chirps are physically identical to pops (Parmentier et al. 2010). In *Dascyllus albisella*, Mann and Lobel (1998) has also indicated that two types of aggressive sounds were produced: a popping sound that was composed of one or two pulses, and a multiple-pulse "chirp" resembling the signal jump sound. However, there was no significant difference between the aggressive pop and the aggressive chirp in average pulse duration, peak frequency or frequency bandwidth (Mann and Lobel 1998).

Despite this overlap, it is possible to differentiate two main kinds of sounds in damselfishes, especially when the same species is able to make both kinds of sounds. It is however quite difficult to provide quantitative and qualitative data for each of the sound types because the acoustic characteristics depend on the species. Both types of sounds can be multiple-pulsed. We propose to call pops, the sounds in which pulses have longer pulse duration, longer pulse period and fewer pulses than chirps (Fig. 1). Pops are made during teeth snapping (Parmentier et al. 2007, Colleye et al. 2012) mainly during aggressive behaviors (chase and defense of the territory), courtship or reproduction. These sounds are found in all the pomacentrids studied so far. Chirps are produced during head shaking and correspond to submissive behavior during agonistic interactions (Schneider 1964, Colleye and Parmentier 2012). The mechanism allowing "chirp" production is currently not known. They were recorded only in *D. aruanus* (Parmentier et al. 2006), *A. akallopisos* (Parmentier et al. 2005) and *A. frenatus* (Colleye and Parmentier 2012). The main reason for this poor list would be that chirps are usually less audible than pops and are more difficult to detect in field studies.

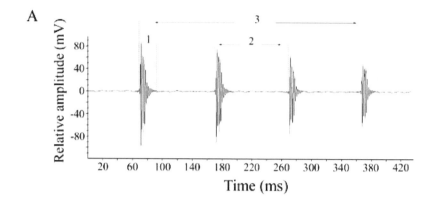

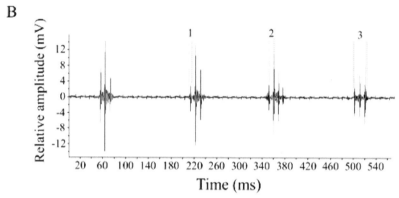

Fig. 1. Oscillograms comparing the temporal patterns between pops (A) and chirps (B) in *Amphiprion frenatus*. In A, there is only one call with four pulses. In, B there are four calls each having different pulses. Modified from Colleye and Parmentier (2012). Note the differences in pulse duration (1) and pulse period (2). The acoustic variable measured in (3) represents the sound duration in the case of submissive sounds (B), and the train duration in the case of aggressive sounds (A).

Details on the study of the sound parameters in teleosts were recently reviewed (Fine and Parmentier 2015). Three major tools used for describing fish sounds are (Fig. 2), an image showing frequency versus time, oscillograms, which depict amplitude against time (Fig. 1) and frequency spectra (Fig. 3), which show amplitude against frequency and indicate dominant frequencies within a sound. A sonogram is simply a series of frequency spectra from short time samples of a signal. Most pomacentrid sounds are a series of short-duration pulses and therefore present as vertical lines (a wide frequency band with a short duration) on a sonogram (Fig. 2). Since damselfish sounds include a series of pulses, one can measure the duration and number of pulses in the series, pulse period (time between the start of one pulse and the next), the related pulse repetition rate (number of pulses per unit time), interpulse interval (the silent period between pulses), pulse duration, and the frequency or power spectrum (an output of the amplitude, typically in dB, against frequency).

In this chapter, we will first focus on the ethological, physiological and morphological knowledge concerning the acoustic communication in damselfish.

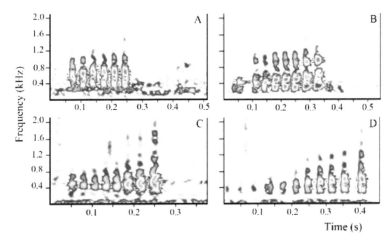

Fig. 2. Spectrogram of dip sounds in *Dascyllus albisella* (A), *D. flavicaudus* (B), *D. aruanus* (C) and *D. trimaculatus* (D). Despite the figure, note that the species do not show a species-specific distribution of the pulses in a call. Each spectrogram shape could apply to any of the four *Dascyllus* species. Modified from Parmentier et al. (2009b).

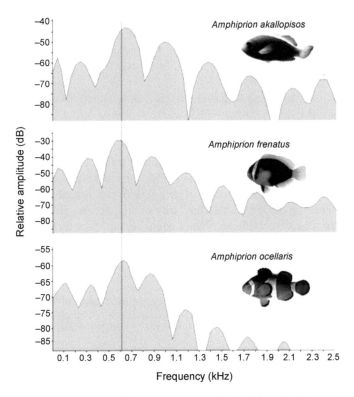

Fig. 3. Power spectrum in *Amphiprion akallopisos, A. frenatus* and *A. ocellaris*. Sound comparisons of these three species based on three specimens having the same size (61–63 mm SL) revealed that their dominant frequency is not significantly different. See Colleye et al. (2011) for further details.

In the second part of the chapter, we will provide additional information concerning the sound production of some species and review the different species for which the sounds are (at least) partly analyzed.

Signal Characters

Acoustic signals may carry much information. Generally speaking, sounds in fishes consist of trains of pulses that can be characterized by different parameters such as: sound duration (ms), pulse duration (ms), number of pulses in a train, pulse period (ms), the interpulse interval (measured as the time from the end of one pulse to the beginning of the next one) and dominant frequency (Hz). All these characters do not carry the same kind of information and the physical properties of the acoustic environment can affect the cues in different ways during sound propagation.

Among all studied parameters, pulse period is least affected by propagation when compared to peak frequency, pulse duration, interpulse interval, and the coefficient of variation of pulse amplitudes within a call (Mann and Lobel 1997). Pulse period varied by only 2% compared to its mean value, whereas the interpulse interval varied by about 10% and the other parameters by 40%. These results suggest that the pulsed sound functions over short distances and that the pulse period provides the most reliable basis for signal identification because it degrades the least with propagation through the environment (Mann and Lobel 1997). On the other hand, frequency spectrum and amplitude do not appear to be critical for species recognition (Ha 1973). This hypothesis is supported by different studies concerning different clades: *Stegastes*, *Dascyllus* and *Amphiprion*.

The bicolor damselfish *Stegastes partitus*, the beaugregory damselfish *Stegastes leucostictus*, the dusky damselfish *Stegastes dorsopunicans* and the threespot damselfish *Stegastes planifrons* are sympatric damselfishes from the coral reefs of southern Florida and the Caribbean. These species appear to have a common reproductive season, and congeners often maintain residences and territories within a few meters of each other (Myrberg et al. 1978). It is not unreasonable to assume that the sounds produced by the members of each species can be heard by the members of all others, based on the remarkable similarity of their hearing abilities (Myrberg and Spires 1980). In such conditions, sounds could be important for purposes of species recognition in order to avoid misidentification during courtship. The characteristics of the sounds of these *Stegastes* species showed that pulse period was different between species, with significant overlap of other metrics (Table 1; Spanier 1979).

Table 1. Acoustic characteristics of the dip sounds in four sympatric species of *Stegastes* (Spanier 1979).

	Number of pulses	Pulse length (ms)	Pulse period (ms)	Frequency range (Hz)
S. partitus	2–4	9.8	38	300–1200
S. leucostictus	3–6	15	31.9	180–1080
S. planifrons	3–6	11.6	42.1	200–900
S. dorsopunicans	4–9	11.7	40.4	200–1100

It was discovered that the males of these species would respond to playbacks of their sounds by performing courtship dips and producing sound (Myrberg et al. 1978, Spanier 1979). To study whether different species discriminated between the calls of other species, calls that varied in the number of pulses and the pulse period were played back to each species (Myrberg and Spires 1972, Myrberg et al. 1978, Spanier 1979). Two sets of calls were used. The first contained those calls having the most prevalent number of pulses for each species (Table 1). The second type contained 4-pulse calls from all species. The effectiveness of sounds was ascertained by counting the number of dips the male made in response to the sound playbacks. Species showed different abilities in recognizing their specific sounds.

The realization of signal jumps in reaction to playbacks indicated that each species was able to respond to the sounds of heterospecifics. However, they all responded significantly more to their own typical dip calls than to those of the other species when the typical number of pulses and pulse period were used, showing the species-specificity of the call. However, having responses to heterospecific sounds also indicate that there is an overlap in call structure between the species. The species-specific nature of the call was confirmed in an additional experiment (Spanier 1975) where the pulse period of the beaugregory (*S. leucostictus*) was artificially increased by 9 ms to produce a pulse interval equivalent to that of the dusky damselfish (*S. dorsopunicans*). This change was sufficient to cause the dusky damselfish to respond to the modified sound.

Another way to study which characters may be important in identification is to determine in closely related species which parameters are clearly different. The sounds of the four species of *Dascyllus* were first compared in 2009 (Parmentier et al. 2009b). The temporal characteristics of *D. aruanus* sounds differed widely from those of the other three species (*D. albisella, D. flavicaudus* and *D. trimaculatus*) and allowed the formation of two groups, corresponding to the phylogenetic branching (Bernardi and Crane 1999, McCafferty et al. 2002). For the purpose of this chapter, sounds in *Dascyllus carneus* and *D. reticulatus* were also analyzed (see further in the text). This comparison is based only on the pulse period (Fig. 4) because this character is least affected by signal propagation and thus would be a reliable carrier of information.

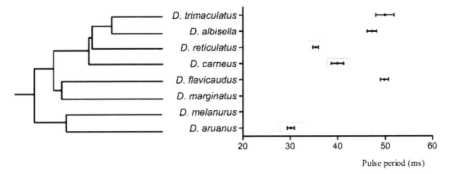

Fig. 4. Confrontation between the *Dascyllus* phylogeny (simplified figure from Bernardi and Crane 1999) and the variation of pulse period in 6 *Dascyllus* species (data from Parmentier et al. 2009b and from analysis made for the purpose of this chapter). Results are represented as means ± 95% confidence intervals. Dotted line rectangles show that the pulse periods of some species are completely isolated from the rest of the taxa.

According to Mann and Lobel (Mann and Lobel 1997), pulse period in *D. albisella* varied by only 2% as compared to its mean value, whereas other parameters (pulse length, frequency) varied by 40% or more of their mean values. Interestingly, the pulse period of *D. carneus* and *D. reticulatus* is significantly (Kruskal Wallis, $p < 0.05$) shorter than that of *D. trimaculatus*, *D. albisella* and *D. flavicaudus* and longer than that of *D. aruanus* (Fig. 4). It means that this character (pulse period) is sufficient to discriminate calls of both these fish species from the other previously studied *Dascyllus* species.

The courtship dip mating display associated with sound production in *Dascyllus aruanus* differs from other species. In *D. aruanus*, the male stops swimming, makes a forward rotating movement by raising its caudal fin, and then swims rapidly downward. In successive signal jumps, the fish stops swimming between each dip, and rises in the water column at the same time as raising its tail before the next jump. The movement resembles a sinusoidal curve. In the other three species, the swimming fish turns abruptly at a right angle and begins the dive to the side and not forwards as in *D. aruanus*. In all four species, the sounds are made not only during the descent, but also during the rise (Parmentier et al. 2009b). Based on an analysis of the sound parameters, *D. trimaculatus* groups with *D. albisella* (McCafferty et al. 2002), rather than one or both these fishes grouping with *D. flavicaudus*. This grouping of acoustic characteristics matches the phylogenetic relationships among these species (Parmentier et al. 2009b).

In the comparison of *D. trimaculatus*, *D. aruanus* and *D. flavicaudus*, significant differences in temporal characteristics (pulse period and interpulse interval) were also found between populations of sympatric species from Moorea in French Polynesia (*D. trimaculatus*, *D. aruanus*, and *D. flavicaudus*) and from Toliara in Madagascar (*D. trimaculatus* and *D. aruanus*) supporting the existence of dialects (Parmentier et al. 2009b). For example, the pulse period and pulse length of *D. trimaculatus* and *D. aruanus* are longer in Moorea than in Tulear. Playback experiments are however required in order to test the fish behavior and know if these statistically supported differences are important for the fish identification.

Interestingly, the temporal acoustic parameters concerning the fourth species (*D. albisella*), that lives in Hawaii and does not co-occur with other *Dascyllus* species, overlap all other *Dascyllus* and cannot be clearly distinguished. In the regions where they live in sympatry, it appears that *Dascyllus* species restrict the variability in their sounds. This could be evidence of adaptation with character displacement of sonic characteristics where different species co-occur (Parmentier et al. 2009b). Next to the sounds, the difference in the courtship dance during sound production can also aid species discrimination between sympatric species. In *D. carneus*, *D. albisella*, *D. flavicaudus* and *D. trimaculatus*, the swimming fish turns abruptly at a right angle and begins the dive to the side. In *D. aruanus*, the male stops swimming, makes a forward rotating movement by raising the caudal fin, and then swims rapidly downward (Parmentier et al. 2009b). In contrast to *D. aruanus*, these species change their coloration during the jump: the anterior part of their body becomes chocolate brown. However, the large white band behind the eye can become greyish in the biggest *D. aruanus* during the dips (pers. com.). Other modalities, such as vision (i.e., recognition of species-specific postures and coloration), probably augment any discrimination process in reproduction.

Another way to avoid any overlap in the communication channel would be to produce sounds at different periods of the day (Fig. 5). Study of the daily cycle in *D. flavicaudus* showed that this species made sounds mainly during the day with a

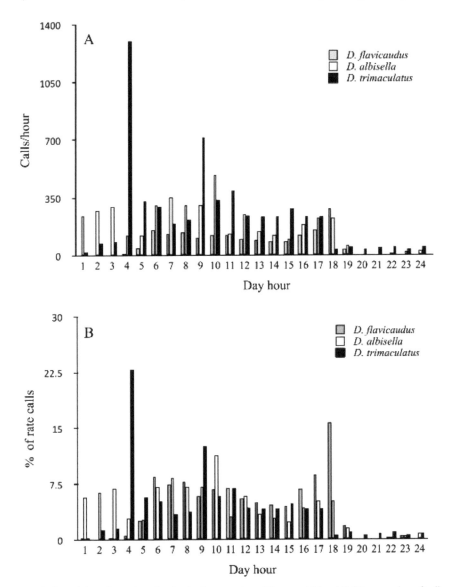

Fig. 5. Daily activity of sound production in three species of Pomacentridae. (A) Mean number of calls produced by hour and (B) ratio of the acoustic activity along the day. Data for *Dascyllus flavicaudus* are from Parmentier et al. (2010) and data for *Dascyllus albisella* are from Mann and Lobel (1997). Data concerning *D. trimaculatus* were acquired during November 2013 in Moorea. The number of recorded specimens is not the same, meaning we cannot assert from this graph that a species is calling more or less than another one.

peak of activity at sunset and a second, higher one, at sunrise (Parmentier et al. 2010). This pattern differs from that of *Dascyllus albisella*, which produces sounds mainly in the morning (Mann and Lobel 1995). Sound production was also detected at night in this species and peaked just before spawning. *Dascyllus albisella* and *D. flavicaudus* are however not sympatric. In November 2013, we studied the daily rhythm of *D. trimaculatus* in order to compare it with its sympatric species *D. flavicaudus*. Recording of daily cycles in *D. trimaculatus* followed the method applied in *D. flavicaudus* (Parmentier et al. 2010). Recordings were made with a Digital Spectrogram Recorder (DSG, Loggerhead Instruments Inc.). The DSG recorder is a long-term, low-power recorder of acoustic signals. The system was scheduled to record for 10 min every 60 min. DSG had been placed during four days next to a group of 8 adult specimens. Sounds were digitized at 22 kHz (16-bit resolution), low-pass filtered at 2 kHz and analyzed using AvisSoft-SAS Lab Pro 4.33 software. Manual analysis consisted simply of identifying and counting the *D. trimaculatus* sounds. The question was to know if sympatric and closely related species share the acoustic daily activity in the same way or not. It appears that *D. trimaculatus* produced sounds mainly around 04:00 h, whereas the main peak of sound production was 18:00 h in *D. flavicaudus* (Fig. 5A). Both species produced few sounds during the night and called most during the day. *Dascyllus trimaculatus* produced more calls than *D. flavicaudus*. However, many factors (spawning, fish number in the colony, etc.) could affect the call rate (Oliver and Lobel 2013), so we also calculated the percentage of call rate during the day for each species (Fig. 5B). This confirms both peaks of calling activity, but it also shows that the time devoted to diurnal calls is roughly the same in the two species. It is worth mentioning that *Dascyllus* species are very prolific callers, they can repeat their sounds hundreds to thousands of times a day (Lobel and Mann 1995, Parmentier et al. 2010), so it is possible that a female could sample many calls before making a decision to spawn with a given male.

The next set of experiments supporting the importance of temporal parameters in sounds, and more precisely the pulse period, can be found in clownfish. Contrary to *Dascyllus* and *Stegastes*, fish from this taxa do not make sounds during courtship or reproduction but mainly during aggressive and concomitant submissive behaviors (Colleye and Parmentier 2012). Agonistic sounds were recorded and compared in fourteen clownfish species (Colleye et al. 2011). Pulse duration and dominant frequency did not help in differentiating the species, because there is a size-related intraspecific variation in dominant frequency and pulse duration: smaller individuals produce higher frequency and shorter duration pulses than larger ones, whatever the sexual status (Colleye et al. 2009). Surprisingly, the relationship between the fish size and both dominant frequency and pulse duration is not species-specific: all the specimens of the 14 species are situated on exactly the same slope. It means that the size of any *Amphiprion* can be predicted by both acoustic features and that the sound-producing mechanism is highly conservative. According to previously described methods (Colleye et al. 2011), we have analyzed the pulse period of an additional species (*Amphiprion sandaracinos*) for the purpose of this book chapter and it confirms this relationship. Such detailed data has not been collected for different species of the same genus in fishes. However, the same kind of relationship between size and dominant frequency was also found in *Dascyllus albisella* (Lobel and Mann 1995) and in *Stegastes partitus* (Myrberg et al. 1993). In the clownfish species, the number of pulses also broadly overlaps among species and does not help in differentiating between species. Once more, the pulse period appears to be useful because it displays the most variation between species and the least variability within species (Fig. 6), even if it shows overlap among sympatric species (Colleye et al. 2011). Again, these results have to be carefully considered because other environmental characteristics can help in differentiating between species: the different clownfish species live in different parts of the world, are not all found in the same sea anemone species, and can have different diet and different coloration patterns (see Chapter XIII). Interestingly, several species (*A. sandaracinos, A. akallopisos* and *A. perideraion*), that have lost their vertical bands, that are phylogenetically closely related (Santini and Polacco 2006, Litsios et al. 2012) and that can cohabit in individual sea anemones (i.e., *A. sandaracinos* with *A. chrysopterus* in the region of Madang, or *A. perideraion* with *A. clarkii* in the region of Okinawa) are all characterized by a smaller pulse period (Fig. 6) than other *Amphiprion* species. As previously stated, non-overlapping in this character may have been important in the taxon diversification. However, pulse period is not systematically significantly different among sympatric species: *A. clarkii, A. frenatus* and *A. ocellaris* have the same pulse period range (Colleye et al. 2011) while living in sympatry on the fringing reef around Sesoko island (Hattori 1991). These three species inhabit different host species, being *Heteractis crispa* for *A. clarkii, Entacmaea quadricolor* for *A. frenatus* and *Stichodactyla gigantea* for *A. ocellaris* (Hattori 1991, 1995), which suggests that overlap in pulse period among these species is of minor importance. Comparing *Dascyllus* and *Amphiprion* species (Figs. 4 and 6) shows that the shortest pulse period in anemonefishes is longer than the longest pulse period of *Dascyllus* species.

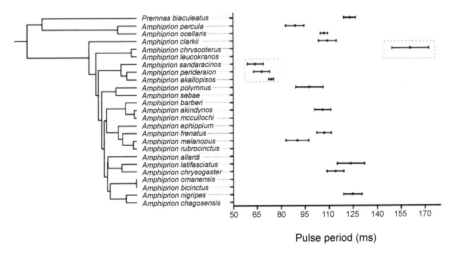

Fig. 6. Confrontation between the clownfish phylogeny (simplified figure from Litsios et al. 2012) and variation of pulse periods in 15 clownfish species (most data from Colleye et al. 2012). Results are represented as means ± 95% confidence intervals. Dotted line rectangles show that the pulse periods of some species are completely isolated from the rest of the taxa. Interestingly three closely related species have the shortest pulse period.

In the clownfish taxa, the dominant frequency of the calls could however be important at the intraspecific level. Within each clownfish species, the sex is controlled socially and there is a size-based dominance hierarchy: the breeding female is the largest individual, the breeding male is the second largest and the non-breeders get progressively smaller as the hierarchy descends (Fricke 1979, Buston 2003a, Buston and Cant 2006). The size hierarchy forms a queue to attain dominant status; individuals only ascend in rank when a higher rank individual disappears, and the smallest fish in the group is always the most recent recruit (Chapters IV and XII). The size-related variation in dominant frequency implies that smaller individuals produce higher frequency sounds than larger ones. Consequently, these sonic features might be useful cues for individual recognition within the group and may convey information on the social rank of the emitter within the group (Colleye et al. 2009, Colleye and Parmentier 2012).

Relationship between Sound Production and Spawning

Most of the sound production in *Dascyllus* is carried out by males during advertising and courtship behavior (Lobel and Mann 1995, Parmentier et al. 2010). The distance these calls are detectable is likely on the order of <15 m (Mann and Lobel 1997). Thus, it is a local communication signal and the receiving fish would have to be close by to detect the sound. *Dascyllus albisella* can make up to 3000 calls per day just prior to spawning (Mann and Lobel 1995). If each dip involves 1 m of swimming, this would mean that the fish would swim 3 km in one day. The rate of calling changes over the spawning cycle, with the highest levels occurring just before spawning (Mann and Lobel 1995). Since each of these sounds is accompanied by a courtship dip, these calls could act as an honest signal of male quality. This is an important assessment for females to make, since the males will guard their eggs in the nest for several days prior to their hatching into planktonic larvae (Oliver and Lobel 2013).

Different metrics of male quality including dominant frequency, inter-pulse interval, pulse duration, pulse number or calling rate were correlated with the measure of mating success in *D. albisella* (Oliver and Lobel 2013). It showed that females would choose mates based on the courtship rate of males. These experiments concerned both visual (dips) and acoustic cue meaning playback experiments should determine the relative contributions of the acoustic and visual modalities to the success of the courtship display (Oliver and Lobel 2013). However, it remains that acoustic call structure does not seem implied in the mating success, meaning that the phenotypic differences between males cannot be clearly explained in the courtship behavior framework.

Mechanism and Hypothesis about Damselfish Sound Production

Despite the numerous studies on sound production in pomacentrids, the nature of the sound-producing mechanism has remained unresolved for a long time, only resting on few assumptions. Some authors claimed that sound was produced by rapid up-and-down movements of the opercula and by the movements of the mouth bones related to taking food (Verwey 1930, Takemura 1983). Others implied that sound was produced by grating pharyngeal teeth, and could then be amplified by the swimbladder (Luh and Mok 1986, Chen and Mok 1988, Rice and Lobel 2003). Sounds produced by *Abudefduf luridus* were thought to involve a swimbladder mechanism (Santiago and Castro 1997), but the authors do not specify whether they believe this mechanism involves extrinsic muscles attached to the swimbladder.

Aggressive sounds emitted by the clownfish *Amphiprion clarkii* result from rapid mouth closing movements (Parmentier et al. 2007). This fast jaw slam is enabled by the cerato-mandibular (c-md) ligament joining the lateral side of the hyoid bar to the medial side of the mandible (Chapter XIV). Briefly, once the mouth is opened, the ligament acts as a cord, forcing the mouth to close and the teeth to snap resulting in the production of sound.

The mechanism can be summarized in 4 phases and highlights that both opening and closing of the mouth can be produced through a single set of movements (Olivier et al. 2014). (1) In the initial phase, the mouth is closed, the neurocranium is lowered and the hyoid apparatus is not depressed. In this situation, the cerato-mandibular ligament is loose and cannot apply any traction on the lower jaw. (2) During the aperture phase, there is an elevation of the neurocranium which mechanically involves the lower jaw and branchial basket depression, a phenomenon well known in fish feeding (Osse 1969, Van Wassenbergh et al. 2005). (3) Rather than accentuating this movement, a higher amplitude elevation of the head actually forces the mandible to rotate around its quadrate articulation and the mouth to close rapidly (5 ms) in a slam. This movement is due to the backward movement of the branchial basket that causes the c-md ligament to tighten and the traction on the lower jaw (Parmentier et al. 2007). The teeth collisions caused by rapid jaw closure correspond only to the onset of the sound. (4) The swimbladder was shown to be a highly damped sound source prevented from prolonged vibrations and could not in this case be the resonator (Colleye et al. 2012). The acoustic radiator results probably from a vibrational wave due to buccal jaw snapping and is likely transferred to the rib cage via different functional units of the skeleton such as the suspensorium, the neurocranium and the vertebral column. Mobile ribs vibrate and drive the oscillations of the swimbladder wall (Colleye et al. 2012).

Since the c-md ligament is a synapomorphic trait of the damselfish family (Stiassny 1981) and because the sounds were recorded in basal clades (Lepidozyginae) such as in *Plectroglyphidodon lacrymatus* (Parmentier et al. 2006) and different *Stegastes* species (Myrberg and Spires 1972, Ha 1973, Spanier 1975, 1979, Myrberg et al. 1978, 1986, 1993), it was assumed all the pomacentrid species are able to make sounds. However, more than 120 species were dissected and some of them did not have the ligament. Stochastic mapping of this synapomorphic trait on a time-calibrated phylogeny of damselfishes suggests that the c-md ligament disappeared at least three times (within Chrominae, Abudefdufinae and Pomacentrinae). More surprisingly with regard to our hypothesis, some species lacking the c-md ligament are not muted and can produce pulsed sounds. Most of the time, the sound duration in pomacentrids is correlated to the number of pulses indicating that calls have a relatively constant pulse period (Lobel and Mann 1995, Parmentier et al. 2006) and that the mechanism of sound production can be repeated at regular intervals. Sound parameters were compared between

11 species having the c-md ligament and two species (*Chromis viridis* and *Chromis atripectoralis*) lacking the c-md ligament (Frédérich et al. 2014). Since all studied species displayed the same type of sound spectra and oscillograms, their sounds seem to be produced in a similar way. *Chromis viridis* and *C. atripectoralis* can generate pulsed sounds by mouth closing but they have to use a different motor pattern than clownfishes and the other species having the c-md ligament (Frédérich et al. 2014). Interestingly, both *Chromis* which do not possess the c-md ligament, show the highest standard deviation around the mean value of pulse period (SD > 36 ms) in comparison with all the other studied species (SD ≤ 30 ms). The inconstancy of pulse periods in *C. viridis* and *C. atripectoralis* seems to be related to the absence of the c-md ligament. Indeed, mouth opening and closing in these species could be achieved by different muscles (e.g., epaxial, sternohyoideus, adductor mandibulae muscles) complicating the synchronization and the temporal pattern. The coordination of such a system is expected to be more variable than the c-md ligament system allowing the opening and rapid closing of oral jaws by the lonely continuous backward movement of the hyoid bar (Parmentier et al. 2007). A high variation in the pulse period is also shown or suggested in different *Abudefduf* species (Santiago and Castro 1997, Lobel and Kerr 1999, Maruska et al. 2007). See for example Fig. 5 in Rice and Lobel (2003). We have dissected *Abudefduf sordidus*. In this species, there is no ligament between the hyoid bar and the lower jaw as is the case in most of pomacentrids. However, there is a tendon on the geniohyoideus that inserts on the angular of the lower jaw. This muscle could play a role in the production of sounds. Beyond the functional aspect of sound mechanism, this inconstancy in pulse period can be perceived as a species-specific acoustic cue in *Chromis* and *A. sordidus* as well as a more constant length of pulse period in others. This variation between species with and without the c-md ligament is a source of diversity in the acoustic repertoire of the family (Frédérich et al. 2014).

The teeth collision (with or without c-md ligament) mechanism that we have described should be the basic mechanism of dip and aggressive sounds of at least the species belonging to *Dascyllus*, *Stegastes*, *Pomacentrus*, *Chromis*, *Abudefduf*, *Plectroglyphidodon*. This mechanism is also used for the production of agonistic sounds in the *Premnas* and *Amphiprion* species.

Clownfish species are also able to produce another kind of sound (called submissive sounds) emitted when fish make head shaking movements (Parmentier et al. 2005, Colleye and Parmentier 2012). Generally speaking, submissive sounds are completely different from aggressive ones. They are always composed of several pulses forming units produced alone or in series, whereas aggressive sounds are composed of a single pulse unit that can be emitted alone or in series (Fig. 1B). They also exhibit shorter pulse periods and shorter pulse durations than aggressive sounds (Colleye and Parmentier 2012).

The importance of sound production in damselfish could be due to their way of life. Most of them establish permanent or temporary territories. Sounds generally occur simultaneously with aggressive actions either during pair-encounters or are produced by territorial residents as they encounter intruders. That vocalizations play an important role in territorial defense has been experimentally demonstrated in both avian and piscine species (Myrberg 1997). So long as territorial individuals could produce sounds, they maintained their territorial boundaries; however, muted individuals were unable to deter intruders from entering their shelter sites, despite appropriate visual displays (Myrberg 1997).

By comparing in detail, the movements related to sound production and those related to biting in *Amphiprion* and *Stegastes*, we can reasonably postulate that sound production in pomacentrid results from the exaptation of feeding movements. Exaptation refers to a functional character previously shaped by natural selection for a particular function and that has been coopted for a new use (Gould and Vrba 1982, Larson et al. 2013). Pops are used in different combinations during different behaviors, but are constructed on the basis of the same mechanism involving the c-md ligament. The parsimony principle implies that this ancestral call was made of only one pulse.

Among the behaviors of all the species that we have examined, one is related to fighting and is mainly made of only one pulse, corresponding to a single jaw slam. The origin of the sound could be found in biting, because fighting sounds usually occur before the display of aggressive behavior with biting (Parmentier et al. 2010). Moreover, single pulse sounds can also be heard in *Plectroglyphidodon* and *Stegastes* species giving teeth strokes while simply grazing algae on their territories (Chapter XIV). This grazing activity also results from a mechanical single slam. We hypothesize that sounds in pomacentrids were first produced incidentally as a by-product of foraging and/or fighting activities. Single sounds were then selected because they resulted in successful territory and nest defense. Currently, one or two pulses are still used to deter conspecifics and heterospecifics. Because dip or visiting calls are made of trains of pulse, it is quite easy to postulate they should result from the repetition of the same motor pattern.

Sonic Behavior

Sounds can be used in different behavioral contexts that are not well defined in all damselfish. Here we review already studied species and we also include data concerning seven species that we have video recorded during field missions to Moorea Island (French Polynesia) and Madagascar. *Pomacentrus pavo, Plectroglyphidodon lacrymatus, Chromis viridis, Chromis atripectoralis, Stegastes nigricans* and *Stegastes punctatus* were recorded in Moorea during January and February 2009. *Dascyllus carneus* were recorded in Madagascar during June 2011. *Dascyllus reticulatus* were recorded in Dongsha atoll (Taiwan) during May 2015. The way we have recorded and analyzed the sounds follows the previous studies on *Dascyllus* and different clownfish species (Parmentier et al. 2010, Colleye and Parmentier 2012). Recording sessions, each lasting from 1 to 4 h, were made at a depth of between 1 and 5 m. Recordings of sound production were made using a SONY HDD video camera placed in a housing (HC3 series, Ocean Images, Cape Coral, Fl) and coupled with an external hydrophone (High Tech. Inc., HTI-96) with a flat response of 20 Hz to 20 kHz. Recordings were made by placing the housing in front of the coral patch. Sounds were extracted in .wav files using the AoA audio extractor setup freeware. Sounds were digitized at 44.1 kHz (16-bit resolution), low-pass filtered at 1 kHz and analyzed using AvisSoft-SAS Lab Pro 4.33 software. Only the sounds with a good signal to noise ratio were used in the analysis. The following sound parameters were measured: sound duration; number of pulses in a sound; pulse period (measured as the average peak-to-peak interval between consecutive pulses in the entire sound; pulse length (measured as the time from the beginning of one pulse and its end); dominant frequency.

Abudefduf

Abudefduf species produce aggressive sounds towards both conspecific and heterospecific intruders while nest guarding, preparing a nest substrate or during courtship while trying to attract a female for spawning. Sounds are mainly produced by males that can be distinguished by territorial behavior and by adoption of courtship/spawning coloration. *Abudefduf* also seem to differ from other genera in the way they form a nuptial parade. The male does not produce signal jumps but performs vigorous horizontal swimming, looping and then zigzags to motivate the female to follow him back to the nest (Lobel and Kerr 1999, Maruska et al. 2007). These courtship displays are not associated with sound production in *Abudefduf abdominalis* (Maruska et al. 2007) but they are in *A. sordidus* (Lobel and Kerr 1999, Lobel and Lobel 2013) and *A. luridus*[1] (Santiago and Castro 1997). In *A. abdominalis* and *A. vaigiensis*, two sounds produced during agonistic encounters with conspecific and heterospecific individuals were an aggressive short pulse (52–88 ms, <500 Hz) identified by 1 to 2 pulses per sound and longer pulse trains identified by more than 2 pulses (Maruska et al. 2007, Tricas and Boyle 2014). In both species, Tricas and Boyle (Tricas and Boyle 2014) described an additional agonistic high-frequency single pulse sound of short duration (18–23 ms) and high peak frequency (805–1162 Hz).

Circadian rhythms were only studied in *A. luridus*: the major proportion of sounds was recorded around sunrise and sunset (Santiago and Castro 1997). As in *Chromis viridis* and *C. atripectoralis*, *Abudefduf* species present a characteristic that allows distinguishing their calls easily from the other pomacentrids: their calls do not exhibit a consistent repeated pattern at the level of the pulse period (Lobel and Kerr 1999, Maruska et al. 2007). In *A. luridus*, it is not possible to establish a relationship between the number of pulses emitted and the pulse period (Santiago and Castro 1997).

Amphiprion and *Premnas*

Clownfishes live in social groups composed of a breeding pair and between zero to four non-breeders. Within each group, numerous agonistic interactions occur and they appear to play an important role by maintaining size differences between individuals adjacent in rank (Fricke 1979, Buston 2003b). Larger fishes chase smaller ones, which means that the smallest one is the recipient of numerous charges (Fricke 1979). All clownfish species have evolved ritualized threats and submissive postures that presumably serve to circumvent physical injury during intraspecific quarreling (Allen 1972). During threat postures, resident fish can face, charge and chase an intruder. Several authors (Schneider 1964, Allen 1972, Fricke 1974) have also highlighted the existence of a typical behavior (commonly called "head shaking") as a reaction

---

[1] According to the display description they give, we think these authors wrongly assume they observed aggressive behaviours.

to aggressive interactions. This behavior consists of a lateral quivering of the body that begins at the head and continues posteriorly. Clownfishes were also reported to produce sounds during both agonistic and submissive interactions (Schneider 1964, Allen 1972, Chen and Mok 1988, Parmentier et al. 2005).

Agonistic sounds are produced by individuals of different sexual status (females, males and non-breeders) who display charge-and-chase reactions when another hetero- or conspecific approaches the sea anemone in which they dwell (Colleye et al. 2009). Sound production in clownfishes can be traced back as early as 1930 when Verwey stated that *A. akallopisos* and *A. polymnus* could produce sounds (Verwey 1930). Studies were thereafter expanded to include different *Amphiprion*, and it has been noted that they produced sounds while swimming, feeding, associating with anemones, and particularly when fighting for an anemone (Schneider 1964, Takemura 1983, Parmentier et al. 2005). Further insight into the description of sounds was provided since two distinct sounds were differentiated based on their duration, frequency range and repetition of pulses (Allen 1972). Aggressive sounds are mainly produced by dominants during charges, chases and threat displays between conspecifics during agonistic interactions (Colleye et al. 2009), whereas submissive sounds (chirps) are always emitted when subordinates exhibit head shaking movements in reaction to aggressive displays by higher-ranking individuals (Colleye and Parmentier 2012). Therefore, both types of sounds seem to be an integral part of the agonistic behavior in clownfishes. Currently, 14 species (including *Premnas biaculeatus*) have been recorded and the sounds analyzed (Colleye et al. 2011). For the purpose of this chapter we have also recorded the sounds of *Amphiprion sandaracinos*.

In addition to these behaviors, it was also reported that clownfishes might produce sounds during courtship. Courtship in clownfishes is generally stereotyped and ritualized, and is typically accompanied by different activities such as nest cleaning, spawning and nest care (Allen 1972). Basically, studies that describe the courtship sounds of clownfishes are limited in number. To date, sound production during reproductive periods has been reported in three clownfish species: *A. ocellaris, A. frenatus* and *A. sandaracinos* (Takemura 1983). However, these observations need to be carefully considered since, according to the author, the sounds were hardly heard and sometimes they do not seem to be directly related to spawning behavior. Moreover, the behavioral relevance of these data are somewhat doubtful since the three species would emit sounds with high frequency components of more than 2 kHz during reproduction, a frequency these fish cannot hear (Parmentier et al. 2009a). These sounds could just be a by-product of the nest cleaning activities. In addition, spawning events were observed and recorded in *A. akindynos, A. melanopus* and *A. percula* living in tank, in *A. clarkii* living in semi-natural condition and in *A. perideraion* living in the field. All these observations correspond to a total of 13 complete spawning events. Overall, the absence of sound production throughout all activities of the reproductive period was complete (Colleye and Parmentier 2012). Unlike other pomacentrids, sounds are not produced for mate attraction in clownfishes. It is likely an evolutionary outcome related to their peculiar way of life: these fishes form small social groups including only one mating pair, inhabit a restricted territory (the sea anemone), spend most of the time in close vicinity of their host and rarely interact with other species on the reef. On the other hand, sounds seem to be important in order to reach and to defend the competition for breeding status.

Chromis

Chromis damselfishes are planktivorous species that usually live in aggregations and schools (Pinnegar et al. 2007, Frédérich et al. 2009), but solitary fish can also be found defending small areas around rocky ledges and crevices. These territorial individuals chase away other *Chromis* and other species. The number of territorial individuals increases greatly during the spawning period (Myrberg et al. 1967). In different species, males spawn repeatedly with different females. They synchronously establish territories, prepare nests and court females through visual and acoustic display (Abel 1961, Picciulin et al. 2001, 2004). Females lay demersal eggs that are guarded and fanned by males until hatching. When eggs have concluded hatching, males abandon the nests and rejoin the feeding school (Picciulin et al. 2004).

In *Chromis chromis*, sounds are composed of a single pulse and are associated with aggressive behavior and the dip (Picciulin et al. 2001). Pulse train sounds were recorded for threespot chromis *Chromis verater* during agonistic interactions with a conspecific (Tricas and Boyle 2014).

Sounds with between 1 to 22 pulses (pulse duration 8 ms, pulse period 7 ms) were produced in tanks during agonistic interactions in *Chromis viridis* (Amorim 1996). The sounds of two sister-species, *Chromis viridis* and *Chromis atripectoralis* (Froukh and Kochzius 2008), were recorded in the field (Frédérich et al. 2014). Different behaviors were observed. Sounds associated with conspecific chases were recorded in *C. atripectoralis* but not in *C. viridis*. Sounds produced during conspecific chases had 4 to 6 pulses with a pulse period of 36 ± 7 ms. During conspecific fighting, both fish rotated around a common axis while attempting to bite the opponent. Sounds associated with chases were restricted to 1 to 2 pulses in both species in the majority of the cases. In *C. viridis*, sounds can be produced during the ascending or descending (81.3% of the cases) phases of the signal jump. During the signal jump, males become yellow, and black areas appear on the pectoral fins. Dip sounds have between 3 and 13 pulses (7 ± 3, n = 32) with a duration of 8 ± 3 ms (n = 196), and the period is highly variable (81 ± 51 ms, n = 183). In *C. atripectoralis*, the color pattern becomes dull during signal jumps. Dip sounds in this species have between 1 and 17 pulses (6 ± 5, n = 18) with a duration of 10 ± 8 ms (n = 103) and a highly variable pulse period (47 ± 20 ms, n = 183). The peak frequency was 661 ± 205 Hz (n = 196) in *C. viridis* and 790 ± 296 Hz (n = 107) in *C. atripectoralis*, but the fish sizes were not measured. Both species also have an unknown characteristic: dip sounds made during signal jumps can be composed of several pulse trains that are randomly emitted during the movements. These sequences appeared in 31% of the signals in *C. viridis* and 37.5% in *C. atripectoralis*. Pulse duration and peak frequency of sounds from *Chromis viridis* of Moorea were similar to sounds previously described from tank recordings (Amorim 1996). However, there is a large difference in the pulse period (81 vs. 8 ms) between both sets of data.

Sounds are also likely to be emitted by other *Chromis* species in which at least signal jumps have been observed: *C. multilineata* (Myrberg et al. 1967), *C. notata*

(Ochi 1985), *C. cyanea* (Albrecht 1969), *C. caeruleus, C. verater, C. ovalis* (Swerdloff 1970) and *C. iomelas* (pers. obs.). No signal jump was observed in *Chromis dispilus*. However, they have a comparable visual signal consisting of rapid alternate expansion and relaxation of the caudal fin, causing a distinctive flashing of the white margin along the inner edge of the tail (Russell 1971).

Dascyllus

Dascyllus species were the subject of numerous studies around the world and constitute probably one of the most complete descriptions of sounds related to different kinds of behavior. *Dascyllus albisella* was recorded in Johnston atoll and Hawaii (Lobel and Mann 1995), *D. flavicaudus* in Rangiroa and Moorea (Parmentier et al. 2009b, Parmentier et al. 2010), *D. aruanus* and *D. trimaculatus* in Rangiroa, Moorea and Madagascar (Parmentier et al. 2009b) and, finally, *D. carneus* in Madagascar (pers. obs.).

Different kinds of sounds were reported within the same *Dascyllus* species. In *D. albisella*, males produced pulsed sounds during the courtship behavior known as the signal jump, when visited by females (during pseudospawning), mating, and aggression towards heterospecifics and conspecifics, and nest preparation (Mann and Lobel 1998). Females made only aggressive sounds in this species. Sounds associated with fighting; mating/visiting, chasing and signal jumps were also recorded in *D. flavicaudus* (Parmentier et al. 2010). In both species, characteristics of the sounds related to different behaviors show significant differences (Lobel and Mann 1995, Mann and Lobel 1998, Parmentier et al. 2010), highlighting that acoustic features should be sufficient to infer the corresponding behavior.

Two studies provided data on the calling rate. In *D. flavicaudus*, daily recordings showed that sound production rates were higher at sunrise and sunset than during the day and that no sound was produced during the night (Parmentier et al. 2010). In *D. albisella*, sound production peaked each day at dawn. However, sound production was detected at night and was most intense just before spawning. The highest rates of sound production occurred on the day before and the day of egg-laying (Mann and Lobel 1995). In this species, sound production rates are also higher during their reproductive season (April) than during the non-reproductive season (October).

Specimens of *D. carneus* and *D. reticulatus* were recorded for the first time in Madagascar and Dongsha atoll respectively (see also above). In calling males we did not notice deep color changes in *D. reticulatus*, but it was marked in *D. carneus*: the anterior part of its body became chocolate brown during the signal jump. In *D. carneus*, the pulse length was (on average ± SD) 17 ± 4 ms (n = 185), the pulse period 40 ± 8 ms (n = 150) and mean dominant frequencies 753 ± 188 Hz. In *D. reticulatus*, the pulse length averaged 16 ± 3 ms (n = 110), the pulse period 35 ± 3 ms (n = 160) and the mean dominant frequency was 812 ± 110 Hz (n = 144).

Plectroglyphidodon

Plectroglyphidodon spp. are territorial species having a major influence on the algal communities within their territories through the exclusion of other herbivorous taxa from their territories (Brawley and Adey 1977) and/or farming activities corresponding to the selective removal of undesirable algae, active site selection, and fertilization (Ceccarelli et al. 2005, Hoey and Bellwood 2010, Emslie et al. 2012). To date, sounds have been recorded from *Plectroglyphidodon lacrymatus*, where sounds were originally recorded from three fish in an aquarium in Madagascar (Parmentier et al. 2006). Sounds were produced only when an observer approached the tank. The fish that was apparently responsible for making the sound faced the observer and spread its pectoral fins, showing an aggressive behavior, probably in relation to the defense of the territory. Sounds were produced in trains of two to five pulses (mean duration of each pulse: 56 ms), with a mean pulse period of 179 ms. Recordings of *P. lacrymatus* living in adjacent territories in Moorea enable the description of other sonic behaviors. Sounds were also recorded during signal jumps, conspecific chases and heterospecific fighting. During signal jumps, *P. lacrymatus* rise in the water column and then rapidly swim downwards while producing a pulsed sound. During chasing or fighting, calls are made consisting of one to two pulses. Dip sounds were composed of three to nine pulses with a pulse duration of 15 ± 3 ms (n = 137) and pulse periods of 75 ± 12 ms (n =107). The large difference in the pulse duration between both populations (Madagascar and Moorea) could be due to the resonating tank effect in Toliara. The difference between the pulse periods of both regions could be due to behavior (agonistic vs. signal jump) but is most probably related to geographic distribution (Parmentier et al. 2009b).

In Hawaii, *Plectroglyphidodon johnstonianus* produce sounds directed towards neighboring conspecifics and heterospecifics (Tricas and Boyle 2014). Different kinds of sounds were described in this species. Single pulse sound or pulses produced in a train seem to correspond to sounds described in other damselfish species. The authors also reported for the first time "a half pulse sound waveform" corresponding to a distinctive single, strong, and rapid negative peak followed by a slower positive half cycle. These half pulses can be emitted in trains or not. Two kinds of growls are also described. They would occur as a series of contiguous pulses. Both growls can be distinguished on the basis of pulse rate and pulse amplitude (Tricas and Boyle 2014). Future studies are however required because few sounds were recorded.

Pomacentrus

Sounds were recorded in *Pomacentrus pavo* for the purpose of this chapter. These data however have to be considered carefully because the number of observations was limited, it concerned individuals coming from two different groups. *Pomacentrus pavo* live in small groups around coral patches but males cluster during the reproduction (pers. obs.). During reproduction, males isolate and defend small territories that are crevices within coral patches or shelters in the sand. In contrast to the nesting behavior of *Dascyllus* and *Amphiprion* which nest on the substrate, *P. pavo* nest in it. Courtship consists of signal jumps that can be accompanied by sounds in some cases. In this case, the sound is single pulsed and is made at the lower end of the signal jump. A second jump can directly follow and, in this case, the sound is again made at the lower end of the dive. Similar movements were observed in *Pomacentrus nagasakiensis* and grunting sounds were heard during female enticement (Moyer 1975). These sounds were however not described. Single pulses were also emitted during fights with conspecifics and sounds of two to three pulses were made during chases. There were two kinds of chases: males towards conspecifics, and males attempting to attract a female to their nests that chased and dipped in alternation. On three occasions, multiple-pulsed sounds (from 6 to 8 pulses) were emitted when the female entered and visited the nest. We could not observe the behavior inside the nest and do not know if it corresponds to spawning or to the visiting sound described in *Dascyllus albisella* (Mann and Lobel 1998) or in *Abudefduf abdominalis* (Maruska et al. 2007). We did not observe modifications of the color pattern during the signal jumps, the agonistic interactions or the female visit. The pulse duration was (mean ± SD) 17 ± 4 ms (n = 89). There was a difference between the pulse lengths of fighting (8 ± 2 ms, n =3) and both signal jump (18 ± 4 ms, n = 32) and chases (17 ± 4 ms, n = 31). However, the small number of fighting pulses should prompt us to be cautious. Pulse periods from sounds produced during signal jumps were significantly longer (217 ± 58 ms, n = 14; Mann-Whitney p < 0.05) than during visiting (133 ± 37 ms, n = 20).

Stegastes

Stegastes species hold permanent territories and culture filamentous algae on dead corals. This farming involves selective weeding of algae in order to maintain algal communities that are distinct from the surrounding undefended substratum (Ceccarelli et al. 2005, Ceccarelli 2007). This taxon is particularly well known for active sound production in many different behavioral contexts such as courtship, territorial defense (against conspecifics and heterospecifics), chases and "keep-out" signals (Myrberg and Spires 1972, Ha 1973, Spanier 1975, 1979, Myrberg et al. 1978, Myrberg 1997). *Stegastes* have been used for different pioneering ethological experiments that were conducted on the Caribbean species *S. dorsopunicans*, *S. planifrons*, *S. leucostictus* and *S. partitus* (Spanier 1975, Myrberg et al. 1978, Spanier 1979). These experiments allowed the identification of signal characteristics that can help in species recognition. Different kinds of sounds have also been recorded in *Stegastes lividus* in Taiwan (Mok, pers. com.) and in the Hawaiian gregory *S. marginatus* during algal turf feeding and breeding territories (Tricas and Boyle 2014). In a tank, *S. rectifraenum* calls were also associated with shelter defense or biting filamentous algae. Pulse durations were significantly longer when biting filamentous algae but no significant difference was found between dominant frequency of calls and bites (Olivier et al. 2014). In the sympatric species *Stegastes nigricans* and *Stegastes punctatus* of French Polynesia, differences were also found in case of the Caribbean species. The sounds of *S. nigricans* were significantly (Mann-Whitney, p < 0.01) different from that of the *S. punctatus* on the basis of the pulse period (mean ± SD, 48 ± 11 ms, n = 47 vs. 63 ± 15 ms, n = 51) but not the pulse duration (Mann-Whitney, p = 0.055). Both species also significantly differed in the dominant frequency of sounds, probably because the specimens of *S. nigricans* (342 ± 160 Hz, n = 76) in Opunohu Bay (Moorea) were generally smaller than the *S. punctatus* (244 ± 21 Hz, n = 67).

Other Species

Sounds were also reported but not analyzed in *Hypsypops rubicundus* (Limbaugh 1964, Fish and Mowbray 1970), *Microspathodon chrysurus* (Emery 1973), and *Chrysiptera leucopoma* (Graham 1992). Few sounds were also recorded in *Dischistodus prosopotaenia* in Dongsha Atoll, Taïwan (pers. com.). In Rangiroa, we failed to record *Chrysiptera glauca*: we observed courtship behaviors but were unable to record any sounds. During courtship, males change coloration, their color pattern looking like the great white shark's color pattern: the back became darker and the belly brighter. The caudal fin forms a sail that seems to be used by the male to attract females to the nest.

Conclusion

Sound production in damselfish is probably used by all the species. The Pomacentridae provide a powerful model for the study of evolution of sound production, because there are so many extant species that produce sound and they live mainly in tropical and subtropical regions where the water is clear and behavior is readily observed. There is diversity in both intra and inter-specific calls, since some species are able to produce at least six different calls. Moreover, the association between the calls and the ability of some species to change their color pattern suggest a greater variety of messages than expected. Future studies should be conducted to describe sounds in more species in order to assess the role of acoustic communication in the evolutionary history of the taxa. It is important to note many factors such as temperature, size, and background noise can affect acoustic parameters in pomacentrids, making comparisons difficult (Demski et al. 1973, Mann and Lobel 1997, Feher et al. 1998, Connaughton et al. 2000, Colleye et al. 2009, Papes and Ladich 2011). Moreover, many other fundamental studies are needed on the sound production mechanism of chirps, the ability of inter-specific communication, and the use of call partitioning at the level of the reef.

We draw these general conclusions, but there is clearly a lot of room for more research:

1) All pomacentrids should be able to make sounds for communication purposes, but sounds are not produced during all behaviors in all species.
2) Sounds mainly consist of trains of pulses in which the number is higher when emitted towards conspecifics.
3) Courtship sounds are associated with stereotyped movements.
4) Dominant frequency (and most probably pulse duration) is related to the fish size in all species and more data should be collected to know if all the family species can be found on the same slope or not.
5) Sounds of sister species show many overlapping characteristics, indicating that this cue alone would not be sufficient to discriminate species.
6) Dialects are found within the same species.
7) Pop sounds result from jaw snapping, but another kind of mechanism has to be found for explaining the emission of chirps.
8) The call rate would be a determining factor in mating success, meaning sonic phenotypic differences related to size are not the evolutionary driving force.

We hypothesize that damselfish sound production originally evolved starting with the single pop associated with feeding/aggression. The coupling of sound production and exaggerated swimming could have evolved through sexual selection by females for male quality. In *Amphiprion* where there is one female per anemone, there is no exaggerated courtship dip associated with spawning.

References

Abel, E.F. 1961. Freiwasserstudien über das Fortpflanzungsverhalten des Monchfishes *Chromis chromis*, einem Vertreter der Pomacentriden im Mittelmeer. Z. Tierpsychol. 18: 441–449.

Albrecht, H. 1969. Behaviour of four species of Atlantic damselfish from Columbia, South America (*Abudefduf saxatilis*, *A. taurus*, *Chromis multilineata*, *C. cyanea*; Pisces Pomacentridae). Z. Tierpsychol. 26: 662–676.

Allen, G.R. 1972. The Anemonefishes: Their Classification and Biology. T.F.H. Publications, Neptune City, N.J.

Amorim, M.C.P. 1996. Sound production in the blue-green damselfish, *Chromis viridis* (Cuvier, 1830) (Pomacentridae). Bioacoustics 6: 265–272.

Amorim, M.C.P. 2006. Diversity of sound production in fish. pp. 71–104. *In*: F. Ladich, S.P. Collin, P. Moller and B.G. Kapoor (eds.). Communication in Fishes. Science Publishers, Enfield.

Bernardi, G. and N. Crane. 1999. Molecular phylogeny of the humbug damselfishes inferred from mtDNA sequences. J. Fish Biol. 54: 1210–1217.

Brawley, S. and W. Adey. 1977. Territorial behavior of threespot damselfish (*Eupomacentrus planifrons*) increases reef algal biomass and productivity. Environ. Biol. Fishes 2(1): 45–51.

Buston, P. 2003a. Size and growth modification in clownfish. Nature 424: 145–146.

Buston, P. 2003b. Social hierarchies: size and growth modification in clownfish. Nature 424(6945): 145–146.

Buston, P. and M. Cant. 2006. A new perspective on size hierarchies in nature: patterns, causes, and consequences. Oecologia 149(2): 362–372.

Ceccarelli, D.M. 2007. Modification of benthic communities by territorial damselfish: a multi-species comparison. Coral Reefs 26(4): 853–866.

Ceccarelli, D.M., G.P. Jones and L.J. McCook. 2005. Foragers versus farmers: contrasting effects of two behavioural groups of herbivores on coral reefs. Oecologia 145: 445–453.

Chen, K.-C. and H.-K. Mok. 1988. Sound production in the Anemonefishes, *Amphiprion clarkii* and *A. frenatus* (Pomcentridae), in captivity. Jpn. J. Ichthyol. 35: 90–97.

Colleye, O. and E. Parmentier. 2012. Overview on the diversity of sounds produced by clownfishes (Pomacentridae): importance of acoustic signals in their peculiar way of life. PLoS ONE 7(11): e49179.

Colleye, O., B. Frédérich, P. Vandewalle, M. Casadevall and E. Parmentier. 2009. Agonistic sounds in the skunk clownfish *Amphiprion akallopisos*: size-related variation in acoustic features. J. Fish Biol. 75(4): 908–916.

Colleye, O., P. Vandewalle, D. Lanterbecq, D. Lecchini and E. Parmentier. 2011. Interspecific variation of calls in clownfishes: degree of similarity in closely related species. BMC Evol. Biol. 11(1): 365.

Colleye, O., M. Nakamura, B. Frédérich and E. Parmentier. 2012. Further insight into the sound-producing mechanism of clownfishes: what structure is involved in sound radiation? J. Exp. Biol. 215(13): 2192–2202.

Connaughton, M., M. Taylor and M.L. Fine. 2000. Effects of fish size and temperature on weakfish disturbance calls: implications for the mechanism of sound generation. J. Exp. Biol. 203: 1503–1512.

Demski, L.S., J.W. Gerald and A.N. Popper. 1973. Central and peripheral mechanisms of teleost sound production. Am. Zool. 13: 1141–1167.

Emslie, M.J., M. Logan, D.M. Ceccarelli, A.J. Cheal, A.S. Hoey, I. Miller and H.P.A. Sweatman. 2012. Regional-scale variation in the distribution and abundance of farming damselfishes on Australia's Great Barrier Reef. Mar. Biol. 159(6): 1293–1304.

Feher, J., T. Waybright and M.L. Fine. 1998. Comparison of sarcoplasmic reticulum capabilities in toadfish (*Opsanus tau*) sonic muscle and rat fast twitch muscle. J. Muscle Res. Cell Motil. 19(6): 661–674.

Fine, M.L. and E. Parmentier. 2015. Mechanisms of sound production. pp. 77–126. *In*: F. Ladich (ed.). Sound Communication in Fishes. Springer, Wien.

Fish, M.P. and W.H. Mowbray. 1970. Sounds of Western North Atlantic Fishes. A Reference File of Biological Underwater Sounds. The John Hopkins Press, Baltimore.

Frédérich, B., G. Fabri, G. Lepoint, P. Vandewalle and E. Parmentier. 2009. Trophic niches of thirteen damselfishes (Pomacentridae) at the Grand Récif of Toliara, Madagascar. Ichthyol. Res. 56(1): 10–17.

Frédérich, B., D. Olivier, G. Litsios, M.E. Alfaro and E. Parmentier. 2014. Trait decoupling promotes evolutionary diversification of the trophic and acoustic system of damselfishes. Proc. R. Soc. B-Biol. Sci. 281(1789).

Fricke, H.W. 1974. Öko-ethologie des monogamen Anemonefisches *Amphiprion bicinctus* (Freiwasseruntersuchung aus dem Roten Meer). Z. Tierpsychol. 36: 429–512.

Fricke, H.W. 1979. Mating system, resource defense and sex change in the anemonefish *Amphiprion akallopisos*. Z. Tierpsychol. 50: 313–326.

Froukh, T. and M. Kochzius. 2008. Species boundaries and evolutionary lineages in the blue green damselfishes *Chromis viridis* and *Chromis atripectoralis* (Pomacentridae). J. Fish Biol. 72(2): 451–457.

Gould, S.J. and E.S. Vrba. 1982. Exaptation-a missing term in the science of form. Paleobiology 8: 4–15.

Graham, R. 1992. Sounds fishy. Australia's Geographic Magazine 14: 76–83.

Ha, S.J. 1973. Aspects of sound communication in the damselfish *Eupomacentrus partitus*. PhD dissertation, University of Miami.

Hattori, A. 1991. Socially controlled growth and size-dependent sex change in the anemonefish *Amphiprion frenatus* in Okinawa, Japan. Jpn. J. Ichthyol. 38: 165–177.

Hattori, A. 1995. Coexistence of two anemonefish, *Amphiprion clarkii* and *A. perideraion*, which utilize the same host sea anemone. Environ. Biol. Fishes 42: 345–353.

Hoey, A.S. and D.R. Bellwood. 2010. Damselfish territories as a refuge for macroalgae on coral reefs. Coral Reefs 29(1): 107–118.

Lagardère, J.P., G. Fonteneau, A. Mariani and P. Morinière. 2003. Les émissions sonores du poisson-clown mouffette *Amphiprion akallopisos*, Bleeker 1853 (Pomacentridae), enregistrées dans l'aquarium de la Rochelle. Ann. Soc. Sci. Nat. Charente-Marit. 9: 281–288.

Larson, G., P.A. Stephens, J.J. Tehrani and R.H. Layton. 2013. Exapting exaptation. Trends Ecol. Evol. 28(9): 497–498.

Limbaugh, C. 1964. Notes on the life history of two Californian pomacentrids: garibaldis, *Hypsypops rubicunda* (Girard), and blacksmiths, *Chromis punctipinnis* (Cooper). Pac. Sci. 18: 41–50.

Litsios, G., C. Sims, R. Wuest, P. Pearman, N. Zimmermann and N. Salamin. 2012. Mutualism with sea anemones triggered the adaptive radiation of clownfishes. BMC Evol. Biol. 12(1): 212.

Lobel, L.K. and P.S. Lobel. 2013. Junkyard damselfishes: spawning behavior and nest site selection. Paper presented at the 2013 AAUS/ESDP Curaçao Joint International Scientific Diving Symposium, Curaçao, Dauphin Island, AL.

Lobel, P.S. and D.A. Mann. 1995. Spawning sounds of the damselfish, *Dascyllus albisella* (Pomacentridae), and relationship to male size. Bioacoustics 6: 187–198.

Lobel, P.S. and L.M. Kerr. 1999. Courtship sounds of the Pacific Damselfish, *Abudefduf sordidus* (Pomacentridae). Biol. Bull. 197: 242–244.

Lobel, P.S., I.M. Kaatz and A.N. Rice. 2010. Acoustical behavior of reef fishes. pp. 307–387. *In*: K.S. Cole (ed.). Reproduction and Sexuality in Marine Fishes: Patterns and Processes. University of California Press, Berkeley.

Luh, H.K. and H.K. Mok. 1986. Sound production in the domino damselfish, *Dascyllus trimaculatus* (Pomacentridae) under laboratory conditions. Jpn. J. Ichthyol. 33: 70–74.

Mann, D. and P.S. Lobel. 1995. Passive acoustic detection of sounds produced by the damselfish, *Dascyllus albisella* (Pomacentridae). Bioacoustics 6: 199–213.

Mann, D. and P.S. Lobel. 1997. Propagation of damselfish (Pomacentridae) courtship sounds. J. Acoust. Soc. Am. 101: 3783–3791.

Mann, D. and P.S. Lobel. 1998. Acoustic behaviour of the damselfish *Dascyllus albisella*: behavioural and geographic variation. Environ. Biol. Fishes 51: 421–428.

Maruska, K.P., K.S. Boyle, L.R. Dewan and T.C. Tricas. 2007. Sound production and spectral hearing sensitivity in the Hawaiian sergeant damselfish, *Abudefduf abdominalis*. J. Exp. Biol. 210(22): 3990–4004.

McCafferty, S., E. Bermingham, B. Quenouille, S. Planes, G. Hoelzer and K. Asoh. 2002. Historical biogeography and molecular systematics of the Indo-Pacific genus *Dascyllus* (Teleostei: Pomacentridae). Mol. Ecol. 11: 1377–1392.

Moyer, J.T. 1975. Reproductive behavior of the damselfish Pomacentrus nagasakiensis at Miyake-jima, Japan. Jpn. J. Ichthyol. 22: 151–163.

Myrberg, A.A. and J.Y. Spires. 1972. Sound discrimination by the bicolour damselfish, *Eupomacentrus partitus*. J. Exp. Biol. 57: 727–735.

Myrberg, A.A. and J.Y. Spires. 1980. Hearing in damselfishes: an analysis of signal detection among closely related species. J. Comp. Physiol. 140(2): 135–144.

Myrberg, A., E. Spanier and S. Ha. 1978. Temporal patterning in acoustic communication. pp. 137–179. *In* : E.S. Reese and F.J. Lighter (eds.). Contrasts in Behaviour. Wiley, New York.

Myrberg, A.A.J. 1972. Ethology of the bicolour damselfish *Eupomacentrus partitus* (Pisces: Pomacentridae): a comparative analysis of laboratory and field behaviour. Anim. Behav. 5: 197–283.

Myrberg, A.A.J. 1997. Underwater sound: its relevance to behavioural functions among fishes and marine mammals. Mar. Freshw. Behav. Phy. 29: 3–21.

Myrberg, A.A.J., D.B. Bradley and A.R. Emery. 1967. Field observations on reproduction of the damselfish, *Chromis multilineata* (Pomacentridae), with additional notes on general behavior. Copeia 1967: 819–827.

Myrberg, A.A.J., M. Mohler and J. Catala. 1986. Sound production by males of a coral reef fish (*Pomacentrus partitus*): its significance to females. Anim. Behav. 34: 913–923.

Myrberg, A.A.J., S.J. Ha and M.J. Shamblott. 1993. The sounds of bicolor damselfish (*Pomacentrus partitus*): predictors of body size and a spectral basis for individual recognition and assessment. J. Acoust. Soc. Am. 94: 3067–3070.

Ochi, H. 1985. Termination of parental care due to small clutch size in the temperate damselfish, *Chromis notata*. Environ. Biol. Fishes 12(2): 155–160.

Oliver, S. and P. Lobel. 2013. Direct mate choice for simultaneous acoustic and visual courtship displays in the damselfish, *Dascyllus albisella* (Pomacentridae). Environ. Biol. Fishes 96(4): 447–457.

Olivier, D., B. Frédérich, M. Spanopoulos-Zarco, E. Balart and E. Parmentier. 2014. The cerato-mandibular ligament: a key functional trait for grazing in damselfishes (Pomacentridae). Front. Zool. 11(1): 63.

Osse, J.W.M. 1969. Functional morphology of the head of the perch (*Perca fluviatilis* L.): an electromyographic study. Neth. J. Zool. 19: 289–392.

Papes, S. and F. Ladich. 2011. Effects of temperature on sound production and auditory abilities in the striped raphael catfish Platydoras armatulus (Family Doradidae). PLoS ONE 6: 1–10.

Parmentier, E., J.-P. Lagardere, P. Vandewalle and M.L. Fine. 2005. Geographical variation in sound production in the anemonefish *Amphiprion akallopisos*. Proc. R. Soc. B-Biol. Sci. 272: 1697–1703.

Parmentier, E., P. Vandewalle, B. Frédérich and M.L. Fine. 2006. Sound production in two species of damselfishes (Pomacentridae): *Plectroglyphidodon lacrymatus* and *Dascyllus aruanus*. J. Fish Biol. 68: 1–13.

Parmentier, E., O. Colleye, M. Fine, B. Frédérich, P. Vandewalle and A. Herrel. 2007. Sound production in the clownfish *Amphiprion clarkii*. Science 316: 1006.

Parmentier, E., O. Colleye and D.A. Mann. 2009a. Hearing ability in three clownfish species. J. Exp. Biol. 212: 2023–2026.

Parmentier, E., D. Lecchini, B. Frédérich, C. Brie and D. Mann. 2009b. Sound production in four damselfish (Dascyllus) species: phyletic relationships? Biol. J. Linnean Soc. 97(4): 928–940.

Parmentier, E., L. Kéver, M. Casadevall and D. Lecchini. 2010. Diversity and complexity in the acoustic behaviour of *Dacyllus flavicaudus* (Pomacentridae). Mar. Biol. 157(10): 2317–2327.

Picciulin, M., M. Constantini, A.D. Hawkins and E.A. Ferrero. 2001. Sound emission of the Mediterranean Damselfish *Chromis chromis* (Pomacentridae). Bioacoustics 12: 236–237.

Picciulin, M., L. Verginella, M. Spoto and E.A. Ferrero. 2004. Colonial nesting and the importance of the brood size in male parasitic reproduction of the Mediterranean damselfish *Chromis chromis* (Pisces: Pomacentridae). Environ. Biol. Fishes 70(1): 23–30.

Pinnegar, J.K., N.V.C. Polunin, J.J. Videler and J.J. de Wiljes. 2007. Daily carbon, nitrogen and phosphorus budgets for the Mediterranean planktivorous damselfish *Chromis chromis*. J. Exp. Mar. Biol. Ecol. 352(2): 378–391.

Rice, A.N. and P.S. Lobel. 2003. The pharyngeal jaw apparatus of the Cichlidae and Pomacentridae: function in feeding and sound production. Rev. Fish Biol. Fish. 13(4): 433–444.

Russell, B.C. 1971. Underwater observations on the reproductive activity of the demoiselle *Chromis dispilus* (Pisces: Pomacentridae). Mar. Biol. 10(1): 22–29.

Santiago, J.A. and J.J. Castro. 1997. Acoustic behaviour of *Abudefduf luridus*. J. Fish Biol. 51(5): 952–959.

Santini, S. and G. Polacco. 2006. Finding Nemo: molecular phylogeny and evolution of the unusual life style of anemonefish. Gene 385: 19–27.

Schneider, H. 1964. Bioakustische Untersuchungen an Anemonenfischen der Gattung *Amphiprion* (Pisces). Z. Morph. Okol. Tiere 53: 453–474.

Spanier, E. 1975. Sound recognition by damselfishes of the genus, *Eupomacentrus*, from Florida waters. Ph.D. Thesis, University of Miami, 145 p.

Spanier, E. 1979. Aspects of species recognition by sound in four species of damselfishes, genus *Eupomacentrus* (Pisces: Pomacentridae). Z. Tierpsychol. 51: 301–316.

Stiassny, M.L.J. 1981. The phyletic status of the family Cichlidae (pisces, perciformes): a comparative anatomical investigation. Neth. J. Zool. 31: 275–314.

Swerdloff, S.N. 1970. Behavioral observations on Eniwetok damselfishes (Pomacentridae: *Chromis*) with special reference to the spawning of *Chromis caeruleus*. Copeia 1970: 371–374.

Takemura, A. 1983. Studies on the Underwater Sound - VIII. Acoustical behavior of clownfishes (*Amphiprion* spp.). Bulletin of the Faculty of Fisheries. Nagasaki University 54: 21–27.

Tricas, T. and K. Boyle. 2014. Acoustic behaviors in Hawaiian coral reef fish communities. Mar. Ecol. Prog. Ser. 511: 1–16.

Van Wassenbergh, S., A. Herrel, D. Adriaens and P. Aerts. 2005. A test of mouth-opening and hyoid-depression mechanisms during prey capture in a catfish using high-speed cineradiography. J. Exp. Biol. 208: 4627–4639.

Verwey, J. 1930. Coral reef studies. The symbiosis between damselfishes and sea anemones in Batavia Bay. Treubia 12: 305–355.

Hearing in Damselfishes

Orphal Colleye,* Eric Parmentier[a] and *Loïc Kéver*[b]

Background

The underwater environment is full of biotic and abiotic sounds, which can have natural or anthropogenic sources (Slabbekoorn et al. 2010, Ladich and Schulz-Mirbach 2013). Animal sound sources are known to be highly diverse since the ability to produce sounds for communication purposes has been recorded in many species of marine mammals, fishes and in different invertebrates (Fish 1964, Steinberg et al. 1965, Cato 1993, Popper et al. 2001, Ladich et al. 2006). In such a context, the ability to detect, discriminate and identify surrounding sounds is crucial for accomplishing different behaviors in which acoustic communication is involved such as reproduction, courtship, territorial defense, predator avoidance or prey detection (Fay and Popper 2000).

Some of the earliest discussions regarding the hearing abilities in fishes were reported by Pliny the Elder and dated from the first century CE. Yet, it was only 2,000 years later that the first official anatomical description of fish ears was reported by Weber (1820), and then by Retzius (1881) who provided the first anatomical comparison of fish (and all vertebrate) ears. In the beginning of the 20th century, investigators clearly demonstrated that fishes were able to detect sounds (Parker 1903). A few years later, Karl von Frisch and his students (von Frisch 1923, von Frisch and Dijkgraaf 1935) carried out several studies highlighting that fishes use their ears for hearing and provided the first quantitative measures of hearing sensitivity and signal discrimination in fishes. Currently, there is an increasing interest in fish hearing with

Laboratoire de Morphologie Fonctionnelle et Evolutive, AFFISH - Research Center, University of Liège, Quartier Agora, Allée du six Août 15, Bât. B6C, 4000 Liège (Sart Tilman), Belgium.
[a] Email: E.Parmentier@ulg.ac.be
[b] Email: Loic.Kever@ulg.ac.be
* Corresponding author: O.Colleye@ulg.ac.be

biologists giving more attention to the hearing abilities of teleost fishes (Fish 1964, Steinberg et al. 1965, Cato 1993, Popper et al. 2005, Ladich et al. 2006). Besides informing scientists about the hearing abilities and inner ear morphology in fishes, these studies have also given further insight into the different mechanisms of hearing and the evolution of fish inner ears (Popper and Fay 1999, Popper et al. 2005, Higgs and Radford 2013).

Given that there already exist extensive reviews dealing with fish hearing (e.g., Fay and Megela Simmons 1999, Bass and Ladich 2008, Webb ct al. 2008, Ladich and Fay 2013), we will just provide the readers with some generalities about this topic. Instead, the emphasis will be placed on what is known about hearing abilities in pomacentrids, which are well known to be sound producers (see Chapter X).

The Inner Ear Structures in Fishes

Basically, the fish inner ear consists of three semicircular canals, three otolithic endorgans (namely the saccule, lagena and utricle), and, in some species, a relatively small macula (or papilla) neglecta (Popper and Fay 1999, Ladich and Popper 2001, Popper et al. 2005; see Fig. 1). Each of the otolithic endorgans and the semicircular canals have sensory epithelia that contain sensory hair cells similar to those found in other vertebrates (Coffin et al. 2004, Manley and Ladher 2008). These sensory hair cells correspond to the transducing elements of the ear by converting mechanical energy into a signal that can stimulate the nervous system (Popper and Lu 2000).

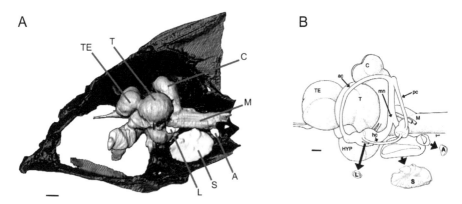

Fig. 1. The inner ear morphology of Pomacentridae. (A) Left lateral view of a three-dimensional reconstruction of the brain, the right half of the neurocranium, and the right otoliths of *Chromis chromis* based on μCT-scans. (B) Left lateral view of the inner ear and brain of the Hawaiian sergeant damselfish, *Abudefduf abdominalis*. The position of the sensory macula is outlined within each otolithic endorgan, while the removed left otoliths are illustrated below (large arrows). Dashed lines represent the location of the crista ampullaris of each semicircular canal. The dotted line represents the position of the macula neglecta (mn). Slightly modified from Maruska et al. (2007).

A, asteriscus otolith of lagena; ac, anterior canal; C, cerebellum; hc, horizontal canal; HYP, hypothalamus; L, lapillus otolith of utricle; M, medulla; pc, posterior canal; S, sagitta otolith of saccule; T, tectum; TE, telencephalon. Scale bar, 1 mm.

Each otolithic endorgan sensory epithelium contains thousands, tens of thousands or even hundreds of thousands of sensory hair cells depending on the species and the size of fish (Lombarte and Popper 1994). These sensory hair cells are grouped in an epithelium called macula (Webb et al. 2008). The sensory epithelium lies close to a dense calcareous otolith and is separated from it by a thin otolithic membrane that mechanically couples them together (Popper et al. 2003, 2005). Hair cell stimulation results from the relative motion between the sensory epithelium and the otolith (Popper et al. 2005).

The semicircular canals serve a vestibular function to encode angular accelerations (Popper et al. 2003). Although the precise role of each otolithic endorgan is not known, they are likely to have two different functions. First, they are involved in determining head position using gravity, as observed in terrestrial vertebrates (Platt 1983). Secondly, the saccule, lagena and utricle are involved in sound detection (Popper et al. 2003). However, the relative contributions of each to sound detection may vary in different species. For example, the connection between the swim bladder and saccule in otophysans implies that this otolithic endorgan may be the primary detector of sounds in these fishes (Ladich and Wysocki 2003, Rogers and Zeddies 2008), whereas the utricle appears to be the major sound detection endorgan in clupeiform fishes, at least for higher frequency sounds (Mann et al. 2001, Higgs et al. 2004, Plachta et al. 2004).

How do Fish Hear?

The ancestral and most common mode of hearing in fishes involves sensitivity to acoustic particle motion via direct inertial stimulation of the otolithic endorgan(s) (Fay and Popper 1974, Popper and Fay 2011). All fishes use an accelerometer-like system for hearing. Indeed, the structure of the otolithic endorgans looks like an accelerometer, with a mass moving in a relative manner to some kind of receptor (Popper and Fay 2011). The otolith is the mass whereas the sensory hair cells are the receptors. The fish body and sensory epithelia have approximately the same density as water, so they move with approximately the same amplitude and phase as the particle motion component of the sound field (Popper and Fay 2011). On the other hand, the otolith moves with a different amplitude and phase than the rest of the body, by being about three times denser. Given that ciliary bundles are connected to the otolith through direct contact or through the otolithic membrane, they undergo a shearing force that induces their deflection (Popper and Fay 2011), which creates a signal that stimulates the auditory system (Popper et al. 2005). Auditory information is transferred from the hair-cell sensory epithelia via primary afferent neurons to several hindbrain auditory nuclei, ascends to the midbrain torus semicircularis, and then to the forebrain processing regions (see McCormick 1999). The hearing sensitivity of fishes using an accelerometer-like ear is good, with some fishes able to detect particle displacement of about 0.1 nm at 100 Hz (Fay 1988, Fay and Edds-Walton 1997). This displacement sensitivity corresponds to about 0 dB re: 1 µPa in the far field (Popper and Fay 2011).

It is obvious that most fishes rely on the detection of particle motion through direct stimulation of the ear (referred to as the "direct path", see Fay and Popper 1974, 1975). However, some fish species are also able to detect sound pressure.[1] These fishes use an "indirect" path of sound stimulation whereby the swim bladder or other gas bubble is set into motion by the pressure component of the sound field since their gas is of different density and compressibility than water (Fay and Popper 1974, 1975). Then, the volume oscillations of the bubble reradiate the signal, which emits a particle motion component that can stimulate the inner ear through inertial stimulation of the otolith endorgans. In some fish species, an intimate mechanical connection (i.e., an otophysic connection) between the swim bladder and inner ear may greatly improve the transmission of the reradiating signal. Numerous examples of these mechanisms of sound detection are reported in Webb et al. (2008).

What do Fishes Listen to?

Although it was clearly shown that hearing is adapted for receiving conspecific communication sounds in many species (e.g., Myrberg and Spires 1972, Zelick et al. 1999), it cannot be denied that the evolution of ears and auditory systems provides fishes with the 'sense' of their environment. Underwater soundscapes probably contain important information about the objects and events in the environment so that the receiving fish may behave appropriately with respect to them (Fay 2009). This overview of the acoustic environment has been called the "auditory scene" (Bregman 1990). In order to make use of this "auditory scene", animals must be able to discriminate between sounds that are and are not of biological relevance (Fay and Popper 2000). Most damselfish species, for example, live in the coral reef environment, which is a noisy place. Reef sounds appear to be used for the orientation and migration during the reef colonization at the end of the pelagic larval phase (see Chapter III). Therefore, they must have the ability to discriminate between the frequency and intensity of sounds, to determine the direction of sound source and to detect signals in the presence of other sounds, which is necessary for their survival and persistence.

What about Hearing in Pomacentrids?

Although the ability to produce sounds in different behavioral contexts has been extensively documented in damselfishes, only few studies have concentrated on sound detection. Basically, Myrberg et al. (1978) was the first to highlight that pomacentrids are able to hear acoustic signals. Using playback experiments to test the responsiveness of different *Stegastes* species to courtships sounds, it was then shown that fishes are able to detect these signals by carrying out a behavior called the signal jump (Myrberg

---

[1] One potentially useful analogy can be made with the 'rock concert effect'. At a rock concert with large speakers, people having a front row seat can feel air particles moving over their body and hear the sound, which corresponds to the near field where there is large particle velocity and sound pressure. On the other hand, people being far away from the stage can only hear the sound and not feel the air particles, which is the far field where particle motion has fallen off dramatically. Thereby, they can no longer feel movement of the air, but they can still hear the sound pressure (Mann 2006).

et al. 1978). Subsequent to this study, several investigations were conducted in the larval and adult individuals of different damselfish species using either a behavioral approach (Myrberg and Spires 1980, Kenyon 1996) or an electrophysiological method called AEP (Auditory Evoked Potential, e.g., Egner and Mann 2005, Wright et al. 2005, 2010, Maruska et al. 2007, Parmentier et al. 2009a). Additionally, Maruska and Tricas (2009) determined the encoding properties of single auditory neurons in the damselfish brain in response to simple tone bursts and playbacks of more complex conspecific sounds. They also examined how hormones, such as the gonadotropin-releasing hormone, can affect central auditory processing in damselfishes (Maruska and Tricas 2011).

Generally speaking, these studies aimed at measuring hearing abilities in order to determine whether acoustic signals might be involved in inter- and intraspecific communication during reproductive and agonistic behaviors, to demonstrate that settling larvae may use the sense of hearing to localize reef sounds or habitat, or to point out the possibility of changes in auditory abilities based on the season or breeding condition. Consequently, there now exist audiograms for different pomacentrid species such as *Stegastes* spp. (Myrberg and Spires 1980, Kenyon 1996), *Abudefduf* spp. (Egner and Mann 2005, Maruska et al. 2007), *Amphiprion* spp. (Parmentier et al. 2009a), *Chromis chromis* (Wysocki et al. 2009), and the larvae of *Pomacentrus nagasakiensis* (Wright et al. 2005) and *P. amboinensis* (Wright et al. 2010). Unfortunately, this dataset does not encompass audiograms of all damselfishes for which sound production has already been recorded (e.g., Parmentier et al. 2009b). For a better understanding of the present chapter and to get more insight into the comparisons among species, the hearing thresholds of additional species (see Fig. 2) have been measured following the previous published AEP protocol (Kenyon et al. 1998). The presentation of sound stimuli and the determination of thresholds followed the detailed description given by Parmentier et al. (2009a).

Considering hearing abilities based on AEP recording methods, it appears that all damselfish species have relatively high sound pressure thresholds (>100 dB re 1 µPa), and low-pass shaped audiograms (Fig. 2). The frequency range over which they are able to detect sounds is between 100 and 2000 Hz, with their best hearing sensitivities varying among species but always being at low frequencies (<400 Hz). The *Abudefduf* audiograms show a region of highest sensitivity between 100 and 400 Hz (Egner and Mann 2005, Maruska et al. 2007). *Chromis* species have their best hearing sensitivity around 200 Hz (Wysocki et al. 2009, this study), whereas the *Dascyllus* and *Amphiprion* species are most sensitive to frequencies below 200 Hz (Parmentier et al. 2009a, this study). Regarding pre-settlement larvae, audiograms for *Pomacentrus nagasakiensis* and *P. amboinensis* have shown their best sensitivity at 100 Hz and 200 Hz, respectively (Wright et al. 2005, 2010). On the other hand, all *Stegastes* species have a frequency of best sensitivity at 500 Hz and a lower sound pressure threshold around 80 dB re 1 µPa (Fig. 2), but these data are based on behavioral measures (Myrberg and Spires 1980).

Audiogram comparisons show differences of several tens of dB in hearing threshold among damselfish species (Fig. 2), but this must be interpreted cautiously. Using different methodologies (i.e., behavioral vs. electrophysiological measures) and experimental conditions can produce dramatically different results (see Ladich

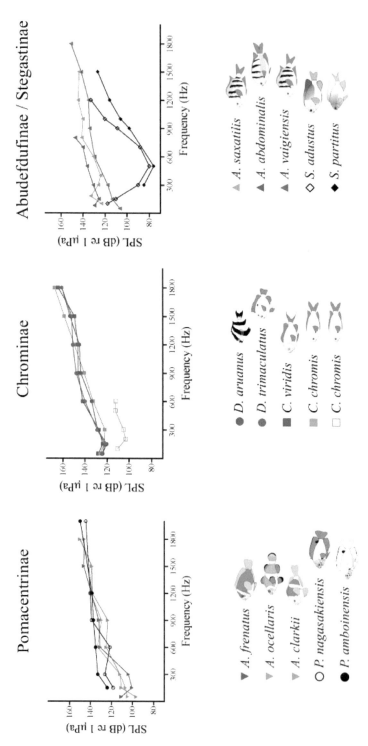

Fig. 2. Comparison of Sound Pressure Level (SPL) audiograms of various pomacentrid species with *Abudefduf vaigiensis* (Red filled triangle), *Dascyllus aruanus* (blue filled circle), *D. trimaculatus* (red filled circle), *Chromis viridis* (blue filled square) and *C. chromis* (green filled square) from this current study; green open square – *C. chromis* (Wysocki et al. 2009); blue inverted triangle – *Amphiprion frenatus*, green inverted triangle – *A. ocellaris*, orange inverted triangle – *A. clarkii* (Parmentier et al. 2009a); open circle – *Pomacentrus nagasakiensis* (Wright et al. 2005); black filled circle – *P. amboinensis* (Wright et al. 2010); green filled triangle – *Abudefduf saxatilis* (Egner and Mann 2005), blue filled triangle – *A. abdominalis* (Maruska et al. 2007); open diamond – *Stegastes adustus* (syn. *Eupomacentrus*, Myrberg and Spires 1980), filled diamond – *S. partitus* (syn. *Pomacentrus*, Kenyon 1996).

and Fay 2013 for review). By comparing behavioral and AEP thresholds in several fish species, Ladich and Fay (2013) emphasized that AEP measures tend to produce detection thresholds that are higher than behavioral values at low frequencies (<1 kHz), whereas they produce lower thresholds than the behavioral ones at higher frequencies. However, this is not a universal relationship for all fish species. Although this tendency was observed for the goldfish *Carassius auratus* and the oyster toadfish *Opsanus tau*, it was not the case for other fish species such as the common carp *Cyprinus carpio*, the Red sea bream *Pagrus major* and the oscar *Astronotus ocellatus* (Ladich and Fay 2013). Overall, these differences between behavioral and AEP thresholds are likely because the fish sensory system is more sensitive than the electrodes used to measure it. Physiological techniques provide less sensitive results than behavioral testing, but they are more appropriate for working with larval fish. Other reasons can explain such differences. First, it is quite difficult with the AEP technique to create short tone bursts at lower frequencies with good precision in the frequency domain. Short tone bursts with a greater rapidity of onset result in a greater efficacy at generating AEPs at higher frequencies (Kenyon et al. 1998, Ladich and Fay 2013). Second, the signals used in AEP studies are usually short (about 20 ms in duration) whereas most of the behavioral studies used long duration signals (several seconds) (see Ladich and Fay 2013). For example, detection thresholds in behavioral studies are known to be higher when signal duration decreases in the goldfish (Fay and Coombs 1983) and in the Atlantic cod *Gadus morhua* (Hawkins 1981).

Similarly, differences in detection thresholds were also observed by comparing AEP recordings to single auditory neuron recordings in the brain of the Hawaiian sergeant major damselfish *Abudefduf abdominalis* (Maruska and Tricas 2009). Single neuron recordings are actually more sensitive by 10–20 dB than AEP recordings; this suggests that damselfishes may be more sensitive than the AEP traces indicate (Maruska and Tricas 2009). Moreover, the fact that damselfishes have low frequency audiograms does not mean they are not able to detect high frequencies: there exist populations of single neurons in their auditory brain that are best tuned to different frequencies. There are also filtering mechanisms that sharpen the frequency response along the ascending auditory pathway (Maruska and Tricas 2009).

Analysis of audiograms may also reveal surprising results when comparing hearing thresholds measured with the same recording technique and for the same species (see *C. chromis* in Fig. 2). Such differences in hearing thresholds may reflect how chronic exposure to high noise levels negatively impacts the hearing abilities of fishes living in different environments (Smith et al. 2003, Wysocki and Ladich 2005, Wysocki et al. 2007). Individuals studied by Wysocki et al. (2009) exhibited a better hearing sensitivity but they were collected in the field, whereas our individuals were reared in captivity, which implies the use of noisy equipment (e.g., aerators, air and water pumps and filtration systems). Similar long-term effects of increased ambient noise on hearing loss have already been reported in the goldfish (Gutscher et al. 2011).

Although some of the inter-species differences clearly appear to be related to methodological differences in audiograms, there are still quite large differences between genera that need to be explained (Fig. 2). Thus, it would be interesting to find out the factors behind such differences by carrying out comparative physiological and morphological studies on several pomacentrid species. So far, no swim bladder diverticulae or other peripheral adaptations have been described in pomacentrids. Yet, rostral extensions of the swim bladder towards the otic capsule have already been shown to enhance hearing abilities in other species. By extending the swim bladder towards the cranium, the oscillations of the swim bladder wall may be transmitted to the otolithic endorgans. Such adaptations are known in many teleost taxa (see Braun and Grande 2008), with some of the best known examples found in holocentrids, chaetodontids and cichlids (Coombs and Popper 1979, Webb 1998, Schulz-Mirbach et al. 2012).

Recently, Ladich and Fay (2013) reported that it is controversial whether fishes without an otophysic connection between the swim bladder and ears detect sound pressure or can only respond to acoustic particle motion. In this regard, it is interesting to note that most of those studies dealing with hearing abilities in damselfishes did not provide responses to particle acceleration. Although some experiments to confirm sensitivity to particle acceleration have been carried out by Myrberg and Spires (1980) in the dusky damselfish *Stegastes audustus*, and by Wysocki et al. (2009) who indicated relatively high particle acceleration thresholds in *C. chromis*, the other damselfish species have not been studied in this respect. Thus, auditory thresholds for fish species without an otophysic connection should be measured in terms of both particle motion (i.e., displacement, velocity, or acceleration) and sound pressure in order to determine what acoustic quantity is most appropriate (Ladich and Fay 2013). It is likely that these fishes detect particle acceleration in their natural habitat, but not sound pressure, and audiometric data measured in sound pressure thresholds may not be appropriate (Ladich and Fay 2013). In addition, the test fish is well within the near field during the AEP recordings due to the enclosed acoustic environment, which includes lateral line input, especially for lower frequencies (see Higgs and Radford 2013). Therefore, AEP results at lower frequencies should be interpreted as multimodal responses since both the ear and lateral line are likely involved in the detection of acoustic stimuli (Higgs and Radford 2013).

Despite these caveats, all of the research on hearing in damselfishes has provided interesting results. Besides giving an insight into their auditory thresholds, it has also highlighted different aspects such as the occurrence of changes in hearing abilities during ontogeny, the matching in the frequency domain between sound production and hearing abilities in some species, the influence of fish size on the hearing range over which some of these species can detect sounds, and the possibility of changes in hearing thresholds and frequency tuning for seasonally breeding damselfish species based on their internal hormonal/physiological state. The next parts of the present chapter will be structured according to these different points.

Ontogenetic changes in the auditory sensitivity of damselfishes

Kenyon (1996) provided the first behavioral evidence that auditory thresholds change during ontogeny by recording the hearing abilities in four juvenile size-groups of the bicolor damselfish *Stegastes partitus*. Although the frequency range over which fishes are able to detect sounds was similar (except some small juveniles unable to detect sounds at 1200 and 1500 Hz), the audiograms showed a clear downward progression (i.e., more sensitive) with increasing size, which highlights that hearing abilities rapidly improved with age. A similar trend was observed for three individual juveniles of the cocoa damselfish *S. variabilis* as the auditory sensitivity increased with standard length (Kenyon 1996). The general pattern exhibited by these damselfish species indicates that young juveniles show flat and untuned audiograms, with the appearance of a best frequency and an increasing sensitivity with age (Kenyon 1996). Although the underlying physiological processes of such changes are unclear, they may be due to a variety of factors such as neural maturation, hair cell proliferation and changes in swim bladder acoustics (Kenyon 1996). More recently, Simpson et al. (2004) emphasized the possible role of hearing during the early life stages by noticing that embryos of two clownfish species *Amphiprion ephippium* and *A. rubrocinctus* were able to detect sounds. They observed an increasing heart rate in response to noise exposure. They also demonstrated an increase in auditory sensitivity with age in terms of sound level detection and hearing range even after a few days post-fertilization (Simpson et al. 2005). Moreover, Wright et al. (2005) investigated the auditory abilities of pre- and post-settlement larvae in *P. nagasakiensis* using an electrophysiological method (AEP). Audiograms of the two ontogenetic stages followed similar trends in shape: pre- and post-settlement larvae have similar hearing abilities between 100 and 2,000 Hz, except at 100 and 600 Hz where pre-settlement larvae show thresholds 8 dB higher than those of post-settlement. All these observations suggest that morphological mechanisms of sound detection did not change considerably but the ear may have become more sensitive, and this slight increase in sensitivity may be due to changes in hair cell number increasing with growth (Wright et al. 2005).

Such studies dealing with hearing abilities in coral reef fish larvae are interesting due to their bipartite lifecycle (see Chapter III). Larvae grow up in the pelagic environment during their dispersal pelagic phase before locating a suitable habitat associated with coral reef environment for settling (Leis 1991, Leis and McCormick 2002). The influence of reef sound on fish colonization has become a subject of particular interest in the last few years, since Stobutzki and Bellwood (1998) first hypothesized that fish larvae could use sound as a cue for nocturnal orientation. Field studies clearly demonstrated that settlement stage coral reef fish larvae are able to hear and to localize reef sounds. They conducted experiments showing that fish larvae are significantly more attracted by noisy light traps than silent ones (Leis et al. 2003, Simpson et al. 2004, Tolimieri et al. 2004, Leis and Lockett 2005). However, light traps do not correspond to natural conditions given that fishes usually colonize the reef in the nighttime without the use of any light. Thereby, double choice experiments (i.e., choice chamber experiments) have been used in order to avoid the bias related to this method (see Tolimieri et al. 2004, Simpson et al. 2010, Holles et al. 2013). Recently, the importance of sound in habitat localization has also been shown at Lizard Island (Australia) using juveniles: they are significantly more attracted to patch reefs with lagoon or fringing reef sound broadcast than to patch reefs with no playback sound (Radford et al. 2011a). Some differences were observed at the family level with pomacentrid species being preferentially found on patch reefs broadcasting fringing reef sounds compared to other treatments (Radford et al. 2011a). Although it has been clearly demonstrated that acoustic stimuli are involved in attracting fishes to the reef, the distance over which fish larvae could detect sounds is still debated (Leis et al. 2011). To estimate the distance of detection of reef sounds, it is necessary to know the hearing abilities of the fish, the reef sound intensity and spectral composition, and the propagation loss with distance. The fact is current models detailing sound propagation do not perfectly reflect reality since propagation loss is complex and site and/or time dependent (Cato 1978). The current estimations are about 500 m to 50 km, based on fish hearing abilities and ambient sound level (Mann et al. 2007, Wright et al. 2010, Radford et al. 2011b). Recently, Wright et al. (2005, 2010) provided AEP data supporting the hypothesis that the *Pomacentrus* larvae can respond physiologically to such auditory cues. Therefore, this is all about matching between auditory threshold and sound source level.

Coupling between sound production and hearing ability in damselfishes

Damselfishes constitute an excellent model to carry out comparative studies on sound production and hearing abilities, and to test hypotheses on the function and evolution of acoustic communication in a soniferous group of fishes. Indeed, they produce different context-dependent sounds mainly related to reproductive and territorial behaviors, and are the subject of numerous acoustic studies (see Chapter X). Myrberg et al. (1978) were first to observe that four *Stegastes* species were able to elicit signal jumps in the test males of other species by playing back courtship sounds. However, there was significantly more responsiveness by males to sounds from their own species than to sounds from congeners. They postulated that inconsequential courtships were due to a misidentification, which appeared to be related to the broad and overlapping ranges in frequency spectra and temporal information contained in their sounds. The same overlaps in temporal and spectral characteristics were also observed in four *Dascyllus* species (Parmentier et al. 2009b).

In teleost fishes, temporal features such as pulse duration, rate and number can all have a communicative value during sound production (see Chapter 10). Temporal sound features have been effectively and reliably encoded in auditory neurons in some fishes (Suzuki et al. 2002, Wysocki 2006). More precisely, Maruska and Tricas (2009) showed that single auditory neurons in the brain of the damselfish *A. abdominalis* also encode temporal information from playbacks of natural sounds. According to Myrberg et al. (1978), damselfishes could be capable of distinguishing temporal differences of sounds at 5–10 ms. By analyzing the auditory evoked potentials in response to sounds with varying periods in different species, Wysocki and Ladich (2002, 2003) showed that temporal resolution ability was below 1.5 ms. Further studies are still needed to determine the real abilities of pomacentrids to discriminate different temporal characteristics. However, the wide overlap between species strongly suggests that recognition based on temporal sonic features would require sampling of multiple calls to perform reliable species discrimination (Parmentier et al. 2009b).

Damselfish sounds are percussive and the temporal properties (i.e., interpulse interval, number of pulses per sound) appear to be the most important in species recognition (see Chapter X). Parmentier et al. (2009b) created a graph analogous to a bar code in order to visualize the differences among pomacentrid sounds. Actually, this graph was made on the basis of a succession of several pulses and interpulse intervals (see Fig. 3). Using this model, *D. aruanus* appears to be clearly separated from the three other *Dascyllus* species. Likewise, *Stegastes leucostictus* is clearly different from the three other *Stegastes* species (see Parmentier et al. 2009b, Fig. 3). For example, there exist important variations in the pulse length in *Dascyllus* species, while the most important variations among *Stegastes* species are found at the level of the interpulse interval (Spanier 1979). This observation emphasizes that the calling signal may not have the same kind of evolution in different taxa. As a result, fishes belonging to different taxa can use a similar code resulting from different evolutionary paths (Parmentier et al. 2009b). However, it is important to remind in this context that other sense organs such as vision (colour pattern and swimming movements) and chemoreception (pheromones) might convey additional cues involved in species recognition and partner selection because they occur during courtship.

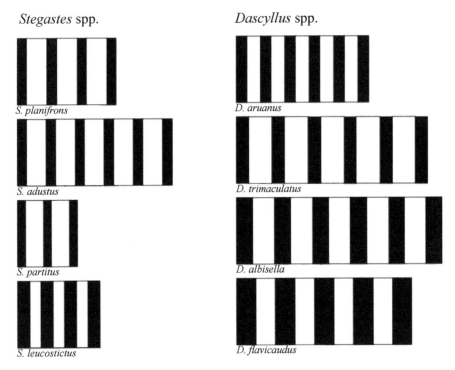

Fig. 3. Bar code corresponding to the typical dip sounds produced by different damselfish species of the *Stegastes* and *Dascyllus* genera. The black bars correspond to a pulse unit and its width to the pulse duration. The white bars correspond to the interpulse interval (IPI) and its width to the interpulse duration. Modified from Parmentier et al. (2009b).

Using this bar code model and the knowledge about the auditory temporal resolution, Parmentier et al. (2009b) postulated that the fish brain might have a type of punch card image corresponding to the bar code. The closer the temporal code to the punch card, the greater the responsiveness will be. Although this model is speculative, it is likely to explain how *Stegastes* species can respond to sounds from closely related species (Myrberg and Spires 1972, Myrberg et al. 1978) and why females differ in how they select a spawning partner. Given that some damselfish species produce courtship sounds hundreds of times a day (Mann and Lobel 1995), it is possible that females could sample many calls before making a decision to spawn with a given male (Parmentier et al. 2009b).

Regarding the spectral characteristics, it is interesting to note that there often exists a relation between the dominant sound frequencies produced and the ones damselfishes are able to detect. For example, the maximum sensitivity at 500 Hz matches well the dominant frequency of the courtship sounds in all *Stegastes* species (Myrberg and Spires 1980). The AEP-pressure audiogram shows best sensitivity at 200 Hz for *C. chromis*, matching the dominant frequency of their sounds (Picciulin et al. 2002). Whatever the type of sounds produced (i.e., courtship, nest preparation or aggressive sounds), the acoustic signals closely match hearing abilities in the frequency domain of *A. abdominalis* (see Fig. 9 in Maruska et al. 2007). For this same species, Maruska and Tricas (2009) demonstrated that the best frequency of midbrain auditory neurons matches the pulse repetition rate associated with natural courtship sounds, which highlights that the auditory midbrain is well-suited for encoding natural sounds produced for communication purposes. Moreover, it is important to take into consideration the reproductive condition and breeding season when examining the overlap between the dominant frequency of sounds and hearing abilities, since sound production abilities and auditory perception can show seasonal and steroid-induced changes (Maruska and Tricas 2011).

By contrast, it is also interesting to note that the best hearing sensitivity of a given species does not necessarily correspond to the dominant frequency of the calls it can produce. *Dascyllus aruanus* and *D. trimaculatus* show their best hearing sensitivity at 150 Hz (Fig. 2), whereas the dominant frequency of their courtship sounds related to signal jumps was around 300–500 Hz (Parmentier et al. 2009b). Likewise, small specimens of three clownfish species were most sensitive to frequencies below 200 Hz, which is lower than the dominant frequency of their own calls. However, those juveniles were shown to have their best sensitivities close to the sound frequencies produced by larger conspecifics (Parmentier et al. 2009a). The more the fish size increases, the more the dominant frequency matches the best auditory frequency (see Fig. 4 in Parmentier et al. 2009a). This observation is of significant importance in clownfishes due to the size-based dominance hierarchy displayed by these fish (see Chapter 12). All this suggests that juveniles might use size-related dominant frequencies as an acoustic cue for identifying larger conspecifics. Future research would need to be carried out with larger clownfishes in order to discover whether this clade presents differences in auditory threshold with increasing size (Parmentier et al. 2009a). If the interest of the larvae and juveniles is in localizing adults on the reef in order to find their habitats, adults may need to detect the presence of smaller fish, which constitute potential new mates or competitors for sex change (Parmentier et al. 2009a). The effect of fish size on auditory sensitivity has also been highlighted in the sergeant major damselfish *A. saxatilis* (Egner and Mann 2005). The most sensitive frequency has always been found to be at around 100 Hz, whatever the fish size.

However, larger fish were found to respond more readily to the higher frequencies at a significant level (Egner and Mann 2005), which is likely to be the frequencies produced by smaller individuals (see Chapter X). Interestingly, such a capacity to discriminate frequency differences has already been shown in damselfish: females of *Stegastes partitus* preferentially respond to lower frequency chirps of a larger male than those from a smaller fish (Myrberg et al. 1986).

Conclusion

As a whole, the present chapter strongly suggests that there is still a need to go further into the understanding of the process of hearing in damselfishes, especially regarding the factors that could explain the interspecific differences observed in hearing thresholds. In this regard, it would be interesting to measure audiograms for numerous species with different body shapes in order to determine whether differences in hearing abilities could be related to the diversity of swim bladder morphology (i.e., size, shape and/or distance to the neurocranium), as it has already been shown in cichlids (e.g., Schulz-Mirbach et al. 2012). Given that all pomacentrids do not produce the same diversity of sounds (e.g., *Dascyllus* spp. and *Amphiprion* spp., see Chapter X), it is likely that relationships exist between the auditory neuronal pathways (i.e., initial encoding of sounds by the primary afferent neurons, see Maruska and Tricas 2009) and the temporal and spectral sound characteristics produced by fishes. Finally, regarding the methodology used, future audiometric studies should determine the fish sensitivity to direct particle motion by measuring audiograms in terms of particle acceleration with specific methods (i.e., shaker table, accelerometer, see Radford et al. 2012). For that matter, it is important to keep in mind that tone stimuli typically used in AEP experiments are probably not the best signal to study fish hearing: single auditory neuron recordings (e.g., Maruska and Tricas 2009) have shown that damselfish are more sensitive to playbacks of natural sounds as compared to the tone bursts.

Acknowledgements

The authors would like to thank the two reviewers (K.P. Maruska and T.C. Tricas) for their helpful comments and constructive criticism of a previous version of the manuscript.

References

Bass, A.H. and F. Ladich. 2008. Vocal-acoustic communication: from neurons to behavior. pp. 253–278. *In*: J.F. Webb, R.R. Fay and A.N. Popper (eds.). Fish Bioacoustics. Springer Science + Business Media, LLC, New York.

Braun, C.B. and T. Grande. 2008. Evolution of peripheral mechanisms for the enhancement of sound reception. pp. 99–144. *In*: J.F. Webb, R.R. Fay and A.N. Popper (eds.). Fish Bioacoustics. Springer Handbook of Auditory Research, New York.

Bregman, A.S. 1990. Auditory Scene Analysis: The Perceptual Organization of Sound. MIT Press, Cambridge, MA.

Cato, D.H. 1978. Features of Ambient Noise in Shallow Waters around Australia Vol. 2. Sonobuoy Working Party, 26–29 September 1978, London.

Cato, D.H. 1993. The biological contribution to the ambient noise in waters near Australia. Acoust. Aust. 20: 76–80.

Coffin, A., M. Kelley, G.A. Manley and A.N. Popper. 2004. Evolution of sensory hair cells. pp. 55–94. *In*: G.A. Manley, A.N. Popper and R.R. Fay (eds.). Evolution of the Vertebrate Auditory System. Springer-Verlag, New York.

Coombs, S. and A.N. Popper. 1979. Hearing differences among Hawaiian squirrelfishes (Family Holocentridae) related to differences in the peripheral auditory system. J. Comp. Physiol. 132: 203–207.

Egner, S.A. and D.A. Mann. 2005. Auditory sensitivity of sergeant major damselfish *Abudefduf saxatilis* from post-settlement juvenile to adult. Mar. Ecol. Prog. Ser. 285: 213–222.

Fay, R.R. 1988. Hearing in Vertebrates, A Psychophysics Databook. Hill-Fay Assoc., Winnetka, Illinois.

Fay, R.R. 2009. Soundscapes and the sense of hearing of fishes. Integr. Zool. 4: 26–32.

Fay, R.R. and A.N. Popper. 1974. Acoustic stimulation of the ear of the goldfish (*Carassius auratus*). J. Exp. Biol. 61: 243–260.

Fay, R.R. and A.N. Popper. 1975. Modes of stimulation of the teleost ear. J. Exp. Biol. 62: 379–387.

Fay, R.R. and S.L. Coombs. 1983. Neural mechanisms in sound detection and temporal summation. Hear. Res. 10: 69–92.

Fay, R.R. and P.L. Edds-Walton. 1997. Directional response properties of saccular afferents of the toadfish, *Opsanus tau*. Hear. Res. 111: 1–21.

Fay, R.R. and A. Megela Simmons. 1999. The sense of hearing in fishes and amphibians. pp. 269–318. *In*: R.R. Fay and A.N. Popper (eds.). Comparative Hearing: Fish and Amphibians. Springer-Verlag, New York.

Fay, R.R. and A.N. Popper. 2000. Evolution of hearing in vertebrates: the inner ears and processing. Hear. Res. 149: 1–10.

Fish, M.P. 1964. Biological sources of sustained ambient sea noise. pp. 175–195. *In*: W.N. Tavolga (ed.). Marine Bio-Acoustics. Pergamon Press Inc., New York.

Gutscher, M., L.E. Wysocki and F. Ladich. 2011. Effects of aquarium and pond noise on hearing sensitivity in an otophysine fish. Bioacoustics 20: 117–136.

Hawkins, A.D. 1981. The hearing abilities of fish. pp. 109–133. *In*: W.N. Tavolga, A.N. Popper and R.R. Fay (eds.). Hearing and Sound Communication in Fishes Springer, New York.

Higgs, D.M. and C.A. Radford. 2013. The contribution of the lateral line to "hearing" in fish. J. Exp. Biol. 216: 1484–1490.

Higgs, D.M., D.T.T. Plachta, A.K. Rollo, M. Singheiser, M.C. Hastings and A.N. Popper. 2004. Development of ultrasound detection in American shad (*Alosa sapidissima*). J. Exp. Biol. 207: 155–163.

Holles, S., S.D. Simpson, A.N. Radford, L. Berten and D. Lecchini. 2013. Boat noise disrupts orientation behaviour in coral reef fish. Mar. Ecol. Prog. Ser. 485: 295–300.

Kenyon, T.N. 1996. Ontogenetic changes in the auditory sensitivity of damselfishes (Pomacentridae). J. Comp. Physiol. A 179: 553–561.

Kenyon, T.N., F. Ladich and H.Y. Yan. 1998. A comparative study of hearing ability in fishes: the auditory brainstem response approach. J. Comp. Physiol. A 182: 307–318.

Ladich, F. and A.N. Popper. 2001. Comparison of the inner ear ultrastructure between teleost fishes using different channels for communication. Hear. Res. 154: 62–72.

Ladich, F. and L.E. Wysocki. 2003. How does tripus extirpation affect auditory sensitivity in goldfish? Hear. Res. 182: 119–129.

Ladich, F. and T. Schulz-Mirbach. 2013. Hearing in cichlid fishes under noise conditions. PLoS one 8: 1–7.

Ladich, F. and R.R. Fay. 2013. Auditory evoked potential audiometry in fish. Rev. Fish. Biol. Fisheries 23: 317–364.

Ladich, F., S.P. Collin, P. Moller and B.G. Kapoor. 2006. Communication in Fishes, Volume 1. Science Publishers, Enfield.

Leis, J.M. 1991. The pelagic stage of reef fishes: the larval biology of coral reef fishes. pp. 183–230. *In*: P. Sale (ed.). The Ecology of Fishes on Coral Reefs. Academic Press, San Diego.

Leis, J.M. and M.I. McCormick. 2002. The biology, behavior, and ecology of the pelagic, larval stage of coral reef fishes. pp. 171–200. *In*: P.F. Sale (ed.). Coral Reef Fishes: Dynamics and Diversity in a Complex Ecosystem. Academic Press, San Diego.

Leis, J.M. and M.M. Lockett. 2005. Localization of reef sounds by settlement-stage larvae of coral-reef fishes (Pomacentridae). Bull. Mar. Sci. 76: 715–724.

Leis, J.M., B.M. Carson-Ewart, A.C. Hay and D.H. Cato. 2003. Coral-reef sounds enable nocturnal navigation by some reef-fish larvae in some places and at some times. J. Fish Biol. 63: 1–14.

Leis, J.M., U. Siebeck and D.L. Dixson. 2011. How Nemo finds home: the neuroecology of dispersal and of population connectivity in larvae of marine fishes. Integr. Comp. Biol. 51: 826–843.

Lombarte, A. and A.N. Popper. 1994. Quantitative analyses of postembryonic hair cell addition in the otolithic endorgans of the inner ear of the European hake *Merluccius merluccius* (Gadiformes, Teleostei). J. Comp. Neurol. 345: 419–428.

Manley, G.A. and R. Ladher. 2008. Phylogeny and evolution of ciliated mechanoreceptor cells. pp. 1–34. *In*: A.I. Bausbaum, A. Kaneko, G.M. Shepherd, G.M. Westheimer, P. Dallos and D. Oertel (eds.). The Senses: A Comprehensive Reference. Audition Vol. 3. Academic Press, San Diego.

Mann, D.A. 2006. Propagation of fish sounds. pp. 107–120. *In*: F. Ladich, S.P. Collin, P. Moller and B.G. Kapoor (eds.). Communication in Fishes, Vol. 1. Science Publishers, Enfield.

Mann, D.A. and P.S. Lobel. 1995. Passive acoustic detection of sounds produced by the damselfish *Dascyllus albisella* (Pomacentridae). Bioacoustics 6: 199–213.

Mann, D.A., D.M. Higgs, W.N. Tavolga, M.J. Souza and A.N. Popper. 2001. Ultrasound detection by clupeiform fishes. J. Acoust. Soc. Am. 109: 3048–3054.

Mann, D.A., B.M. Casper, K.S. Boyle and T.C. Tricas. 2007. On the attraction of larval fishes to reef sounds. Mar. Ecol. Progr. Ser. 338: 307–310.

Maruska, K.P. and T.C. Tricas. 2009. Encoding properties of auditory neurons in the brain of a soniferous damselfish: response to simple tones and complex conspecific signals. J. Comp. Physiol. A 195: 1071–1088.

Maruska, K.P. and T.C. Tricas. 2011. Gonadotropin-releasing hormone (GnRH) modulates auditory processing in the fish brain. Horm. Behav. 59: 451–464.

Maruska, K.P., K.S. Boyle, L.R. Dewan and T.C. Tricas. 2007. Sound production and spectral hearing sensitivity in the Hawaiian sergeant damselfish, *Abudefduf abdominalis*. J. Exp. Biol. 210: 3990–4004.

McCormick, C.A. 1999. Anatomy of the central auditory pathways of fish and amphibians. pp. 155–217. *In*: R.R. Fay and A.N. Popper (eds.). Comparative Hearing: Fish and Amphibians. Springer, New York.

Myrberg, A.A. 1978. Underwater sound—its effect on the behaviour of sharks. pp. 391–417. *In*: E.S. Hodgson and R.F. Mathewson (eds.). Sensory Biology of Sharks, Skates and Rays. Government Printing Office, Washington D.C.

Myrberg, A.A. and J.Y. Spires. 1972. Sound discrimination by the bicolour damselfish, *Eupomacentrus partitus*. J. Exp. Biol. 57: 727–735.

Myrberg, A.A. and J.Y. Spires. 1980. Hearing in damselfishes: an analysis of signal detection among closely related species. J. Comp. Physiol. 140: 135–144.

Myrberg, A.A., E. Spanier and S.J. Ha. 1978. Temporal patterning in acoustic communication. pp. 137–179. *In*: E.S. Reese and F.J. Lighter (eds.). Contrasts in Behaviour. John Wiley & Sons, New York.

Myrberg, A.A., M. Mohler and J. Catala. 1986. Sound production by males of coral reef fish (*Pomacentrus partitus*): its significance to females. Anim. Behav. 34: 913–923.

Parker, G.H. 1903. The sense of hearing in fishes. Am. Nat. 37: 185–203.

Parmentier, E., O. Colleye and D.A. Mann. 2009a. Hearing ability in three clownfish species. J. Exp. Biol. 212: 2023–2026.

Parmentier, E., D. Lecchini, B. Frédérich, C. Brié and D.A. Mann. 2009b. Sound production in four damselfish (*Dascyllus*) species: phyletic relationships? Biol. J. Linnean Soc. 97: 928–940.

Picciulin, M., M. Costantini, A.D. Hawkins and E.A. Ferrero. 2002. Sound emissions of the Mediterranean damselfish *Chromis chromis* (Pomacentridae). Bioacoustics 12: 236–238.

Plachta, D.T.T., J. Song, M.B. Halvorsen and A.N. Popper. 2004. Neuronal encoding of ultrasonic sound by a fish. J. Neurophysiol. 91: 2590–2597.

Platt, C. 1983. The peripheral vestibular system in fishes. pp. 89–124. *In*: R.G. Northcutt and R.E. Davis (eds.). Fish Neurobiology. University of Michigan Press, Michigan.

Popper, A.N. and R.R. Fay. 1999. The auditory periphery in fishes. pp. 43–100. *In*: R.R. Fay and A.N. Popper (eds.). Comparative Hearing: Fish and Amphibians. Springer-Verlag, New York.

Popper, A.N. and Z. Lu. 2000. Structure-function relationships in fish otolith organs. Fish. Res. 46: 15–25.

Popper, A.N. and R.R. Fay. 2011. Rethinking sound detection by fishes. Hear. Res. 273: 25–36.

Popper, A.N., M. Salmon and K.W. Horch. 2001. Acoustic detection and communication by decapod crustaceans. J. Comp. Physiol. A 187: 83–89.

Popper, A.N., R.R. Fay, C. Platt and O. Sand. 2003. Sound detection mechanisms and capabilities of teleost fishes. pp. 3–38. *In*: S.P. Collin and N.J. Marshall (eds.). Sensory Processing in Aquatic Environments. Springer-Verlag, New York.

Popper, A.N., J.U. Ramcharitar and S.E. Campana. 2005. Why otoliths? Insights from the inner ear physiology and fisheries biology. Mar. Freshw. Res. 56: 497–504.

Radford, C.A., J.A. Stanley, S.D. Simpson and A.G. Jeffs. 2011a. Juvenile coral reef fish use sound to locate habitats. Coral Reefs 30: 295–305.

Radford, C.A., C.T. Tindle, J.C. Montgomery and A.G. Jeffs. 2011b. Modelling a reef as an extended sound source increases the predicted range at which reef noise may be heard by fish larvae. Mar. Ecol. Progr. Ser. 438: 167–174.

Radford, C.A., J.C. Montgomery, P. Caiger and D.M. Higgs. 2012. Pressure and particle motion detection thresholds in fish: a re-examination of salient auditory cues in teleosts. J. Exp. Biol. 215: 3429–3435.

Retzius, G. 1881. Das Gehörorgan der Wirbelthiere, Vol. 1. Samson and Wallin, Stockholm.

Rogers, P.H. and D.G. Zeddies. 2008. Multiple mechanisms for directional hearing in fish. pp. 233–252. In: J.F. Webb, A.N. Popper and R.R. Fay (eds.). Fish Bioacoustics, Springer, New York.

Schulz-Mirbach, T., B. Metscher and F. Ladich. 2012. Relationship between swim bladder morphology and hearing abilities—a case study on Asian and African Cichlids. PLoS ONE 7: e42292.

Simpson, S.D., M.G. Meekan, R.D. McCauley and A. Jeffs. 2004. Attraction of settlement-stage coral reef fishes to reef noise. Mar. Ecol. Prog. Ser. 276: 263–268.

Simpson, S.D., H.Y. Yan, M.L. Wittenrich and M.G. Meekan. 2005. Response of embryonic coral reef fishes (Pomacentridae: *Amphiprion* spp.) to noise. Mar. Ecol. Prog. Ser. 287: 201–208.

Simpson, S.D., M.G. Meekan, N.J. Larsen, R.D. McCauley and A. Jeffs. 2010. Behavioral plasticity in larval reef fish: orientation is influenced by recent acoustic experiences. Behav. Ecol. 21: 1098–1105.

Slabbekoorn, H., N. Bouton, I. van Opzeeland, A. Coers, C. ten Cate and A.N. Popper. 2010. A noisy spring: the impact of globally rising underwater sound levels on fish. Trends Ecol. Evol. 25: 419–427.

Smith, M.E., A.S. Kane and A.N. Popper. 2003. Noise-induced stress response and hearing loss in goldfish. J. Exp. Biol. 207: 427–435.

Spanier, E. 1979. Aspects of species recognition by sound in four species of damselfishes, genus *Eupomacentrus* (Pisces: Pomacentridae). Z. Tierpsychol. 51: 301–316.

Steinberg, J., W. Cummings, B. Brahy and J.Y. Spires. 1965. Further bioacoustics studies off west coast of Bihimi area, Bahamas. Bull. Mar. Sci. 15: 942–963.

Stobutzki, I.C. and D.R. Bellwood. 1998. Nocturnal orientation to reefs by late pelagic stage coral reef fishes. Coral Reefs 17: 103–110.

Suzuki, A., J. Koslozki and J.D. Crawford. 2002. Temporal encoding for auditory computation: physiology of primary afferent neurons in sound-producing fish. J. Neurosci. 22: 6290–6301.

Tolimieri, N., O. Haine, A. Jeffs, R.D. McCauley, J.C. Montgomery. 2004. Directional orientation of pomacentrid larvae to ambient reef noise. Coral Reefs 23: 184–191.

von Frisch, K. 1923. Ein Zwergwels der kommt, wenn man ihm pfeift. Biol. Zentralbl. Leipzig 43: 439–446.

von Frisch, K. and S. Dijkgraaf. 1935. Können Fische die Schallrichtung wahrnehmen? Z. Vergl. Physiol. 22: 641–655.

Webb, J.F. 1998. Laterophysic connection: a unique link between the swim bladder and the lateral-line system in *Chaetodon* (Perciformes: Chaetodontidae). Copeia 4: 1032–1036.

Webb, J.F., R.R. Fay and A.N. Popper. 2008. Fish Bioacoustics. Springer Science + Business Media, LLC, New York.

Weber, E.H. 1820. De Aure et Auditu Hominis et Animalium. Pars 1. De Aure Animalium Aquatilium, Gerhard Fleischer, Leipzig.

Wright, K.J., D.M. Higgs, A.J. Belanger and J.M. Leis. 2005. Auditory and olfactory abilities of pre-settlement larvae and post-settlement juveniles of a coral reef damselfish (Pisces: Pomacentridae). Mar. Biol. 147: 1425–1434.

Wright, K.J., D.M. Higgs, D.H. Cato and J.M. Leis. 2010. Auditory sensitivity in settlement-stage larvae of coral reef fishes. Coral Reefs 29: 235–243.

Wysocki, L.E. 2006. Detection of communication sounds. pp. 177–206. In: F. Ladich, S.P. Collin, P. Moller and B.G. Kapoor (eds.). Fish Communication. Science Publishers, Enfield.

Wysocki, L.E. and F. Ladich. 2002. Can fishes resolve temporal characteristics of sounds? New insights using auditory brainstem responses. Hear. Res. 169: 36–46.

Wysocki, L.E. and F. Ladich. 2003. The representation of conspecific sounds in the auditory brainstem of teleost fishes. J. Exp. Biol. 206: 2229–2240.

Wysocki, L.E. and F. Ladich. 2005. Hearing in fishes under noise conditions. J. Assoc. Res. Otolaryngol. (JARO) 6: 28–36.

Wysocki, L.E., J.W. Davidson, M.E. Smith, A.S. Frankel, W.T. Ellison, P.M. Mazik, A.N. Popper and J. Bebak. 2007. Effects of aquaculture production noise on hearing, growth, and disease resistance of rainbow trout *Oncorhynchus mykiss*. Aquaculture 272: 687–697.

Wysocki, L.E., A. Codarin, F. Ladichand and M. Picciulin. 2009. Sound pressure and particle acceleration audiograms in three marine fish species from the Adriatic Sea. J. Acoust. Soc. Am. 126: 2100–2107.

Zelick, R., D.A. Mann and A.N. Popper. 1999. Acoustic communication in fishes and frogs. pp. 363–411. *In*: R.R. Fay and A.N. Popper (eds.). Comparative Hearing: Fish and Amphibians. Springer-Verlag, New York.

Clownfishes

Orphal Colleye,[a,*] *Eri Iwata*[b] and *Eric Parmentier*[a]

Background

Clownfishes[1] are brightly colored fishes that became literally famous to the general public in 2003 due to the release of "Finding Nemo", an American computer-animated film produced by Pixar Animation Studios. Yet, it was in the mid-20th century with the advent of scuba diving that clownfishes began to be known worldwide. Numerous investigations conducted by naturalists and marine scientists have contributed to various underwater discoveries, and among them the fascinating natural history of clownfishes. These fishes are especially well known for their outstanding symbiosis with tropical sea anemones that was first reported in 1868 (Collingwood 1868). This intimate relationship has become a textbook example for mutualistic interactions (Fautin and Allen 1997, Ollerton et al. 2007, Ricciardi et al. 2010). A great deal of attention has been given to the nature of this symbiosis and the immunity mechanisms which enable the fish to live unharmed among the stinging tentacles of its host. Admittedly, this is the most glamorous aspect of the general biology of clownfishes, but the considerable emphasis placed on this topic has tended to obscure other equally interesting areas of research such as their evolutionary history, their social structure and sex change, their reproductive behavior and their acoustic communication.

[a] Laboratoire de Morphologie Fonctionnelle et Evolutive, AFFISH - Research Center, University of Liège, Quartier Agora, Allée du six Août 15, Bât. B6C, 4000 Liège (Sart Tilman), Belgium.
Email: E.Parmentier@ulg.ac.be
[b] College of Science and Engineering, Iwaki Meisei University, 5-5-1 Chuoudai, Ihino, Iwaki, Fukushima 970-8032, Japan.
Email: asealion@iwakimu.ac.jp
* Corresponding author: O.Colleye@ulg.ac.be

[1] The term clownfishes instead of anemonefishes will be used throughout this chapter for *Amphiprion* and *Premnas* because anemonefishes may also refer to other non-pomacentrid fishes that occasionally inhabit actinians (see Randall and Fautin 2002).

Since many aspects related to the biology and ecology of clownfishes were extensively studied by G.R. Allen (1972), the main focus of the present chapter is to give an overview of the most recent observations made on this amazing group of fishes.

Diagnosis of Clownfishes

The following combination of characters is sufficient for differentiating a clownfish (Allen 1972, Nelson 2006; see Fig. 1 and Chapter I): (1) suborbitals and the operculum (i.e., hard bony flap covering the gills and composed of three bones: opercular, interopercular and subopercular) have serrated margins (hence the genus name *Amphiprion*: from the Greek *amphi*, on both sides, and *priōn*, saw), (2) the dorsal fin is composed of 10 spines (rarely 9 or 11) and usually 14–20 soft rays, (3) the buccal jaw teeth are uniserial and usually caniniform (some species have incisiform teeth), (4) the snout is mostly naked, (5) the color pattern usually consists of 0–3 white vertical bars on a darker background, which is usually various shades of red, orange, brown or black.

Fig. 1. *Amphiprion clarkii* adult, 102 mm in SL, Shikoku, Japan (Photo by Randall J.E.).

Geographical Distribution

Clownfishes are widely distributed in the tropical Indo-West Pacific regions (35°40'48" N–31°33'00" S; 34°51'00" E–149°34'00" W), living in a coral reef environment (Allen 1972, Fautin and Allen 1997, Allen et al. 2008, 2010). They also occur where warm, tropical waters are carried by currents such as the Red Sea and the East coast of Japan, as far North as the latitude of Tokyo for example (Allen 1972, Fautin and Allen 1997). According to Allen (1972), clownfishes can be grouped into three categories on the basis of their distribution: (1) species which are relatively widespread, (2) species with a limited regional distribution, and (3) species confined to islands or an island group (Fig. 2).

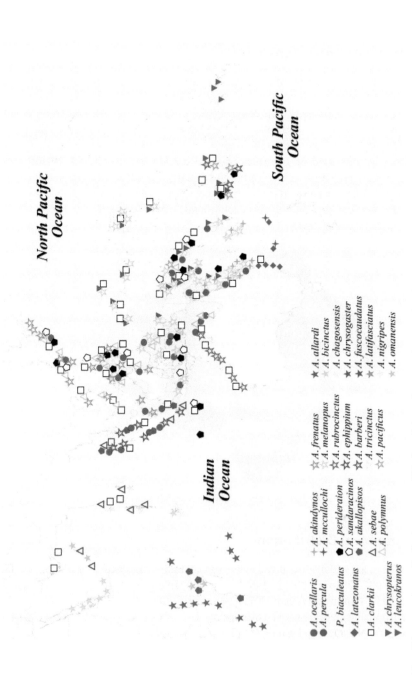

Fig. 2. Map of the approximate distribution ranges of 29 clownfish species. Data about the distribution ranges were obtained from Allen (1972), Fautin and Allen (1997) and Allen et al. (2008, 2010).

The North coast of Papua New Guinea, and more precisely the region of Madang, is probably the most species-rich region of the world with nine sympatric clownfish species (Fautin 1988, Elliott and Mariscal 2001). Such a coexistence of nine species is considered possible because of niche partitioning. Indeed, each species of host sea anemone and clownfish live within a particular range of zones related to their preferred habitat (e.g., nearshore, mid-lagoon, outer barrier, and offshore). Moreover, each species of clownfish lives with a primary species of host; and when two clownfish species use the same host sea anemone, they usually have different distribution patterns among zones (Elliott and Mariscal 2001). For example, *Amphiprion percula* occupies *Heteractis magnifica* in nearshore zones while *Amphiprion perideraion* lives with *H. magnifica* in offshore zones (Elliott and Mariscal 2001).

Outward from this center of abundance, the diversity of clownfish species decreases: Lizard Island on the Great Barrier Reef has up to six different species whereas only one clownfish species is known from the Red Sea and from French Polynesia (Allen 1972). Note that clownfishes are also absent from different localities on the eastern fringe of the Pacific Ocean such as the Hawaiian Islands, Johnston Islands and the Marquesas Islands (Randall 1955). Concomitantly, they are not found off the coast of Central and South America. Different reasons can explain this lack of dispersal across the East Pacific Barrier, which is typical for many widespread Indo-Pacific species (Robertson et al. 2004). Being obligate symbionts, the distribution of clownfishes is intimately linked to the distribution and availability of their host sea anemones. Their range cannot extend beyond that of their hosts, and these actinians are found throughout the tropical and sub-tropical parts of the Indo-West Pacific region (Fautin 1986, 1991). Moreover, host sea anemones usually occur within specific habitat parameters (e.g., reef zonation, substrate, depth; Fautin 1991). They are confined to the photic zone ($\leq$200 m) due to their association with zooxanthellae, and thus they live in sediments surrounding coral reefs as well as on reefs themselves (Fautin 1991). For example, sea anemone individuals occurring in murky environments seldom contain fish (Fautin 1991).

The dispersal and resultant distribution of clownfishes may also be affected by other factors. Due to the very short planktonic period related to their pelagic larval stage (lasting 9–12 days; Thresher et al. 1989), clownfishes are known for high larval retention to natal reefs (Jones et al. 2005, Almany et al. 2007). For example, Jones et al. (2005) marked the newly hatched larvae of the panda clownfish *Amphiprion polymnus* by using a tetracycline immersion method. They showed that one-third of the settled juveniles returned to a 2 hectare natal area, with many settling <100 m from their hatching site. This represents the smallest scale of dispersal known for any marine fish species with a pelagic larval phase (Jones et al. 2005). Likewise, Almany et al. (2007) showed that after a planktonic larval period of 10–12 days, up to 60 percent of the juvenile *Amphiprion percula* settling at Kimbe Island were the offspring of resident adults.

Note that ocean currents could also affect clownfish distribution given that many island areas are relatively isolated and their endemic integrity is probably maintained by localized current gyres (Allen 1972).

Taxonomy and Phylogeny

The clownfishes are composed of 30 known species[2] belonging to the Pomacentridae family (Allen 1972, Ollerton et al. 2007, Allen et al. 2008, 2010). Traditionally, taxonomic studies based on morphological characters have divided the clownfishes into two genera (Allen 1972): the monotypic genus *Premnas* that includes only a single species (*P. biaculeatus*) and the genus *Amphiprion*, containing 29 species. This division into two genera was originally reported by Georges Cuvier, who traditionally used two main characters to differentiate *Premnas* Cuvier 1817 (see summary in Cuvier and Valenciennes 1830) from *Amphiprion* Bloch and Schneider 1801: (1) the presence of two backward projecting spines on the suborbital bones, which is not completely divergent from the typical pattern of serration found in *Amphiprion* and (2) the number of transverse scale rows in excess of 60 (Allen 1972). This second character loses much of its significance since the scale row counts for *A. akallopisos*, *A. nigripes*, *A. ocellaris* and *A. percula* approach the lower limit for *Premnas* (Allen 1972). *Premnas biaculeatus* would also have other typical features such as the relatively large size attained by adults, and a general reduction in the serration of the suborbitals, interopercular and opercular bones (Allen 1972).

The species of *Amphiprion* were further grouped by Allen (1972) according to their morphological similarities into four subgenera (*Actinicola*, *Paramphiprion*, *Phalerebus* and *Amphiprion*) and two species complexes within the subgenus *Amphiprion* (*ephippium*-complex and *clarkii*-complex) (Fig. 3).

Because the development of a specialist condition from a generalist ancestor is commonly used to explain the evolution of specialization of symbiotic organisms (Futuyma and Moreno 1988), it was first thought that the ancestral clownfish became adapted to live with sea anemones and that, over evolutionary time the fishes have radiated into a variety of niches by becoming more specialized for living with particular host species (Elliott et al. 1999). On this basis, Allen (1972) first hypothesized that the members of the subgenus *Amphiprion* (*clarkii*-complex) are related to the basal node (*sensu* Omland et al. 2008) on the phylogenetic tree of clownfishes because they are most similar to other free-living pomacentrids in terms of morphology and behavior: they are relatively deep-bodied, good swimmers (i.e., with truncate or emarginate caudal fin), and less dependent on their host sea anemones for shelter. For example, *Amphiprion clarkii* is the least host specific, occurring in all of the 10 symbiotic actinian species (Fautin and Allen 1997; see Table 1). This clownfish species is a very good swimmer and often takes shelter in coral crevices (instead of its host), when attempts are made to capture it in the field (Allen 1972). Long-term observation *in situ* of this species in association with a soft coral was also reported by Arvedlund and Takemura (2005). Given that most species of the *clarkii*-complex are also able to live in symbiosis with three or more species of host sea anemones (Fig. 3, Table 1), their generalist condition tends to place them ideally at the ancestral node on the tree.

---

[2] The species status of *A. thiellei* and *A. leucokranos* is still debated as they seem to represent natural hybrid crosses between *A. chrysopterus* and *A. sandaracinos* (see Ollerton et al. 2007).

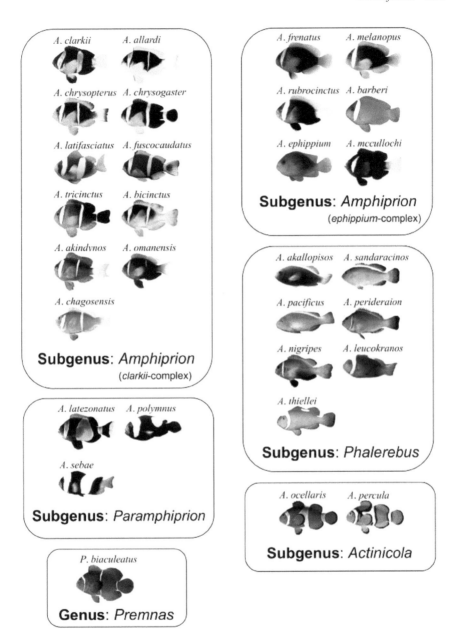

Fig. 3. The five subgenera of *Amphiprion* and the genus *Premnas* (modified from Allen 1991 and Allen et al. 2008, 2010).

Traditional morphological analyses do not always provide enough informative characters to produce robust phylogenies (Hillis 1987, 1995). Molecular phylogenetic studies were conducted within the clownfishes to test the hypothesis that the ancestral condition was the generalist one, and the descendant lineages became specialized to live

Table 1. Mutualistic interactions between clownfishes and their host sea anemones.

Fish species	Sea anemone species									
	Ca	Eq	Md	Hm	Hc	Ha	Hu	Sh	Sg	Sm
P. biaculeatus		x								
A. chagosensis		[x]								
A. frenatus		x								
A. fuscocaudatus										x
A. latezonatus					x					
A. latifasciatus										x
A. mccullochi		x								
A. nigripes				x						
A. pacificus				x						
A. sebae								x		
A. akallopisos				x						x
A. barberi		x			x					
A. ephippium		x			x					
A. omanensis		x			x					
A. polymnus					x			x		
A. rubrocinctus		x							x	
A. sandaracinos					x					x
*A. thiellei**					[x]					[x]
A. allardi		x				x				x
A. leucokranos				x	x					x
A. melanopus		x		x	x					
A. ocellaris				x					x	x
A. percula				x	x				x	
A. perideraion			x	x	x				x	
A. tricinctus		x			x	x				x
A. bicinctus		x		x	x	x			x	
A. chrysogaster				x		x		x		x
A. akindynos		x		x	x	x		x		x
A. chrysopterus		x	x	x	x	x		x		x
A. clarkii	x	x	x	x	x	x	x	x	x	x

Host specificity patterns for 30 clownfish species (*Premnas* (*P*) and *Amphiprion* (*A*)) and ten species of host sea anemones. Ca = *Cryptodendrum adhaesivum*, Eq = *Entacmaea quadricolor*, Md = *Macrodactyla doreensis*, Hm = *Heteractis magnifica*, Hc = *H. crispa*, Ha = *H. aurora*, Hu = *H. malu*, Sh = *Stichodactyla haddoni*, Sg = *S. gigantea*, Sm = *S. mertensii*. *Field records are lacking for *A. chagosensis* and *A. thiellei*, the most probable host is shown between brackets.
Data about association between host anemone species and clownfish species were obtained from Fautin and Allen (1997), Ollerton et al. (2007) and Allen et al. (2008, 2010).

with particular species of host sea anemones. Interestingly, the first studies supported the monophyletic origin of the clownfish species (Elliott et al. 1999, Santini and Polacco 2006), but they rejected the *clarkii*-complex hypothesis as suggested by Allen (1972). Phylogenetic data assigned species such as *A. percula* and *A. ocellaris* to the basal node on the phylogenetic tree (Santini and Polacco 2006). Unlike *A. clarkii*, these clownfish species are considered as specialist because they are more dependent on their hosts for protection, always hiding in the tentacles while being pursued (Elliott et al. 1999). They are also slender-bodied with a rounded caudal fin, which does not place them as effective swimmers (Allen 1972). Therefore, these molecular phylogenetic studies supported the progression from specialist to generalist; host generalization would therefore be considered to be a derived trait as exemplified by *A. clarkii*.

Cooper et al. (2009; see also Chapter II) then presented a phylogenetic tree of Pomacentridae showing that *Amphiprion* and *Premnas* form a distinct group or tribe (Amphiprionini), within the subfamily Pomacentrinae. In this study, the Amphiprionini were recovered as a monophyletic group, which is consistent with previous molecular studies (Elliott et al. 1999, Santini and Polacco 2006). Surprisingly, some of their analyses destabilized the relationship between *Premnas* and *Amphiprion* by rendering *Amphiprion* polyphyletic. Therefore, Cooper et al. (2009) suggested that *Premnas* should be synonymized with *Amphiprion*. However, additional data are still required to support the elimination of *Premnas* since a few phylogenetic studies dealing with damselfishes gave different findings (Quenouille et al. 2004, Santini and Polacco 2006).

Litsios et al. (2012) gave further information about the evolutionary history, showing that clownfishes likely experienced an adaptive radiation driven by ecological speciation. They suggested that the mutualistic interaction with sea anemones might be the key innovation that allowed clownfishes to radiate rapidly in untapped ecological niches. Indeed, these fishes exhibit morphological changes that seem to be correlated to the different ecological niches, as attested by the relation between the morphological characters of clownfishes and the ecological niches associated with the sea anemones (Litsios et al. 2012). As a whole, this case of adaptive radiation gives another example of how species diversification occurs in clownfishes: reproductive isolation between populations might be reinforced by some peculiar characteristics related to their life-history such as the short larval duration and the high larval retention to natal reef (Litsios et al. 2012).

More recently, a new study conducted by Litsios and his collaborators (Litsios et al. 2014) confirmed previous studies by ensuring a monophyletic origin of the group, and by supporting the hypothesis of the progression from specialist to generalist by assigning again species such as *A. percula* and *A. ocellaris* to the ancestral node on the phylogenetic tree (Fig. 4). Central Indo-Pacific and Central Pacific provinces could be considered to be the most likely locations of the origin of clownfishes. Interestingly, the radiation of the clownfishes could have two geographical replicates: most species first arose and diversified in the Indo-Australian Archipelago while a geographically independent clade colonized the eastern shores of Africa and diversified there (Litsios et al. 2014; see Fig. 4).

On a different note, Colleye et al. (2011) showed that acoustic communication (see also Chapter X) is not the main driving force involved in the diversification of

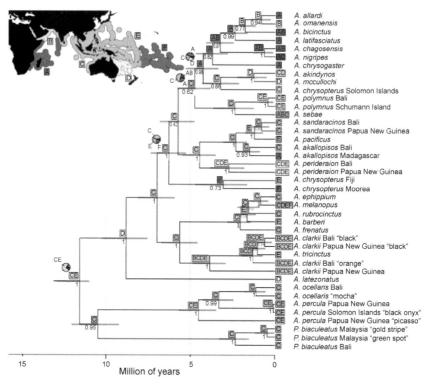

Fig. 4. Majority-rule consensus tree of the clownfishes (genera *Amphiprion* and *Premnas*) inferred in MrBayes. Numbers below nodes indicate Bayesian posterior probabilities. The size of the numbers varies in order to be reader friendly. Localization or phenotype is given for species in which several individuals were sampled (see Appendix S1 in Litsios et al. 2014). Credible intervals for the age estimates are shown on nodes as blue bars. Letters from A to F represent the provinces (see inlayed map) used for the biogeographical reconstructions. (A) Western Indian province; (B) North-western Indian province; (C) Central Indo-Pacific province; (D) South-western Pacific province; (E) Central Pacific province; (F) Polynesian province. Ancestral areas inferred using the BayArea+J model are shown above nodes and current distributions are shown left of the species names. Coloured pie charts show the probability of each area on specific nodes. Outgroup species were only used to root the tree and are not shown (from Litsios et al. 2014). © 2014 John Wiley & Sons Ltd.

clownfishes. Calls cannot act as pre-zygotic isolating mechanisms leading to speciation because they are not produced to find mates and consequently are less subject to variations due to partner preference, which restricts the constraints of diversification.

The Clownfish-host Sea Anemone Interaction

The basic question for a long time revolved around the exact nature of this association. Either it is a mutualistic one whereby both partners make a profit, or it is a form of commensalism in which the fish obtains the benefits while the sea anemone is neither harmed nor gains any advantage (see Parmentier and Michel 2013 for more details on the different types of symbiosis). Although many biologists have considered the interaction as a strictly one-sided relationship and even if the fish seems to gain more benefits than the sea anemone; the symbiosis between clownfish and sea anemone is an extremely rare example of mutualism in which both partners protect one another from predators (Ollerton 2006). Additionally, this association is mediated in part by the zooxanthellae, which act as a third mutualistic partner by living in the sea anemone tissues (see Fautin 1991).

The primary benefit obtained by the fishes is undoubtedly that of protection. Clownfishes are never encountered in nature without their host sea anemones: this is an obligate association for them (Allen 1972, Fautin and Allen 1997). Many species are relatively poor swimmers with a truncate or rounded caudal fin. They rapidly fall prey to larger fishes when isolated from their host (Mariscal 1970a, Allen 1972, Fautin and Allen 1997). Using data on mortality and social structure, Buston and Garcia (2007) constructed a matrix model and estimated that the life expectancy of a female *A. percula* (i.e., the oldest individual in a group) might be up to 30 years, which is twice as long as that of any other damselfish and six times greater than the longevity expected for a fish of that size (Buston and Garcia 2007). Such a long life expectancy in nature emphasizes the protection offered by the host sea anemone.

Although not deemed an essential component of their association with their invertebrate host, clownfishes may also obtain at least part of their nourishment by feeding on the half-digested food rejected by the sea anemone (Verwey 1930, Mariscal 1966, Allen 1972).

Benefits or detriments to the sea anemone are less obvious. Several of the host species are regularly found in nature without symbiotic fish (Verwey 1930, Allen 1972, Fautin 1991), indicating they are less dependent than the fishes upon the association. However, the resident *Amphiprion* provides a degree of protection for its host from various coelenterate-feeding fishes such as butterflyfishes (Allen 1972, Godwin and Fautin 1992, Holbrook and Schmitt 2005). Another minor benefit is that clownfishes may rid their hosts of parasitic copepods from the surface of the tentacles (Mariscal 1970b). Waste ammonia excreted by the clownfish could also be used by the zooxanthellae (Porat and Chadwick-Furman 2004, Roopin et al. 2008). On the other hand, Scott and Francisco (2006) revealed that resident clownfishes may feed upon the gametes being released by their host, which demonstrates a deleterious effect to the sea anemone, which likely only occurs when normal planktonic food is in short supply. This example shows that the relationship on rare occasions may turn into parasitism depending on environmental conditions.

Additionally, it was recently demonstrated that certain behaviors displayed by the fish when it interacts with its host could be beneficial for one or both partners at night (Szczebak et al. 2013). Indeed, clownfishes appear to carry out behaviors such as wedging and switching by involving rapid caudal and pectoral fin movement. Szczebak et al. (2013) experimentally demonstrated that clownfishes engage in more flow-modulating behaviors in presence of their host than when they are kept alone, which leads to an increase of gas exchange across sea anemone tissues due to the enhanced ambient water flow. By modulating water flow among sea anemone tentacles through increased activity, it is likely that clownfish behaviors (i.e., wedging and switching) improve both sea anemone and clownfish oxygen uptake at night (Szczebak et al. 2013).

Note that another species of pomacentrid, the threespot damselfish *Dascyllus trimaculatus*, sometimes shelters among the stinging tentacles of large sea anemones

(Mariscal 1972, Fautin and Allen 1997). Unlike the situation for clownfishes, it is not an obligate symbiont because it is not dependent on the host sea anemone for survival. Only juveniles are known to live with actinians, but they eventually become independent with increased growth (Mariscal 1972, Fautin 1986, Randall and Fautin 2002).

Host utilization and specificity

Of nearly 1000 species of sea anemones distributed worldwide, clownfishes are known to develop mutualistic interactions with only 10 different species (Table 1). A large variation in host usage exists within the group (Ollerton et al. 2007). About one-half of clownfish species are specialized, living with one or just a few species of sea anemones while the remainder are generalists capable of living with several host species (Allen 1972, Fautin and Allen 1997, Elliott and Mariscal 2001). Only *A. clarkii* naturally occurs with all 10 host actinians. Conversely, *Entacmaea quadricolor*, *Heteractis crispa*, *H. magnifica* and possibly *Stichodactyla mertensii* host from 12–14 fish species, whereas *Cryptodendrum adhaesivum* and *H. malu* host only one species of fish (Table 1). According to Fautin (1991), at least four components are involved in host selection and specificity: (1) the conditioned or innate preferences of the fish, which is the shorter-lived and mobile partner, and thus responsible for the pattern, (2) a shared pattern of environmental and ecological requirements because only species that occur in the same geographical area and have similar ecological preferences (sand or reef, deep or shallow) can live together, (3) competition by fish for hosts, considering that fishes with a high degree of host specificity are usually the most competitive with regard to colonizing preferred anemones, and (4) random chance.

Arvedlund et al. (1999) highlighted that host recognition and selection are mediated by an imprinting-like mechanism. By testing the different environmental conditions (contact with its natural host sea anemone *E. quadricolor*; contact with *H. malu*, which is not a natural host except for other clownfishes; and without a sea anemone), they clearly demonstrated that olfaction is used by juvenile *A. melanopus* to recognize host sea anemones. However, clownfish host-imprinting appears to be restricted because it was not possible to imprint *A. melanopus* larvae to the non-host sea anemone *H. malu*. They hypothesized that the recognition of chemicals may be innate or alternatively, it may be learned early in life. Indeed, clownfish eggs are incubated close to an actinian and exposed to frequent contact with the tentacles. During incubation, chemicals from the sea anemone may penetrate the egg case and imprint the embryonic fish (Allen 1972, Fautin and Allen 1997).

Interestingly, the importance of olfactory cues in settlement site selection has already been demonstrated for some species (Gerlach et al. 2007, Dixson et al. 2008, 2011). In particular, settling larvae are able to discriminate among reefs using olfactory cues related to the waterborne odors of their home reefs (Gerlach et al. 2007). They are notably able to detect olfactory cues produced by vegetated islands (Dixson et al. 2008, 2011). Juveniles of the orange clownfish *A. percula* prefer water coming from reefs surrounding vegetated islands with rainforest vegetation compared with that from reefs without islands (Dixson et al. 2008).

In some cases, host preference and habitat segregation can occur between the juvenile and adult life stages through competitive exclusion, and this process appears to be related to anemone morphology. Huebner et al. (2012) experimentally demonstrated that juveniles of the two-band clownfish *A. bicinctus* mainly inhabit leathery sea anemones *Heteractis crispa*, whereas breeding adults almost exclusively occupy bulb-tentacle *Entacmaea quadricolor* because they are larger. In the field, unoccupied *E. quadricolor* are colonized more rapidly than *H. crispa*, which is supported by laboratory experiments showing that juveniles and adults prefer *E. quadricolor* (Huebner et al. 2012). However, large adults monopolize and relegate subordinates to *H. crispa* due to competition. Therefore, juveniles may find refuge in the non-preferred host *H. crispa* while they wait for available space in the preferred host *E. quadricolor* (Huebner et al. 2012). Conversely, some clownfish species also appear to partition the host sea anemone due to a size discrepancy: small clownfish species such as *A. leucokranos* and *A. sandaracinos* sometimes share their host with larger *A. chrysopterus* because there is no competitive exclusion (Elliott and Mariscal 2001).

Clownfishes and their host sea anemones have been used as model systems to test theories about both the evolution of host selection and specificity (Arvedlund et al. 1999, Elliott et al. 1999). In common with most other mutualistic interactions, the anemone-clownfish mutualism is nested in structure: the most generalist fishes interact with both generalist and specialist anemones, and the most generalist anemones interact with generalist and specialist fishes; specialist-specialist interactions do not occur (see Ollerton et al. 2007). Across the distributional range of their association, the mutualistic interaction between clownfishes and their hosts is significantly nested in structure (Ollerton et al. 2007): specialized clownfishes tend to use generalist host sea anemones (Table 1). For example, the most specialist-specialist interaction involves *A. sebae* and *Stichodactyla haddoni*, an anemone species known to host six species of clownfishes (Table 1). However, at both regional and local scales, the interaction between clownfish and host sea anemones appears not to be significantly nested (Ricciardi et al. 2010). Due to the dispersal abilities of fish and anemone larvae, or a combination of local conditions with competition, regional species being more generalists are forced to become more specialist (Ricciardi et al. 2010).

How clownfish can inhabit a sea anemone?

For many years, the prime concern was how clownfishes manage to survive in a deadly environment to most animal species. Several hypotheses to account for this invulnerability have been advanced despite reasons for refuting them (Fautin 1991). Among the most plausible of these assumptions are: (1) the skin of clownfishes is impenetrable to nematocysts. However, it may be slightly thinner than that of other damselfishes, and an unacclimated clownfish (i.e., a fish that has never been in contact with its host) can be killed by its own host (Mariscal 1966), and (2) the sea anemone does not discharge its nematocysts in presence of a clownfish. Actually, this cannot explain the invulnerability because an actinian can sting and capture prey while harboring clownfish (Mariscal 1966). Thereby, the fish appears to be responsible for its invulnerability to being stung (Fautin 1991).

It has been demonstrated that the source of protection resides in the mucus coating of the fish (Davenport and Norris 1958). From this point of view, two theories confront each other. Some authors (Davenport and Norris 1958, Schlichter 1972, Schlichter 1976) claimed this mucus originates from the host actinian and is transferred to the fish through acclimation behavior. The fact that the fish frequently comes into contact with its host may serve to maintain its protection (Schlichter 1972). To some extent, the fish is chemically camouflaged in the anemone's clothing (Schlichter 1972, 1976). According to this theory, it is the fish's behavior that allows it to live in this unusual habitat (Schlichter 1968). Conversely, Lubbock (1980) demonstrated that the protective mucus originates from the fish. Appreciable amounts of anemone mucus were not found on resident clownfish, so anemone mucus on fish may be the result of the protection rather than its cause (Lubbock 1980). Moreover, the fish does not appear to be more stimulatory to nematocysts when denaturing its mucus, so the fish mucus could have evolved to lack components that stimulate firing (Lubbock 1980). According to this explanation, the unusual biochemistry of their mucus could enable clownfishes to live in this peculiar habitat. Overall, there is probably some truth on both sides even if the question is still open.

Clownfishes are closely related and all live in symbiosis with sea anemones. They also differ in the number of host species that they can occupy (Table 1). On this last point, adaptation to host may differ among species with the fish behavior (i.e., acclimation) and the biochemical composition of its mucus both probably playing roles. Fautin and Allen (1997) considered that for fish living with many types of hosts such as *A. clarkii*, the behavior is likely to be more important to adaptation, whereas for host-specific fish such as *A. frenatus*, the biochemical composition of the mucus is probably the more significant factor. In a different way, Brooks and Mariscal (1984) suggested that both fish and anemone may be active in forming the symbiosis. The average acclimation time following the prolonged separation of *A. clarkii* from its host *Macrodactyla doreensis* was significantly reduced if the fish was kept with a surrogate sea anemone made of rubber bands glued to a Petri dish (Brooks and Mariscal 1984). Therefore, it appears that the fish produces specific protective mucus due to physical contact with what it perceives to be a host, which is further augmented in the presence of a real actinian (Fautin and Allen 1997).

Social Structure and Sex Change

Clownfishes are commonly known as protandrous (i.e., male to female) sex-changing fishes (see Chapter IV). They live in social groups composed of a breeding pair and a varying number of sexually immature individuals, depending on the species (Fricke 1979, Buston 2003) and host size (Mitchell and Dill 2005). The social structure of clownfishes consists of a size-based dominance hierarchy, in which the body size ratios are well defined between each member of the group (Fricke 1979, Buston 2003). Generally speaking, the hierarchy of the social rank is strictly maintained in a group and is considered to be one of the critical factors that induce sex change or sex differentiation (Iwata and Manbo 2013). For example, sex change is prevented by social dominance of the female over the breeding male (Fricke and Fricke 1977,

Moyer and Nakazono 1978). The largest and socially dominant individual of a group is the breeding female (rank 1) whose gonads are functioning ovaries with remnants of degenerate testicular tissue (Fricke 1979, Godwin and Thomas 1993, Casadevall et al. 2009). The breeding male is the second largest individual (rank 2) with gonads that are functioning testes, but also possesses non-functioning or latent ovarian cells (Fricke 1979, Godwin and Thomas 1993, Casadevall et al. 2009). The non-breeding fish remain as non-reproductive individuals with ambisexual gonads and get progressively smaller as the hierarchy descends (rank 3 and higher ranks). If the dominant female of a group dies or is experimentally removed, then the male becomes the breeding female (its gonads cease to function as testes and the egg producing cells become active). Simultaneously, the largest non-breeding individual turns into the breeding male (Fricke and Fricke 1977, Moyer and Nakazono 1978). The protandry and social system of clownfishes is thought to have evolved as a mechanism to allow for relatively expeditious pair formation given that sea anemones are generally sparsely distributed (Bollinger et al. 2008). They have also been regarded as adaptations to the extreme difficulty of moving between hosts due to the related predation pressure in subtropical and tropical waters (Allen 1972, Fricke and Fricke 1977, Hattori 1991). Under certain conditions immature individuals with ambisexual gonads may differentiate directly into males or females. For example, in temperate waters where the density of sea anemones is high and the population size of the fish is small, non-breeding *A. clarkii* can become females without passing through a functional male stage (Ochi 1989, Bruslea-Sicard et al. 1994).

The relationship between sex differentiation and physical stress, rather than social stress, is well documented in gonochoristic teleost species with thermolabile sex determination (TSD), such as the Japanese flounder *Paralichthys olivaceus* and pejerrey *Odontesthes bonariensis* (Kitano et al. 2007, Hattori et al. 2009). High temperatures result in increased levels of the stress-related hormone cortisol in the blood, and high cortisol levels suppress transcription of the aromatase gene, which leads to masculinization of larvae (Kitano et al. 2007, Hattori et al. 2009). The mechanism of female to male sex change might be explained in the same manner as TSD, but clownfishes, unlike other damselfishes (Godwin 1995, Asoh 2003) and many other non-pomacentrids, are male to female sex-changing fish. In very early stages of social hierarchy formation, a high blood cortisol level was observed in dominant individuals of false clownfish *A. ocellaris*, which later become females, and higher transcription levels of the aromatase genes were detected (Iwata et al. 2012). It suggests the existence of a regulatory mechanism other than TSD that leads to feminization in protandrous fish species. Moreover, clownfishes usually take at least about 45 days for male to female sex change (Godwin and Thomas 1993, Casadevall et al. 2009), and several months or more for sex differentiation in an ambisexual pair (Iwata et al. 2008). Therefore, long-term social interaction in a group of clownfish may be essential in order for sex change to occur.

Within this social structure, subordinates benefit from settling in an anemone and queuing for breeding positions, their chances of success are thus determined by their rank within the group (Buston 2004a). In such a context, dominants gain no benefits from subordinates, which are potential challengers for their position. This asymmetry generates conflict over subordinate group members, and dominants occasionally evict or kill subordinates that are of similar size to themselves (Allen 1972, Buston 2004b). However, clownfishes seem to use a ploy in order to minimize such conflicts, and to avoid a subordinate representing a threat to its immediate dominant (Buston 2003). They adjust their size according to their position in the group hierarchy, maintaining a well-defined size difference with respect to individuals above them in social rank (Buston 2003). For example, it was shown that the growth of individual *A. percula* is regulated so that the body size ratio between each dominant and its immediate subordinate is about 1.26 (Buston and Cant 2006). Likewise, Hattori (2012) developed a mathematical optimization model showing that the body size composition of clownfishes appears to be essential for coexisting within limited shelter space. This body size composition is determined by the carrying capacity of the host and the body size differences among group members adjacent in rank (Hattori 2012). Overall, this model could explain the function of monogamy and protandry in clownfishes when inhabiting single isolated hosts (Hattori 2012). In addition, aggressive displays play a role in maintaining size differences within the hierarchy (Fricke 1979). The dominant female attacks all group members, especially the two larger ones that pose the biggest threat to usurping its position, whereas the functional beta male displays the highest rate of intragroup aggressive acts (Fricke 1979; see Fig. 5A). Therefore, it appears that females and males show more aggressive behavior than ambisexual individuals. In reaction to aggressive acts, group members also exhibit appeasement behavior, which is most frequently exhibited by the functional beta male in response to the alpha female (Fricke 1979). However, this behavior is also displayed by the next lower-ranked members of the group (Fig. 5B).

Sound production is frequently associated with the previously described agonistic interactions and apparently serves an important function within the group (see Colleye et al. 2012). Clownfishes produce aggressive and submissive sounds during interactions between group members. Interestingly, these sounds are strongly related to fish size (Colleye et al. 2012; see Chapter 10), conveying information regarding the identity of the sound-producing fish. Aggressive sounds possibly serve by reinforcing the dominant position of the larger fish in the hierarchy, whereas submissive

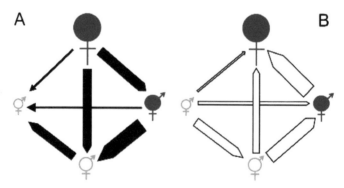

Fig. 5. Aggressive acts in a group. Filled symbol = functional breeding individual. Receiver of aggressive act shown by arrow. (A) Distribution of aggressive acts in % of the total of observed intragroup interactions (= width of arrow). (B) Distribution of appeasement behavior among group members. Treatment of the data is the same as in A (modified from Fricke 1979).

sounds would have the opposite effect. Therefore, acoustic signals appear to be a strategy for preventing conflicts, which otherwise might result in severe physical injury.

Reproductive Behaviors

Clownfishes spawn throughout most of the year at tropical locations, although there may be seasonal peaks of activity (Allen 1972, Ross 1978). Cooler temperatures inhibit spawning and therefore it is generally restricted to spring and summer in subtropical or warm temperate seas (Moyer and Bell 1976, Ochi 1985). In tropical regions, spawning is also strongly correlated to the lunar cycle, usually occurring within six days before and after the full moon (Allen 1972, Ross 1978). Moonlight seems to be useful by maintaining a high level of alertness in the male, who assumes most of the nest guarding duties (Allen 1972, Fautin and Allen 1997). Moreover, moonlight may attract newly hatched larvae to swim towards the surface because they are photopositive (Allen 1972), thereby facilitating their subsequent dispersal by waves and currents (Fautin and Allen 1997).

Courtship is generally stereotyped and ritualized (Allen 1972, Fautin and Allen 1997). Several days prior to spawning, the male selects a nesting site that is usually situated on a rock positioned next to the sea anemone. The spawning period is indicated by the noticeably distended belly of the female and by an increase of cleaning activity by the male, who spends considerable time clearing algae and debris from the nesting site (Allen 1972, Fautin and Allen 1997). In addition, social interactions increase between mates as expressed by chasing, fin-erection, and rapid side by side swimming (Fautin and Allen 1997). More vigorous pecking activity by the male at the surface of the nest indicates that spawning will take place soon (Allen 1972). On the day of spawning, nest-cleaning activities are carried out more rigorously by the female (Fautin and Allen 1997; see Fig. 6A), who also repeatedly presses her belly against the substrate. These activities generally signal the completion of the nesting site. In addition, the whitish cone-shaped ovipositor of the female is clearly apparent just prior to spawning (Allen 1972, Fautin and Allen 1997).

Spawning usually occurs during morning hours, and generally lasts from about 30 minutes to more than two hours (Fautin and Allen 1997). The female starts to lay eggs while swimming a roughly circular path around the nest with her belly just brushing the surface (Allen 1972). The male follows closely behind and fertilizes the spawn. Both fish also nibble the tips of the sea anemone tentacles, preventing the eggs from contacting them (Allen 1972).

Throughout incubation, which lasts 6–8 days, the nest is meticulously guarded and cared for by the male (Fautin and Allen 1997; see Fig. 6B). Other fishes are aggressively chased from the vicinity, especially potential egg-eaters such as wrasses (Fautin and Allen 1997). Two other important parental functions are exhibited, mainly by the male. Egg fanning is the most common behavior and is achieved by fluttering the pectoral fins (Allen 1972). In addition, mouthing behavior is periodically used to remove dead eggs from the nest (Allen 1972; see Fig. 6C).

Recently, Colleye et al. (2012) showed that clownfishes do not use acoustic signals for synchronizing reproductive activities, which might be explained by some aspects

Fig. 6. Breeding in clownfishes. (A) *Amphiprion ocellaris* female cleaning the nest on the day of spawning. (B) Eggs inspection by *Amphiprion percula* male. (C) *Amphiprion percula* male mouthing the eggs (Photos by Allen G.R.).

related to their peculiar way of life. They form small social groups including only one mating pair (Allen 1972, Fricke 1979, Buston 2003). Thereby, the male does not need to have elaborate courtship behavior for attracting a mate. Pair-bonding is very strong and lasts for several years in most species (Allen 1972, Fautin and Allen 1997). Moreover, the relative insignificance of nuptial activity is correlated to the small size of their territories (centered on actinians), which in turn, is correlated to their unusual social hierarchy. However, it seems that visual signals are important for synchronizing reproductive activities. Particularly important is the female's increased nest-cleaning activity on the day of spawning, signaling its readiness to deposit eggs. Similarly, the male possibly regulates its level of nest-caring activity in response to visual stimuli received during nest inspection (Allen 1972). A visual stimulus of this sort would signal the stage of egg development and the need for increased fanning and mouthing activities. Allen (1972) experimentally demonstrated that strong agitation of the eggs is a requisite for hatching. There is also a pronounced increase in the amount of male nest care on day six of incubation (Allen 1972). On that day, the embryos are well developed with one of the most noticeable features being the large eyes with their silvery pupils, likely functioning as an important visual cue (Allen 1972).

Conclusion

Overall, this chapter describes the most recent findings about clownfishes, especially those related to their biogeographical history. It also provides further insights into the mutualistic interaction between sea anemones and clownfishes, showing how important this symbiotic association could be for both partners. Nonetheless, there is still further research required such as identifying the exact mechanism by which clownfishes are able to live unharmed in the habitat offered by their host sea anemones.

Acknowledgements

The authors are greatly indebted to the two reviewers (G.R. Allen and G. Litsios) for their insightful comments and interesting criticisms of the original version of the manuscript. Permission is granted for the use requested Fig. 4 (Glenn Litsios, Peter B. Pearman, Déborah Lanterbecq, Nathalie Tolou and Nicolas Salamin, The radiation of the clownfishes has two geographical replicates, Journal of Biogeography, John Wiley & Sons, Inc.).

References

Allen, G.R. 1972. The Anemonefishes: Their Classification and Biology. T.F.H. Publications Inc., Neptune City, New Jersey.
Allen, G.R. 1991. Damselfishes of the World. Mergus Publishers, Melle, Germany.
Allen, G.R., J. Drew and L. Kaufman. 2008. *Amphiprion barberi*, a new species of anemonefish (Pomacentridae) from Fiji, Tonga, and Samoa. Aqua, Int. J. Ichthyol. 14: 105–114.
Allen, G.R., J. Drew and D. Fenner. 2010. *Amphiprion pacificus*, a new species of anemonefish (Pomacentridae) from Fiji, Tonga, Samoa, and Wallis Island. Aqua. Int. J. Ichthyol. 16: 129–138.
Almany, G.R., M.L. Berumen, S.R. Thorrold, S. Planes and G.P. Jones. 2007. Local replenishment of coral reef fish populations in a marine reserve. Science 316: 742–744.
Arvedlund, M. and A. Takemura. 2005. Long-term observation *in situ* of the anemonefish *Amphiprion clarkii* (Bennett) in association with a soft coral. Coral Reefs 24: 698.
Arvedlund, M., M.I. McCormick, D.G. Fautin and M. Bildsoe. 1999. Host recognition and possible imprinting in the anemonefish *Amphiprion melanopus* (Pisces: Pomacentridae). Mar. Ecol. Prog. Ser. 188: 207–218.
Asoh, K. 2003. Gonadal development and infrequent sex change in a population of the humbug damselfish, *Dascyllus aruanus*, in continuous coral-cover habitat. Mar. Biol. 142: 1207–1218.
Bollinger, E.K., P.V. Switzer, J. Pfammatter and J. Allen. 2008. Group formation and anemone use in captively reared anemonefish (*Amphiprion frenatus*). Ichthyol. Res. 55: 394–398.
Brooks, W.R. and R.N. Mariscal. 1984. The acclimation of anemone fishes to sea anemones: protection by changes in the fish's mucous coat. J. Exp. Mar. Biol. Ecol. 81: 277–285.
Bruslea-Sicard, S., R. Reinboth and B. Fourcault. 1994. Germinal potentialities during sexual state changes in a protandric hermaphrodite, *Amphiprion frenatus* (Teleostei, Pomacentridae). J. Fish. Biol. 45: 597–611.
Buston, P.M. 2003. Size and growth modification in clownfish. Nature 424: 145–146.
Buston, P.M. 2004a. Territory inheritance in clownfish. Proc. R. Soc. Lond. B 271: 252–254.
Buston, P.M. 2004b. Does the presence of non-breeders enhance the fitness of breeders? An experimental analysis in the clown anemonefish *Amphiprion percula*. Behav. Ecol. Sociobiol. 57: 23–31.
Buston, P.M. and M.A. Cant. 2006. A new perspective on size hierarchies in nature: patterns, causes, and consequences. Oecologia 149: 362–372.
Buston, P.M. and M.B. Garcia. 2007. An extraordinary life span estimate for the clown anemonefish *Amphiprion percula*. J. Fish. Biol. 70: 1710–1719.
Casadevall, M., E. Delgado, O. Colleye, S. BerMonserrat and E. Parmentier. 2009. Histological study of the sex-change in the skunk clownfish *Amphiprion akallopisos*. Open Fish. Sci. J. 2: 55–58.
Colleye, O. and E. Parmentier. 2012. Overview on the diversity of sounds produced by clownfishes (Pomacentridae): importance of acoustic signals in their peculiar way of life. PLoS ONE 7: e49179.
Colleye, O., P. Vandewalle, D. Lanterbecq, D. Lecchini and E. Parmentier. 2011. Interspecific variation of calls in clownfishes: degree of similarity in closely related species. BMC Evol. Biol. 11: 365.

Collingwood, C. 1868. Rambles of a Naturalist on the Shores and Waters of the China Sea. John Murray, London.

Cooper, J.W., L.L. Smith and M.W. Westneat. 2009. Exploring the radiation of a diverse reef fish family: phylogenetics of the damselfishes (Pomacentridae), with new classifications based on molecular analyses of all genera. Mol. Phylogenet. Evol. 52: 1–16.

Cuvier, G. and A. Valenciennes. 1830. Histoire naturelle des poissons, Tome 5. F.G. Levrault, Strasbourg.

Davenport, D. and K.S. Norris. 1958. Observations on the symbiosis of the sea anemone *Stoichactis* and the pomacentrid fish, *Amphiprion percula*. Biol. Bull. 115: 397–410.

Dixson, D.L., G.P. Jones, P.L. Munday, S. Planes, M.S. Pratchett, M. Srinivasan, C. Syms and S.R. Thorrold. 2008. Coral reef fish smell leaves to find island homes. Proc. R. Soc. Lond. B 275: 2831–2839.

Dixson, D.L., G.P. Jones, P.L. Munday, M.S. Pratchett, M. Srinivasan, S. Planes and S.R. Thorrold. 2011. Terrestrial chemical cues help coral reef fish larvae locate settlement habitat surrounding islands. Ecol. Evol. 1: 586–595.

Elliott, J.K. and R.N. Mariscal. 2001. Coexistence of nine anemonefish species: differential host and habitat utilization, size and recruitment. Mar. Biol. 138: 23–36.

Elliott, J.K., S.C. Lougheed, B. Bateman, L.K. McPhee and P.T. Boag. 1999. Molecular phylogenetic evidence for the evolution of specialization in anemonefishes. Proc. R. Soc. Lond. B 266: 677–685.

Fautin, D.G. 1986. Why do anemonefishes inhabit only some host actinians? Env. Biol. Fish. 15: 171–180.

Fautin, D.G. 1988. Sea anemones of Madang Province. Sci. New Guinea 14: 22–29.

Fautin, D.G. 1991. The anemonefish symbiosis: what is known and what is not. Symbiosis 10: 23–46.

Fautin, D.G. and G.R. Allen. 1997. Field guide to anemonefishes and their host sea anemones. Western Australian Museum, Perth, Western Australia, 160 pp.

Fricke, H.W. 1979. Mating system, resource defense and sex change in the anemonefish *Amphiprion akallopisos*. Z. Tierpsychol. 50: 313–326.

Fricke, H.W. and S. Fricke. 1977. Monogamy and sex change by aggressive dominance in coral reef fish. Nature 266: 830–832.

Futuyma, D.J. and G. Moreno. 1988. The evolution of ecological specialization. Annu. Rev. Ecol. Syst. 19: 207–233.

Gerlach, G., J. Atema, M.J. Kingsford, K.P. Black and V. Miller-Sims. 2007. Smelling home can prevent dispersal of reef fish larvae. Proc. Natl. Acad. Sci. USA 104: 858–863.

Godwin, J. 1995. Phylogenetic and habitat influences on mating system structure in the humbug damselfishes (*Dascyllus*, Pomacentridae). Bull. Mar. Sci. 57: 637–652.

Godwin, J. and D.G. Fautin. 1992. Defense of host actinians by anemonefishes. Copeia 1992: 903–908.

Godwin, J. and P. Thomas. 1993. Sex change and steroid profiles in the protandrous anemonefish *Amphiprion melanopus* (Pomacentridae, Teleostei). Gen. Comp. Endocrinol. 91: 144–157.

Hattori, A. 1991. Socially controlled growth and size-dependent sex change in the anemonefish *Amphiprion frenatus* in Okinawa, Japan. Jpn. J. Ichthyol. 38: 165–177.

Hattori, A. 2012. Determinants of body size composition in limited shelter space: why are anemonefishes protandrous? Behav. Ecol. 23: 512–520.

Hattori, R.S., J.I. Fernandino, A. Kishii, H. Kimura, T. Kinno, M. Oura, G.M. Somoza, M. Yokota, C.A. Strussmann and S. Watanabe. 2009. Cortisol-induced masculinization: does thermal stress affect gonadal fate in pejerrey, a teleost fish with temperature dependent sex determination? PLoS One 4: e6548.

Hillis, D.M. 1987. Molecular versus morphological approaches to systematics. A. Rev. Ecol. Syst. 18: 23–42.

Hillis, D.M. 1995. Approaches for assessing phylogenetic accuracy. Syst. Biol. 44: 3–16.

Holbrook, S.J. and R.J. Schmitt. 2005. Growth, reproduction and survival of a tropical sea anemone (Actiniaria): benefits of hosting anemonefish. Coral Reefs 24: 67–73.

Huebner, L.K., B. Dailey, B.M. Titus, M. Khalaf and N.E. Chadwick. 2012. Host preference and habitat segregation among Red Sea anemonefish: effects of sea anemone traits and fish life stages. Mar. Ecol. Prog. Ser. 464: 1–15.

Iwata, E. and J. Manbo. 2013. Territorial behaviour reflects sexual status in groups of false clown anemonefish (*Amphiprion ocellaris*) under laboratory conditions. Acta Ethol. 16: 97–103.

Iwata, E., Y. Nagai, M. Hyoudou and H. Sasaki. 2008. Social environment and sex differentiation in false clown anemonefish, *Amphiprion ocellaris*. Zool. Sci. 25: 123–128.

Iwata, E., K. Mikami, J. Manbo, K. Moriya-Ito and H. Sasaki. 2012. Social interaction influences blood cortisol values and brain aromatase genes in the protandrous false clown anemonefish *Amphiprion ocellaris*. Zool. Sci. 29: 849–85.

Jones, G.P., S. Planes and S.R. Thorrold. 2005. Coral reef fish larvae settle close to home. Curr. Biol. 15: 1314–1318.

Kitano, T., N. Yoshinaga, E. Shiraishi, T. Koyanagi and S. Abe. 2007. Tamoxifen induces masculinization of genetic females and regulates P450 aromatase and Müllerian inhibiting substance mRNA expression in Japanese flounder (*Paralichthys olivaceus*). Mol. Reprod. Dev. 74: 1171–1177.

Litsios, G., C.A. Sims, R.O. Wüest, P.B. Pearman, N.E. Zimmermann and N. Salamin. 2012. Mutualism with sea anemones triggered the adaptive radiation of clownfishes. BMC Evol. Biol. 12: 212.

Litsios, G., P.B. Pearman, D. Lanterbecq, N. Tolou and N. Salamin. 2014. The radiation of the clownfishes has two geographical replicates. J. Biogeogr. 41: 2140–2149.

Lubbock, R. 1980. Why are clownfishes not stung by sea anemones? Proc. R. Soc. Lond. B 207: 35–61.

Mariscal, R.N. 1966. The symbiosis between tropical sea anemones and fishes: a review. pp. 157–171. *In*: R.I. Bowman (ed.). The Galapagos: Proceedings of the Symposia of the Galapagos International Scientific Project. University of California Press, Berkeley.

Mariscal, R.N. 1970a. A field and laboratory study of the symbiotic behavior of fishes and sea anemones from the tropical Indo-Pacific. Univ. California Publ. Zool. 91: 1–43.

Mariscal, R.N. 1970b. The nature of the symbiosis between Indo-Pacific anemonefishes and sea anemones. Mar. Biol. 6: 58–65.

Mariscal, R.N. 1972. Behavior of symbiotic fishes and sea anemones. pp. 327–360. *In*: H.E. Winn and B.L. Olla (eds.). Behavior of Marine Animals, Volume 2. Plenum Publishing Corporation, New York.

Mitchell, J.S. and L.M. Dill. 2005. Why is group size correlated with the size of the host se anemone in the false clown anemonefish? Can. J. Zool. 83: 372–376.

Moyer, J.T. and L.J. Bell. 1976. Reproductive behavior of the anemonefish *Amphiprion clarkii* at Miyake-Jima, Japan. Japan. J. Ichthyol. 23: 23–32.

Moyer, J.T. and A. Nakazono. 1978. Protandrous hermaphroditism in six species of anemonefish genus *Amphiprion* in Japan. Japan. J. Ichthyol. 25: 101–106.

Nelson, J.S. 2006. Fishes of the World. John Wiley & Sons, Hoboken, New Jersey.

Ochi, H. 1985. Temporal patterns of breeding and larval settlement in a temperate population of the tropical anemonefish, *Amphiprion clarkii*. Japan. J. Ichthyol. 32: 248–257.

Ochi, H. 1989. Mating behavior and sex change of the anemonefish, *Amphiprion clarkii*, in the temperate waters of Japan. Env. Biol. Fish. 26: 257–275.

Ollerton, J. 2006. "Biological Barter": patterns of specialization compared across different mutualisms. pp. 75–87. *In*: Z. Dubinsky (ed.). Ecosystems of the World, Vol. 25. Elsevier, Amsterdam.

Ollerton, J., D. McCollin, D.G. Fautin and G.R. Allen. 2007. Finding NEMO: nestedness engendered bu mutualistic organization in anemonefish and their hosts. Proc. R. Soc. B 274: 591–598.

Omland, K.E., L.G. Cook and M.D. Crisp. 2008. Tree thinking for all biology: the problem with reading phylogenies as ladders of progress. BioEssays 30: 854–867.

Parmentier, E. and L. Michel. 2013. Boundary lines in symbiosis forms. Symbiosis 60: 1–5.

Porat, D. and N.E. Chadwick-Furman. 2004. Effects of anemonefish on giant sea anemones: expansion behavior, growth, and survival. Hydrobiologia 530: 513–520.

Quenouille, B., E. Bermingham and S. Planes. 2004. Molecular systematics of the damselfishes (Teleostei: Pomacentridae): Bayesian phylogenetic analyses of mitochondrial and nuclear DNA sequences. Mol. Phylogenet. Evol. 31: 66–88.

Randall, J.E. 1955. Fishes of the Gilbert Islands. Atoll Res. Bull. 47: 243.

Randall, J.E. and D.G. Fautin. 2002. Fishes other than anemonefishes that associate with sea anemones. Coral Reefs 21: 188–190.

Ricciardi, F., M. Boyer and J. Ollerton. 2010. Assemblage and interaction structure of the anemonefish-anemone mutualism across the Manado region of Sulawesi, Indonesia. Environ. Biol. Fish. 87: 333–347.

Robertson, D.R., J.S. Grove and J.E. McCosker. 2004. Tropical transpacific shore fishes. Pac. Sci. 58: 507–565.

Roopin, M., R.P. Henry and N.E. Chadwick-Furman. 2008. Nutrient transfer in a marine mutualism: patterns of ammonia excretion by anemonefish and uptake by giant sea anemones. Mar. Biol. 154: 547–556.

Ross, R.M. 1978. Reproductive behavior of the anemonefish *Amphiprion melanopus* on Guam. Copeia 1978: 103–107.

Santini, S. and G. Polacco. 2006. Finding Nemo: molecular phylogeny and evolution of the unusual life style of anemonefish. Gene 385: 19–27.

Schlichter, D. 1968. Das Zusammenleben von Riffanemonen und Anemonenfischen. Z. Tierpsychol. 25: 933–954.

Schlichter, D. 1972. Chemischer Nachweis der Übernahme anemoneneigener Schutzstoffe durch Anemonenfische. Natuwissenschaften 57: 312–313.

Schlichter, D. 1976. Macromolecular mimicry: substances released by sea anemones and their role in the protection of anemone fishes. pp. 433–441. *In*: G.O. Mackie (ed.). Coelenterate Ecology and Behavior. Plenum Press, New York.

Scott, A. and B. Francisco. 2006. Observations of the feeding behaviour of resident anemonefish during host sea anemone spawning. Coral Reefs 25: 451.

Szczebak, J.T., R.P. Henry, F.A. Al-Horani and N.E. Chadwick. 2013. Anemonefish oxygenate their anemone hosts at night. J. Exp. Biol. 216: 970–976.

Thresher, R.E., P.L. Collin and L.J. Bell. 1989. Planktonic duration, distribution and population structure of western and central Pacific damselfishes (Pomacentridae). Copeia 1989: 420–434.

Verwey, J. 1930. Coral reef studies. The symbiosis between damselfishes and sea anemones in Batavia Bay. Treubia 12: 305–355.

Vision and Colour Diversity in Damselfishes

Ulrike E. Siebeck

Introduction

Most damselfishes live in the relatively shallow waters around coral reefs, where the spectrum of downwelling sunlight still includes a wide array of wavelengths. The damselfishes make use of the full spectrum and exhibit a vast number of different colours and patterns. In this chapter, I will first discuss the underwater light environment, as it sets the limits for vision and determines the spectrum of possible colours. I will then go into what we know about the colours of damselfish and finally discuss current knowledge of visual systems in this family.

The underwater light environment

The underwater light environment is highly variable and depends on the depth as well as the quality of the water (Jerlov 1976). Water essentially acts as a monochromator such that with increasing depth, the spectrum of available light is progressively attenuated until a narrow band of wavelengths around 475 nm is all that remains at several hundred metres depth, at least in clear waters (Jerlov 1976). In more turbid waters, or waters rich in organic material (DOM, dissolved organic material), the light spectrum is attenuated more rapidly as compared to clear waters. Both, the quantity and quality of the material suspended in the water column influence the light environment at any given depth. The light spectrum in coastal waters, for example, is generally

School of Biomedical Sciences and Global Change Institute, The University of Queensland, St. Lucia 4072, QLD, Australia.
Email: u.siebeck@uq.edu.au

long-wavelength shifted, often "greener" due to chlorophyll in phytoplankton as compared to the spectrum at the same depth in offshore waters. Also, light is more strongly attenuated with depth in coastal relative to offshore waters (Jerlov 1976). The reason for spectral shifts in different types of waters is due to the wavelength-specific nature of the attenuation process. As light comes in contact with water and other molecules, long wavelength light (e.g., infrared light) is most strongly affected by absorption while short wavelength light (e.g., ultraviolet radiation) is mostly affected by scattering (Jerlov 1976). In the habitat of damselfish living on and around coral reefs, the spectrum of light includes ultraviolet wavelengths through to longer wavelengths, or light which appears red to humans (Fig. 1a).

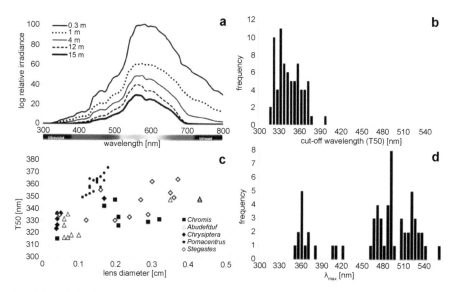

Fig. 1. (a) Relative irradiance at depths to 15 m measured at the Cobia Hole, Lizard Island, Australia. With increasing depth, the irradiance spectrum is reduced in both, quantity and quality of the light. The band of colours indicates approximate colours perceived by human eyes across the spectrum of visible wavelengths. Irradiance at depth was normalised relative to peak irradiance at the surface. All measurements were made through a UV-VIS fibre-optic cable with irradiance head (diameter 200 μ; length 1 m; Ocean Optics, Florida, USA) attached to a USB2000 fibre-optic spectrometer (Ocean Optics, Florida, USA) in a custom underwater housing (Will's Housings, Melbourne, Australia) (unpublished data). (b) Distribution of ocular media transmission cut-offs (T50 values) of damselfish whole eyes. T50 values of the largest specimen in each species measured are shown. (c) T50 values increase significantly with age/lens diameter. The size/T50 relationship is shown for species belonging to the five genera *Chromis, Abudefduf, Chrysiptera, Pomacentrus* and *Stegastes*. (d) Distribution of cone photoreceptor peak sensitivities (λ_{max}) of all measured damselfish. See Table 1 for references for Fig. b, c and d.

Colour Diversity

Background

When light encounters an object, it is reflected, absorbed and/or transmitted depending on the properties of the object. While the colour of an object is determined by the

Table 1. Summary of current knowledge for damselfish colours and visual system properties.

Species	Ocular media T50	Rod	Singles	Double	O	UV	UV Bl	UVh Bl	Bl	G	Y	O	Br	B	W	LV	UV Gr	UV Y	Ref
Abudefduf abdominalis		492	347	464 457, 519															b,c
Abudefduf bengalensis	338													x	x	x			c,m
Abudefduf saxatilis	350–356																		d
Abudefduf septemfasciatus (juv)	329																		d
Abudefduf sexfasciatus (juv)	317				x														a,d
Abudefduf vaigiensis (juv)	315–328																		d
Abudefduf whitleyi	330–349																		c,d
Amblyglyphidodon curacao	348–353															x	x		c,m
Amblyglyphidodon leucogaster	352				x						x			x	x				c,m
Amphiprion akindynos	360																		c,a
Amphiprion melanopus													x			x			m
Amphiprion perideraion	370							x					x		x				c,m
Cheiloprion labiatus (juv)	320																		d
Chromis amboinensis	324															x			d,m
Chromis atripectoralis	348								x	x						x	x		c,d,m

Table 1. contd....

Table 1. contd.

Species	Ocular media	Spectral sensitivities			O	Colours (as classified in Marshall 2000)													Ref
	T50	Rod	Singles	Double		UV	UV Bl	UVh Bl	Bl	G	Y	O	Br	B	W	LV	UV Gr	UV Y	
Chromis cyanea	324–329									x									d,m
Chromis hamui		477	355, 482	470, 514															b
Chromis insolata	327–338																		d
Chromis multilineata	336																		d
Chromis nitida	365																		c,d
Chromis ovalis	370	492	404	473, 518															d,b
Chromis punctipinnis		498	352, 490	490, 537															e
Chromis punctipinnis (juv)		498	417	486, 528															e
Chromis vanderbilti		498		462, 522			x				x			x	x				b,d,f,m
Chromis verater		480	410	471, 514										x					d,b
Chromis viridis		499	367, 493	478, 524	x		x	x											a,g,m
Chrysiptera cyanea (male)							x	x				x							q,m
Chrysiptera parasema								x		x	x								m
Chrysiptera rollandi	334																		d
Chrysiptera rollandi (juv)	322–334				x														a,d
Dascyllus albisella		490	376, 359, 464	467, 510															b
Dascyllus aruanus (juv)	316–319													x	x				d

Species												Note
Dascyllus aruanus	331											c
Dascyllus melanurus	500	357	482	469, 520								g
Dascyllus reticulatus	343								x	x		c,d,h
Dascyllus trimaculatus	499	368	485	471, 512	x				x			a,g
Dascyllus trimaculatus	491	360	490	490, 516	x							a,e
Dischistodus melanotus								x	x	x		Fig. 2
Dischistodus perspicillatus	355								x			c
Dischistodus prosopotaenia (juv)	331											d
Dischistodus prosopotaenia	355					x		x	x			c,m
Hypsypops rubicundus	392											d
Lepidozygus tapeinosoma (juv)	314						x					d
Microspathodon chrysurus	373							x				d
Neoglyphidodon melas (juv)	318					x	x		x			d,m
Neoglyphidodon melas	498		509	488, 532				x				i,c
Neoglyphidodon nigroris (juv)	328						x	x				i
Neopomacentrus azysron	330										x	m
Neopomacentrus azysron (juv)	333											d,c

Table 1. contd....

Table 1. contd.

Species	Ocular media T50	Spectral sensitivities Rod	Singles	Double	O	UV	UV Bl	UVh Bl	Bl	G	Y	O	Br	B	W	LV	UV Gr	UV Y	Ref
Neopomacentrus bankieri					x														a,d
Neopomacentrus cyanomos	331																		d
Plectroglyphidodon dickii					x														a
Plectroglyphidodon johnstonianus		495		474, 518															b
Parma oligolepis					x														a
Pomacentrus amboinensis	320	494	360, 504	485, 523	x	x					x							x	a,d,i,j,k
Pomacentrus bankanensis	319							x				x	x			x			c,m
Pomacentrus coelestis (juv)	317–340																		d
Pomacentrus coelestis	338	491	360, 490	490, 532			x		x									x	e,c
Pomacentrus crysurus	331–367													x	x				c,m
Pomacentrus lepidogenys (juv)	324									x									d,n
Pomacentrus lepidogenys (juv)	346																		d
Pomacentrus melanochir		501	502	502, 560															l
Pomacentrus moluccensis (juv)	320–330										x							x	d

Species	λmax (nm)				Opsins (x)				References
Pomacentrus moluccensis	359–370				x			x	d,j,m
Pomacentrus nagasakiensis (juv)	328						x		d,m
Pomacentrus pavo (juv)	345								d
Pomacentrus wardi (juv)	327–358					x	x	x	d,m
Pomacentrus wardi	365								c
Premnas biaculeatus	345				x		x	x	c,m
Stegastes apicalis	354–368				x	x	x	x	c,m
Stegastes diencaeus	340								d
Stegastes fasciolatus	328	495	363	470, 528					b,c,d
Stegastes gascoynei				x					a
Stegastes leucostictus	355								d
Stegastes partitus	345				x			x	d,m
Stegastes planifrons	342–362								d
Stegastes variabilis	360				x			x	d,m

Species: When data exist for juveniles and adults, the same species appears twice. (juv) juvenile (male), indicates sex in the case of a sexual dichromatic species. **Ocular media:** T50 values are given for each species. If a range is given, multiple specimens have been measured which differ in T50 values due to age/size of the lens (Fig. 1c). **Spectral sensitivities:** λmax values are given for each species and photoreceptor type measured. For each species, a rod, one or two single cones (given in separate columns) and one double cone type (λmax of both members is given as x,y). **Opsins:** x indicates species for which the opsins have been determined, five cone opsin genes (SWS1, SWS2B, RH2A, RH2B and LWS) and one rod opsin gene were found; see Hoffmann et al. 2012. **Colours:** Categories are as defined in Marshall 2000b. UV – ultraviolet, UV Bl – Ultraviolet – blue; UVhBl – ultraviolet hump blue; Bl – blue; G – green; Br – brown; B – black; W – white; LV – light 'violet'; UV Gr – ultraviolet green; UV Y – ultraviolet yellow. See Fig. 2 for example spectra for each category. **References:** [a]Hofmann et al. 2012; [b]Losey et al. 2003; [c]Siebeck and Marshall 2001; [d]Siebeck and Marshall 2007; [e]McFarland and Loew 1994; [f]Marshall et al. 2003; [g]Hawryshyn et al. 2003; [h]Losey 2003; [i]Waller 2005; [j]Siebeck 2010; [k]Siebeck et al. 2006; [l]Loew and Lythgoe 1978; [m]Marshall 2000b; [n]Kasukawa and Oshima 1987.

spectrum of light it reflects, it is important to remember that colour and reflectance are not identical. Reflectance describes a physical property of an object and is the product of this physical property as it interacts with the ambient light environment. The colour perceived by the observer depends on the observer's visual and perceptual systems. The reflectance of an object can be quantified with spectro-radiometric methods (Marshall 2000b) while the perceived colour cannot be easily quantified and varies between different observers. Despite perceptual differences between human observers, we have developed labels for different wavelength spectra so that we can at least talk about our interpretations of reflected light, i.e., colours. In this chapter, I will talk about reflectance as well as colour, and it is important to note that whenever the term colour is used it refers to human sensation only. Reflectance does not translate directly into colour as almost any perceived colour can have more than one underlying spectral power distributions (Wyzecki and Stiles 2000). If we want to understand how fish see colours, we have to take into account the properties of their visual system and ideally also the processing in the fish brain, which is an area of study that is still in its infancy. With information about the illumination and the spectral sensitivity of the visual system, we can model whether two objects are expected to be discriminable based on their reflectance spectra (colour) alone (Vorobyev and Osorio 1998). However, behavioural experiments are required to provide conclusive evidence (Kelber et al. 2003).

Three parameters are commonly used to describe colours: brightness, hue and chroma (Endler 1990). Brightness is defined as the total intensity of light reaching the eye from a colour patch at a given distance. Hue and chroma are associated with the physical properties of colour. Hue is the shade of colour, i.e., red, green, yellow, etc. and is defined by the part of the reflectance spectrum that contains most photons. Chroma is a measure of the saturation of the colour and is a function of how rapidly the intensity changes with wavelength (Endler 1990). A reflectance spectrum with gradual changes will appear less saturated than a spectrum with steep slopes and large differences between the different parts of the spectrum.

Colour patterns found on fish are created by two different mechanisms that can occur in combination, or isolation of one another. Specialised colour cells, or chromatophores (xanthophores: yellow; melanophores: black; leucophores: white; erythrophores: orange/red, and very rarely also cyanophores: blue) in the dermis of fish either contain pigment ("pigment colours") or reflective/refractive structures ("structural colours"; iridophores: iridescent colours, most blues and UV; Cott 1940, Fox and Vevers 1960). Recently, two novel types of chromatophores have been described, erythro-iridophores, which contain both pigment and reflecting platelets (Goda et al. 2011), and chromatophores, which contain fluorescent red pigment (Wucherer and Michiels 2012).

Blue and UV colours are generally of structural origin (for detailed review of blue colours see Bagnara et al. 2007). A true blue pigment colour has so far only been found in callionymid fish (Goda and Fujii 1995). Structural colours are created by interference phenomena, similar to the colours on butterfly wings (e.g., Ghiradella et al. 1972), or as a result of Tyndall (= Ralleigh) scattering. In the case of interference phenomena, stacks of crystals, in fish skin usually guanine, with a high refractive index are interspersed with cell material of low refractive index and are thought to be responsible for the wavelength specific reflection of light. The distance between the layers determines which wavelengths are reflected, the smaller the distance, the shorter the reflected wavelengths (Land 1972, Jordan et al. 2014). In the case of scattering phenomena, the incident light is scattered by fine particles (e.g., guanine) smaller than the wavelengths of the light. Short wavelengths are reflected, while long wavelengths are absorbed by a melanin layer situated behind the scattering layer (Fox and Vevers 1960). The blue colour of the sky is caused by such scattering phenomena. In this case, predominantly UV and blue light is scattered due to molecules in the upper atmosphere and dust particles in the air.

Damselfish colours

Damselfish exhibit a large diversity of colours and colour patterns and have been likened to African cichlids in terms of their shallow water habitats as well as their colour diversity (Hofmann et al. 2012). The colours of only a small proportion of the 394 valid species of damselfish have been described objectively (i.e., independent of the human visual system) using spectrometry (Kasukawa et al. 1987, Oshima et al. 1989, Marshall 2000b, Siebeck et al. 2006, Siebeck 2014; Table 1). The damselfish species measured to date have colours belonging to 13 of the 21 colour categories described by Marshall 2000a (Table 1, Fig. 2). Many damselfish species have highly contrasting colour patterns, such as black and white stripes (e.g., many *Abudefduf* spp. and *Dascyllus* spp.), orange and white stripes (e.g., *Amphiprion* spp.) and blue and yellow patches (e.g., some *Chrysiptera* spp.). On the other hand, there are many other species with more uniform and often dark colours (e.g., some *Pomacentrus*, *Stegastes* and *Chromis* species; see Allen 1991).

Some damselfish have UV-reflective patterns or patches on their bodies (Fig. 3). In most cases, UV reflection is paired with reflection of longer wavelengths, such as a blue, green, yellow and red as well as white (Marshall 2000a,b) so that UV-blind animals can see a pattern/patch but will perceive it to be of a different colour than that seen by UV-sensitive animals. Few fish have been described that have body areas which only reflect UV light and thus, that have patterns that are invisible to UV-blind animals (*Apogon fragilis*, Marshall 2000b, and *Apogon leptacanthus*, Siebeck 2014). *Pomacentrus amboinensis* and *P. moluccensis* both have UV-yellow reflective patterns on a yellow background and therefore their patterns can also only be detected by UV sensitive animals (Siebeck et al. 2010; Fig. 3). As our visual system is not sensitive to UV, we have to use specialised equipment to make these patterns visible. A possible reason of why so few of these pure UV patterns are known is that they are difficult to detect for us, UV-blind creatures, and that they may not always be present as some fish can change colours and may switch their UV colours on/off.

Function of colour

The function of the coloration of fish has attracted much interest over the years (e.g., Longley 1916, Lorenz 1962, Lythgoe 1979, Marshall et al. 2003, Siebeck et al. 2010), but we are still far from an in-depth understanding, probably because specific functions might vary between different species. Also, and, maybe more importantly, we need

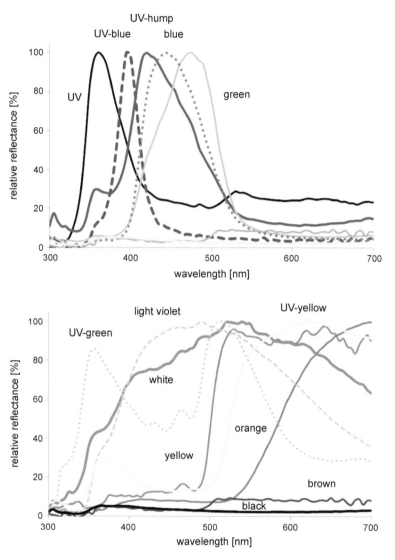

Fig. 2. Reflectance graphs representative of colour categories in Table 1. UV and yellow – *Pomacentrus amboinensis*; UV-blue – *Pomacentrus coelestis*; UV hump, Green, Black and Light violet – *Dascyllus aruanus*; Blue – *Pomacentrus coelestis*; Orange – *Chrysiptera cyanea*, Brown – *Dischistodus melanotus*; White – *Stegastes partitus*; UV-green – *Chromis viridis*; UV-yellow – *Pomacentrus moluccensis*. Reflectance spectra were measured with an Ocean Optics USB2000 fibre-optic spectrometer with UV/VIS reflectance probe attached (Ocean Optics, Dunedin, USA). A Spectralon white standard (99% reflection) was used as reference and an Ocean Optics PX-2 served as the light source.

to address this question from the perspective of the visual system of the intended receivers, rather than our own (Ali et al. 1977, Lythgoe 1979, Muntz 1990, Bennett et al. 1994). Despite the probably species-specific or even behavior-specific nature of

Fig. 3. *Pomacentrus amboinensis* (a) and *P. moluccensis* (b) photographed (Sony DSCF727) in visible light and through a UV-pass filter (Filter Nr. 51124, Oriel) and an infrared-absorbing filter (Filter Nr. 51720, Oriel, CT. USA) with a combined spectral transmission of 350–400 nm. Natural sunlight served as illumination. Facial UV patterns of four different individuals are shown for each species (*P. amboinensis*: a1-a4 and *P. moluccensis*: b1-b4).

colour patterns, several general principles have been proposed, including the possibility that certain colour patterns have no behavioural relevance at all (Marshall 2000a,b).

Camouflage is essential for the survival of fish. Whether or not an animal appears camouflaged to an observer depends on the visual system of the observer as well as on the ambient light environment and the background against which the fish is seen. It is therefore important to remember that animals, which appear conspicuous to us, may in fact be camouflaged to the intended receiver, such as potential predators, or may be camouflaged only when seen against certain backgrounds or even distances from the observer (Marshall 2000a). Extreme examples of camouflage are found in stonefish which resemble stones so closely that it is almost impossible to identify specific features such as the eye even once the general shape of the fish has been recognised. This extreme type of background matching only works for sedentary or slow moving fish and is not found in damselfishes. Another type of perfect camouflage would occur if the reflectance spectrum of a fish in the water column matched the background absolutely when viewed from any direction (Muntz 1990). While this is never completely achieved, several attempts have been made in that direction. Many fishes are darker on their dorsal surfaces than on their ventral ones (countershading) making them less conspicuous against both downwelling light when viewed from below and the dark background when viewed from above (Cott 1940). Most damselfishes are found close to the substrate and are probably rarely viewed from underneath, which may explain why there are not many examples of countershading (but see some *Azurina, Lepidozygus, Chromis* and *Chrysiptera* species). *Chromis viridis* is a good example for the behavioural modulation of camouflage/conspicuousness. Their blue/green colour is a close match to the underwater space light, so that when the fish are feeding above the reef in the water column, the contrast between them and their background is minimal, at least to our eyes (Marshall 2000a) and to the eyes of potential fish predators, which are typically UV blind (Siebeck and Marshall 2001; Fig. 4). In contrast, *C. viridis* stands out against the background illumination when seen by other damselfish, which typically possess a UV cone maximally sensitive to around 360 nm (Fig. 4b). When startled, these fish hide in amongst life coral branches where they are physically protected and the contrast between their blue/green colour and the background can be relatively large (depending on the colour of the coral, Fig. 4g). Another way to reduce conspicuousness is to be compressed and have silvery surfaces that act as mirrors reflecting the entire incident light, which is mostly seen in pelagic fish (Denton 1970, Jordan et al. 2014). Camouflage can also be achieved with disruptive coloration, drawing attention to individual elements of the pattern while at the same time concealing the outline of the fish (Longley 1917, Cott 1940, Muntz 1990). Examples for this would be the many damselfish species with bold black and white stripes (e.g., *Abudefduf* spp., *Dascyllus aruanus, Chrysiptera annulata*).

Certain colour patterns have been shown to convey information to either conspecifics or heterospecifics and are thus involved in **communication**. Many damselfish use visual cues including colour, or colour patterns to recognise conspecifics (e.g., *Dascyllus aruanus*, Katzir 1981 and *Pomacentrus amboinensis*, Siebeck et al. 2010; see Fig. 3), mates (e.g., *Amphiprion bicinctus*, Fricke 1973 and *Stegastes planifrons*, Thresher 1979) and predators (e.g., *Dascyllus marginatus*, Karplus et al. 2006). Many damselfishes change aspects of their colouration, either within a few seconds during behavioural encounters (e.g., during courtship, for review see Thresher 1984), or at intervals throughout their life history (e.g., when they change from juveniles into adults). Such colour changes can happen in response to the environment (e.g., a dark fish will pale if placed into a white bucket), as part of behavioural interactions and communication, as response to the illumination (melanisation or "sun tanning": melanin dispersion in response to UV exposure) and during ontogenetic development. Some types of colour changes are under the neural and/or endocrine control of the animals, e.g., changes during behavioural interactions, while others, such as ontogenetic colour changes, or sun tanning are not (for detailed review on fish colour changes see Leclercq et al. 2010).

Ontogenetic colour change is commonly observed in damselfish. The larval stages of many damselfish species are already pigmented and display colour patterns, however very few studies exist describing larval fish pigmentation and/or coloration (but see Leis and Carson-Ewart 2000, Baldwin 2013). Pomacentrid preflexion larvae have

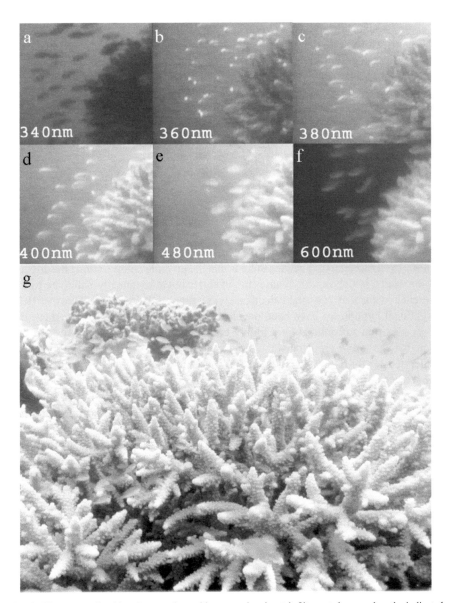

Fig. 4. *Chromis viridis* video frames taken with narrow band notch filters at the wavelengths indicated (a–f). Sun angle is roughly to the side of the angle between the camera and the coral head. Video was taken at 2 m depth near Lizard Island, Australia. For details on the equipment see Losey (2003) and (g) *Chromis viridis* photographed in broad spectrum visible light near Lizard Island, Australia (Pictures a–f: Losey and g: Siebeck).

melanophores in the dermis over the brain, gut and on the ventral and lateral midlines of their caudal fins and become more heavily pigmented during flexion (Leis and Carson-Ewart 2000). Species-specific pigmentation patterns and their development during ontogeny are thought to be useful for species identification even before settlement

(Leis and Carson-Ewart 2000, Baldwin 2013). Many settlement stage pomacentrid species already resemble their adult forms and colours. However, examples of striking post-settlement ontogenetic colour changes exist in the genus *Neoglyphidodon* (e.g., *N. melas* and *N. nigroris*, Randall et al. 1997). In some fish species two colour morphs develop post-settlement. The drivers for such polymorphisms have rarely been investigated but appear to include environmental cues such as the brightness of the habitat (e.g., *Chrysiptera leucopoma*; Frédérich et al. 2010), and the colour of other fish species in the habitat (*Pseudochromis fuscus*; Munday et al. 2003). Another type of more subtle ontogenetic colour change is the presence of an ocellus, or eyespot on juveniles. Ambon damselfish have a prominent ocellus on their dorsal fin, which most, but not all, lose when they mature (Gagliano and Depczynski 2013).

Various theories exist about why juveniles have different colours from adults, such as the 'intraspecific camouflage' hypothesis. The idea is that juveniles gain access to adult territories as they are not recognised by conspecific adults (e.g., Fricke 1973, Booth 1990, Mahon 1994), and the "adult habituation" hypothesis, which applies mainly to territorial fish. The idea there is that the juvenile colours attract the attention of the adults, which initially chase the juveniles but subsequently habituate to their presence so that they can establish themselves within adult territories (Thresher 1978). Additional hypotheses have been developed for eyespots. For example, these dark spots are often surrounded by a highly contrasting white ring and have long been thought to have anti-predatory function. However, the exact mechanism of this is still under debate (for review see Stevens 2005) and may not be the only explanation for the presence of eyespots (Gagliano and Depczynski 2013). *Pomacentrus amboinensis* have eyespots as larvae and juveniles. As they mature, most lose the eyespot, however there are some examples where this is not the case. Interestingly, the male fish, which retained their eyespot, more closely resembled immature females and juveniles than mature males (Gagliano and Depczynski 2013). So it is possible that the eyespot serves to camouflage the sexual identity and maturity of the fish thus making it easier for such individuals to gain access to the territories and possibly also to the females of territorial males.

Permanent sexual dimorphism is rare in damselfish and, if found at all, relatively subtle (e.g., black or white line along the anal fin of males in several species, for review see Thresher 1984), with the exception of *Chrysiptera cyanea*. Only male specimens have a bright orange caudal fin in this species. Temporal sexual dimorphism is found relatively frequently in damselfish as many males change colours during courtship events (Thresher 1984).

Behavioural colour change

Many fish change colour depending on their circumstances. Night colours are often different from day colours and fast colour changes are commonly observed during communication, for example during courtship (for description of courtship colour patterns see Thresher 1984). Behavioural colour change is an area that has not been well studied, with the exception of courtship signals.

For example, *Pomacentrus amboinensis* can shift the peak reflectance of their UV patterns from 370 nm to 400 nm (Siebeck 2014). Seen from the perspective of

UV-blind animals, these fish can effectively switch their patterns on and off by shifting them from visible to UV colours. The function of this is currently unknown but it is possible that the animals modulate the distance over which their patterns can be seen in this way (violet light is less scattered than UV and can thus be detected over larger distances). While we can make UV-reflective areas visible with the help of technology, what a UV-yellow or UV-blue looks like to a UV-sensitive animal is just about impossible to imagine for UV-blind creatures like us. We are not only missing a colour but a whole colour dimension. Consider the world from the perspective of human dichromats (people with extreme red/green colour blindness), or in fact from the perspective of most non-primate mammals. Dichromats see colours completely differently as compared to human trichromats (normal red, green and blue spectral sensitivities) and they would struggle to imagine what a striking contrast a poppy field in flower looks like to a trichromat (Fig. 5).

Fig. 5. Illustration of how a poppy field would appear to (a) a human trichromat (normal colour vision) and (b) dichromat (colour blind). Picture credit: George Losey.

Mechanisms of colour change

Colour changes can be achieved by dispersing or concentrating pigment granules within the chromatophores, or by changing the distance between the crystals in the multilayer stacks within the iridophores (Land 1972). Often, both mechanisms work together to produce the overall colour and colour change (Kasukawa et al. 1985, Kasukawa et al. 1986a, Fujii et al. 1989, Oshima et al. 1989).

Some damselfishes have been observed to rapidly change colour from the UV through to blue (*Pomacentrus amboinensis*, Siebeck 2014) or to blue-green (*Chrysiptera cyanea*, Kasukawa et al. 1986b, and *Pomacentrus coelestis*, Siebeck 2014). The mechanisms of colour change are best studied in *Chromis viridis* (Oshima et al. 1989) and *Chrysiptera cyanea* (Kasukawa et al. 1985, Kasukawa et al. 1986a). Interestingly, the control of this colour change was found to be solely under the sympathetic adrenergic system for *Chrysiptera cyanea* (Kasukawa et al. 1986a) and a combination of both, the nervous and endocrine systems in *Chromis viridis* (Oshima et al. 1989). In other species, the iridophores are not motile and any colour change is due to dispersion and aggregation of pigment in the chromatophores only (e.g., *Pomacentrus lepidogenys*, Kasukawa and Oshima 1987).

Summary – colours

In general, all fish have to solve the apparent trade-off between being simultaneously conspicuous to intended receivers such as potential mates and being inconspicuous to potential predators. As a result, colours have evolved under the constraints of attempting to maximize both, camouflage and conspicuousness, which, at first glance, appears difficult to achieve. It is however possible in cases where the visual systems of the intended observers and those of the relevant predators or other "illegitimate receivers" (Alcock 2009) differ enough so that colours can be used for communication that are difficult to detect by potential predators while being conspicuous to intended receivers. The best-studied example of such a system is the communication with UV patterns, which is often referred to as 'secret' or 'private communication' (Cummings et al. 2003, Siebeck 2004). Many damselfish exhibit UV colours (Marshall 2000b), which have been found to be important for species recognition and territorial defense in *Pomacentrus amboinensis* (Siebeck 2004, Siebeck et al. 2010) and well suited for simultaneous crypsis and communication in *Dascyllus reticulatus* (Losey 2003), and all relevant predators so far investigated have been UV blind (see section on vision below as well as Siebeck and Marshall 2001). Alternatively, temporal colour change can be used during important behavioural interactions, such as found in many damselfish species during courtship behaviour. The risk of attracting the attention of predators is then limited in time and may be managed by risk assessment prior to engaging in the specific behaviour. Yet another way to limit detection by predators, which usually detect their prey over larger distances is to use complex colour patterns, which, seen from a distance, attenuate in a way that the colours blur together and thus provide little contrast seen against the background (Marshall 2000b).

Vision

Introduction

An object is only visible if there is sufficient contrast between it and the background against which it is seen and if sufficient numbers of photons reach the photoreceptor cells. The downwelling light spectrum as well as the light reflected off objects is filtered not only on its passage through water, but also once it reaches the eye. Light has to pass through the ocular media (cornea, lens and vitreous) on its way to the retina, the place where light energy is converted, via a series of steps, into action potentials, the language of the nervous system and the brain. Photoreceptors with different sensitivities and spectral properties in the retina are responsible for the detection and conversion of light. It is the comparison of the output of different photoreceptors tuned to different wavelengths which is analysed in the visual system and eventually leads to the perception of colour. Visual processing by the retina and brain of fishes is an area that is not well studied. The best studied fish are zebrafishes as they serve as model organisms with which a wide range of mostly neuroscience questions can be addressed (Nikolaou et al. 2012).

Ocular media

As light enters the eye, it has to pass through the cornea, lens and vitreous humour. Depending on the filtering properties of these structures, the light spectrum and intensity is changed and/or reduced. While ocular media do not limit the long wavelength end of the spectrum available for vision, large variation can be found in the transmission of short wavelengths (e.g., Siebeck and Marshall 2001). Ultraviolet light is well known for its harmful properties, and many animals, including fish, protect their sensitive visual tissues in the retina by preventing UV from reaching these tissues (Douglas and McGuigan 1989, Siebeck and Marshall 2001, Losey et al. 2003, Siebeck and Marshall 2007). In most cases, the lens is the limiting filter of the eye (Siebeck and Marshall 2001), but UV-absorbing properties have also been found in the cornea (Siebeck and Marshall 2000, Siebeck et al. 2003) and vitreous humour of marine fish (Nelson et al. 2001). Mycosporine-like amino acids (MAAs) and/or carotenoids are responsible for the UV-absorbing properties of the ocular media (Dunlap et al. 1989, Siebeck et al. 2003). MAAs have also been found in the external mucus of many reef fishes, where they are thought to act as natural sunscreens (Zamzow and Losey 2002, Eckes et al. 2008). The specific combination and quantities of the different MAAs found in fishes, such as Palythine (320 nm), Asterina-330 (330), Palythinol (332 nm) and Palythene (360 nm), determine how much UV light the ocular media absorb/ transmit. Measuring ocular media transmission across different species of fish and/ or ontogenetic series within a species of fish is a relatively quick way to establish the potential for UV vision (Siebeck and Marshall 2001, Losey et al. 2003). Transmission properties are generally reported as T50 values, i.e., the wavelength at which 50% of the maximal transmission is reached (see Douglas and McGuigan 1989). Based on the assessment of ocular media transmission properties, around 50% of all measured marine fish have a T50 value < 400 nm and therefore possess the potential for UV vision (Siebeck and Marshall 2001, Losey et al. 2003, Siebeck and Marshall 2007).

All tested damselfish species (Table 1) have ocular media with T50 values in the UV and the limiting filter is the lens. There are 22 additional families, which include at least one species with T50 values below 400 nm, but pomacentrids have the lowest T50 values reported for any marine fish, which makes them suitable subjects for the study of UV vision and communication (Siebeck and Marshall 2001, Losey et al. 2003). The average T50 values (±sd) for the family are 340 nm based on the species measured in Siebeck and Marshall (2001, 2007) and 350 nm based on the 8 Hawaiian damselfish species measured in Losey et al. (2003) (Fig. 1b). There is only one known example of damselfish with a combined ocular media T50 value close to 400 nm, the Garibaldi damselfish, *Hypsypops rubicundus* (T50: 392 nm; Siebeck and Marshall 2007). As this is based on the results of a single specimen (standard length: 19 cm), it is unclear whether this value is typical for the species as a whole, or whether it is typical for the specific age/size of the fish.

The lens transmission properties of various damselfish species change with age, due to the combination of the increase in the path length (caused by lens growth) and a variable amount of UV-absorbing compounds being deposited in the lens (Douglas and McGuigan 1989, Siebeck and Marshall 2007). The general pattern in pomacentrids is that the T50 values increase with age (Regression analysis $F = 42$, $p < 0.0001$; $N = 94$; $t(slope) = 6.9$; $p < 0.0001$; Fig. 1c). The rate at which T50 increases with lens diameter varies between species. A fourfold increase in the lens diameter leads to a 60 nm shift towards longer wavelengths in *Pomacentrus moluccensis* (Siebeck and Marshall 2007). Interestingly, there are families that do not show this ontogenetic long-wavelength shift of T50 values (e.g., serranid species such as the Coral Trout, *Plectropomus leopardus*) and also families with lenses that transmit more UV with age as a result of decreasing T50 values (e.g., some holocentrid species; Siebeck and Marshall 2007). It is not clear how the different patterns are achieved, mostly because the mechanisms for MAA acquisition, potential storage and transport to the eyes are currently unknown. Metazoans do not appear to be capable of producing MAAs *de novo* and must acquire MAAs via the food chain (Mason et al. 1998, Zamzow 2004). There is, however, some indication that, once acquired, fish can modify the structure of the MAA (Kandel 2012). MAAs can be stored for up to a week, but it is unclear where this would take place and how MAAs might be mobilised when required (Zamzow et al. 2013). When fish are exposed to conditions lacking UV wavelengths, MAA levels in the epithelial mucus decrease (Zamzow 2004). It is currently unknown whether the MAAs in the ocular media are also affected by the environmental light conditions of the habitat of the fish.

Retina - spectral sensitivity

Teleost fish have a duplex retina, containing rods (dim vision) and cones (bright light and colour vision). Two different morphological cone types are found in damselfishes, single and double cones (Fig. 6). As the name suggests, double cones consist of two cones that are closely associated with each other. Unlike the twin cones found in other teleosts, double cones have different spectral sensitivities (Engström 1963).

Fig. 6. Photoreceptor mosaic of a *Pomacentrus amboinensis* flat mounted retina. Each central single cone is surrounded by four double cones and 4 corner (single) cones (Siebeck 2002). Scale bar 5 μm.

The discovery of close electric coupling between the two members of double cones has led to the assumption that the spectral sensitivity of the individual members is combined and thus that individual members cannot contribute separately to colour vision (Marchiafava 1985). This assumption has recently been challenged by a study showing that the double cones of the Blackbar Triggerfish, *Rhinecanthus aculeatus* do contribute to colour vision as separate channels (Pignatelli et al. 2010).

The spectral sensitivity of adult fish visual systems is generally well matched to the spectral properties of their habitat (Lythgoe 1979). A typical fish visual system is considered to include a cone type with spectral sensitivity matched to the wavelength of maximum downwelling light and a second cone type with spectral sensitivity off-set from the first (McFarland and Munz 1975, Loew and Lythgoe 1978, Lythgoe 1979, 1984). The spectral sensitivities of the damselfish visual pigments measured so far show a distribution with 3 peaks (525 nm, 490 nm and 360 nm). The peak at the longest wavelength is about 30 nm off-set from the wavelength of maximum transmission (560 nm) as measured at the Cobia Hole, Lizard Island, Australia (Fig. 1a,d). The other two peaks (490 nm and 360 nm) are off-set towards shorter wavelengths. Longer wavelength light (600 nm–700 nm), i.e., the part of the spectrum, which appears red to humans, seems to lie outside the spectral sensitivity of the damselfish measured so far. So overall the visual system of damselfish is well adapted to vision in the brightly lit shallow environment of coral reefs where the downwelling spectrum still includes significant amounts of UV light.

Spectral sensitivity of photoreceptors can be measured using a technique called Microspectrophotometry, MSP. This technique is powerful as the absorbance of individual photoreceptors can be assessed. All of the 18 damselfish species for which

MSP measurements have been carried out were found to have double cones with different spectral sensitivities, at least one type of single cone as well as rods (Table 1). The main drawback of the method is that, while there are millions of photoreceptors in the retina, the distribution of different spectral types is not equal. As the method is slow and depends on the selection of individual photoreceptors for measurement, only a small fraction of all photoreceptors can be measured per individual. UV-sensitive cones, just like the blue cones in the human visual system, are relatively rare and thus may be missed. So while the presence of a certain cone type can be taken as evidence of the spectral sensitivity of the visual system, the lack of a UV cone type as determined by an MSP study does not necessarily mean that UV sensitivity is absent in the fish. One also has to take into account that even if UV cones are absent, UV-sensitivity can be achieved via the secondary absorption maximum (β-band) of one of the cones maximally sensitive to longer wavelengths. However, in the case of β-band absorption, UV wavelengths cannot contribute to colour vision as it cannot be discriminated from longer wavelength light. UV cones have been found for most damselfish that have been measured with MSP (Table 1). There are seven species of damselfish for which no UV cones have been found so far. In two of the six *Chromis* species studied, UV cones have been found. The presence of UV cones in these species, combined with low T50 values makes it likely that the remaining four species also have UV cones. *Plectroglyphidodon johnstonianus, Pomacentrus melanochir* and *Neoglyphidodon melas* are the other three species for which no UV cones have been found. While no ocular media data exist for *P. johnstonianus* and *P. melanochir*, *N. melas* has T50 values in a range suitable for UV vision (350 nm–360 nm). Future studies are required to confirm whether UV cones are indeed lacking in these species, or whether they have been missed.

Molecular genetics has been used to study the evolution of visual pigments and to identify the various opsins (pigment proteins) present in different photoreceptor types (e.g., Bowmaker 2008). In vertebrates, there are four spectrally distinct classes of cone visual pigments and a single rod opsin class. Depending on their specific tuning, these opsin classes produce visual pigments within a range of spectral sensitivities: SWS1 ~360–440 nm; SWS2 ~400–450 nm; RH1 ~480–510 nm; RH2 ~450–530 nm; M/LWS ~510–560 nm (Yokoyama 2008). Many teleost families have extensively duplicated their opsin genes and as a result there are often several opsins within each class (Bowmaker 2008). A recent study on damselfish opsin genes based on ten species of damselfish (Table 1) revealed five cone opsin genes (both members of the RH2 duplication: RH2A/B; LWS; SWS1 and one member of the SWS2 duplication: SWS2B, while SWS2A was assumed to have been lost, which has recently been confirmed by Cortesi et al. 2015) and one rod opsin RH1 (Hofmann et al. 2012). A surprisingly similar pattern of sequence variability was found for damselfish and cichlids, with maximal variability at the long and short-wavelength ends of the spectrum. While sequence changes can lead to an extension of the spectral sensitivity of the visual pigment over longer time-scales, it is possible that differential gene expression can provide an evolutionary labile mechanism which allows the tuning of visual pigment sensitivity in the short term. The ability to fine-tune spectral sensitivities in addition to the long-term evolution of spectral sensitivities (via opsin gene duplication and diversification) may be particularly important for animals, such as fish, allowing them to populate the many spectrally diverse underwater environments.

Colour vision

While colour vision is only possible if multiple spectral photoreceptor types exist, the mere presence of multiple spectral types is not enough to conclude the presence of colour vision (Kelber et al. 2003). Another prerequisite for colour vision is that the output signal of at least two spectral cone types is compared as part of signal processing. The output of each cone is related to photon flux (the number of photons captured per unit time) in a certain wavelength range and it is the relative activation of the different cone types which, at least in humans, is interpreted by the brain as a certain colour. So while MSP measurements can determine the spectral sensitivities of individual photoreceptors, behavioural experiments are required to demonstrate that the quantum catches of these photoreceptors are compared before conclusions can be drawn about colour vision. Kelber and Osorio (2010) suggested that there are in fact four increasingly complex levels of extracting and analyzing spectral information. The most basic level can be achieved by simple animals that are trying to find a suitable light environment. In the simplest case, all that is required is an analysis of light present versus light absent (photokinesis and phototaxis). Very early in development, larval damselfishes have limited light sensitivity which rapidly increases as they develop during their pelagic phase (Job and Bellwood 2000). Younger larvae are found at shallower depths, which put together with their relatively lower light sensitivity, suggests that they select their preferred swimming depth on the basis of brightness. Settlement stage damselfish larvae are regularly captured in light traps, where they make up a large proportion of the entire catch (Doherty 1987). From an early age onwards damselfish clearly have the ability to perform at this basic level. The second level is achieved by animals that show certain colour preferences, i.e., certain colours on a potential mate trigger courtship behaviour, or the blue and yellow colours of a cleanerfish are thought to signal the presence of a cleaning station to potential clients (Cheney et al. 2009). This level has been termed "wavelength-specific" behaviour and one of its characteristics is that no prior experience with the colour is needed to trigger the behaviour. Damselfish are known to change their colours during courtship interactions and it is possible that these courtship colours trigger courting behaviour in a wavelength-specific fashion. However, this has never been directly tested to our knowledge. The third level requires the animal to be able to learn an arbitrary colour, which is taken as evidence that quantum catches are compared across multiple receptor types and a neural representation of the colour exists. There is evidence that *Pomacentrus amboinensis* can learn arbitrary colours (such as blue and yellow) and discriminate between them on the basis of hue alone (Siebeck et al. 2008). This study was the first to demonstrate this level of colour vision in a reef fish. The fourth level is only achieved if colour appearance can be evaluated (e.g., in terms of saturation) and this evaluation can be used for colour categorization. While this has not been directly investigated in reef fishes, and there is some debate as to how it is possible to test colour appearance in non-human animals, there is some evidence that *P. amboinensis* are able to generalize from a learned specific colour, such as blue or yellow, to other blues and yellows (Siebeck et al. 2008). These fish associated novel blues (or yellows) with the category 'rewarded stimuli' and discriminated them from another category of yellow (or blue) 'distracter stimuli'. Future studies are required to fully study potential categorical perception of colours in these fish.

Overall, colour vision in damselfish in general is not well studied despite the multitude of colours and patterns that the fish display. An increasing number of MSP studies are being published, which is helpful when designing behavioural experiments required for testing higher level colour vision. Once spectral sensitivities are known, visual modelling can be used to predict which colours should be discriminable by a certain visual system (Osorio and Vorobyev 2005). These predictions can then be tested using behavioural experiments. It is important to remember that even if, for example, five spectral types of cones are present in a retina, it cannot be assumed that the animal is pentachromatic, unless behavioural experiments testing the underlying assumption (i.e., that the output signals of all receptors are compared) are carried out. It is also possible that different sub-sets of cones contribute to colour vision during different behaviours, or under different light conditions (Neumeyer and Arnold 1989), and that spectral sensitivities change over time. While MSP allows us to measure the spectral sensitivity of various cone types (phenotype), molecular studies allow us to study the underlying mechanisms (genotype) for these spectral sensitivities and with that allow us to investigate the genetic mechanisms underlying the plasticity of the fish visual system (Sabbah et al. 2010).

Polarization vision

The sun radiates unpolarized light. Light becomes partially linearly polarized due to scattering in the atmosphere and water column, and also due to refraction and reflection at the water's surface (Horváth and Varjú 2004, Sabbah et al. 2005). The underwater polarization pattern is variable and depends on a range of variables, such as depth, quality of the water and, most importantly, the position of the sun (Horváth and Varjú 2004, Sabbah et al. 2005, Waterman 2006). Many terrestrial and aquatic animals use polarized light or a solar compass for orientation and recent studies suggest that such visual cues may also be used by damselfish larvae (e.g., *Chromis atripectoralis*) at a time when they must find suitable habitat for settlement (Leis and Carson-Ewart 2003). This idea is supported by a study on juvenile damselfish in which electroretinogram recordings of the species *Chromis viridis*, *Dascyllus trimaculatus* and *D. melanurus* were used to demonstrate that these species possess complex polarization sensitivity (Hawryshyn et al. 2003). Further evidence comes from behavioural tests in *C. viridis*, which demonstrated that these fish are able to discriminate between different e-vector directions independently of brightness cues (Mussi et al. 2005). They were also able to demonstrate that UV light was required for polarization sensitivity in this species, as the fish were no longer able to discriminate between different e-vector orientations when UV light was excluded from the test stimuli. Tests of *in situ* use of polarization cues in settlement stage damselfish are currently underway (Siebeck et al. manuscript in preparation).

Summary - function of vision

Vision is important for many aspects of a damselfish's life. The eyes develop rapidly so that larvae are able to capture food as soon as they have depleted their yolk sac. The sensitivity to light in this early stage appears to set a limit to their swimming depth as they are visual feeders (Job and Bellwood 2000, Job and Shand 2001). At night the vertical distribution of damselfish larvae is more variable compared to during the day when they are found at relatively shallow depths (Leis 1991, 2004). Towards the end of the pelagic phase, damselfish have to find a reef to settle on and there is evidence that they are able to use visual cues for orientation in the pelagic environment (Leis and Carson-Ewart 2003, Berenshtein et al. 2014). While the ability to hold a bearing does not necessarily assist larvae to locate a particular reef, it can enable them to orient in a particular direction, and this is helpful in counteracting a prevailing current (Leis and Carson-Ewart 2003, Leis et al. 2011). At settlement, damselfish select their habitat based on a range of cues, including vision and olfaction (Booth 1992, Gerlach et al. 2007). They are able to use visual cues to discriminate between conspecifics and heterospecifics as new settlers (Brolund et al. 2003, Lecchini et al. 2005) as well as later on in life (Thresher 1979, Katzir 1981, Siebeck et al. 2010). Damselfish are able to visually discriminate between non-predatory fish and predators (Karplus et al. 1982, Karplus et al. 2006), use visual cues for territory selection and defense (Siebeck 2004, Rilov et al. 2007) and there is ample evidence that visual cues are important during courtship events (Thresher 1984, Oliver and Lobel 2013).

Climate change is affecting the chemical as well as the physical properties of the water damselfish live in and thus poses challenges for the survival of the fish exposed to these changes (Munday et al. 2012). Chemical changes (acidification - increased CO_2 levels) have already been shown to alter sensory as well as cognitive abilities in damselfishes (e.g., Nilsson et al. 2012) while physical changes, such as increased particulate matter (due to storms or eutrophication) lead to increased turbidity, loss of visual contrast and possibly also a shift in spectral irradiance (depending on the properties of the suspended particulate matter; for review see Siebeck et al. 2015). The combined effects can have devastating consequences and lead to decreased survival (e.g., reduced ability to detect and recognize predators) and ultimately a loss of species, such as has been described for cichlids exposed to increasingly eutrophic waters (Seehausen et al. 2008). Future studies will reveal if and, if so, how the sensory systems of fish may adapt to the changing conditions.

This review has concentrated on visual cues in isolation of other sensory modalities, which of course is an unnatural scenario as a multitude of various sensory cues are available to any animal going about its business in its habitat. Scientists studying an individual sense usually go to great trouble to exclude potentially confounding cues from other sensory systems so that behavioural responses can be tied to a single sense alone. This is an important step in the process of trying to understand the limitations and abilities of the particular sense. Once this has been accomplished, the next step should be to conduct trade-off experiments in which two or more senses are combined and their relative importance can be evaluated. Damselfish, which have a variety of sensory modalities available to them and have proven to be amenable to experimentation, should provide interesting subjects for such research.

Acknowledgements

I would like to thank George Losey for discussions and valuable feedback on the manuscript and Justin Marshall for discussions about colour categories. Damselfish colour spectra were collected with support from the Sea World Research and Rescue Foundation and the project was supported by the University of Queensland and the Australian Research Council (DP140100431).

References

Alcock, J. 2009. Animal Behavior: An Evolutionary Approach. Sinauer, Sunderland, Massachusetts.

Ali, M.A., R.A. Ryder and M. Anctil. 1977. Photoreceptors and visual pigments as related to behavioural responses and preferred habitats of perches (*Perca* spp.) and pikeperches (*Stizostedion* spp.). J. Fish. Res. Board Can. 34: 1475–1480.

Allen, G.R. 1991. Riffbarsche der Welt. Mergus, Hongkong.

Bagnara, J.T., P.J. Fernandez and R. Fujii. 2007. On the blue coloration of vertebrates. Pigm. Cell. Res. 20(1): 14–26.

Baldwin, C.C. 2013. The phylogenetic significance of colour patterns in marine teleost larvae. Zool. J. Linn. Soc. 168(3): 496–563.

Bennett, A.T.D., I.C. Cuthill and K.J. Norris. 1994. Sexual selection and the mismeasure of color. Am. Nat. 144(5): 848–860.

Berenshtein, I., M. Kiflawi, N. Shashar, U. Wieler, H. Agiv and C.B. Paris. 2014. Polarized light sensitivity and orientation in coral reef fish post-larvae. PLoS ONE 9(2): e88468.

Booth, C.L. 1990. Evolutionary significance of ontogenetic colour-change in animals. Biol. J. Linn. Soc. 40(2): 125–163.

Booth, D.J. 1992. Larval Settlement-patterns and preferences by domino damselfish *Dascyllus albisella* Gill J. Exp. Mar. Biol. Ecol. 155(1): 85–104.

Bowmaker, J.K. 2008. Evolution of vertebrate visual pigments. Vision Res. 48(20): 2022–2041.

Brolund, T.M., L.E. Nielsen and M. Arvedlund. 2003. Do juvenile *Amphiprion ocellaris* (Pisces: Pomacentridae) recognize conspecifics by chemical or visual cues? J. Mar. Biol. Assoc. UK 83(5): 1127–1136.

Cheney, K.L., A.S. Grutter, S.P. Blomberg and N.J. Marshall. 2009. Blue and yellow signal cleaning behavior in coral reef fishes. Curr. Biol. 19(15): 1283–1287.

Cortesi, F., Z. Musilová, S.M. Stieb, N.S. Hart, U.E. Siebeck, M. Malmstrøm, O.K. Tørresen, S. Jentoft, K. Cheney, N.J. Marshall, K.L. Carleton and W. Salzburger. 2015. Ancestral duplications and highly dynamic opsin gene evolution in percomorph fishes. Proc. Natl. Acad. Sci. USA 112: 1493–1498.

Cott, H.B. 1940. Adaptive Colouration in Animals. Methuen, London.

Cummings, M.E., G.G. Rosenthal and M.J. Ryan. 2003. A private ultraviolet channel in visual communication. Proc. R. Soc. Lond. Ser. B 270(1518): 897–904.

Denton, E.J. 1970. On the organisation of reflecting surfaces in some marine animals. Phil. Trans. R. Soc. Lond. B 258: 285–313.

Doherty, J.P. 1987. Light-traps: selective but useful devices for quantifying the distributions and abundances of larval fishes. Bull. Mar. Sci. 41(2): 423–431.

Douglas, R.H. and C.M. McGuigan. 1989. The spectral transmission of freshwater teleost ocular media—an interspecific comparison and a guide to potential ultraviolet sensitivity. Vision Res. 29(7): 871–879.

Dunlap, W.C., D.M. Williams, B.E. Chalker and A.T. Banaszak. 1989. Biochemical photoadaptation in vision: UV-absorbing pigments in fish eye tissues. Comp. Biochem. Physiol. B. Biochem. Mol. Biol. 93(3): 601–607.

Eckes, M.J., U.E. Siebeck, S. Dove and A.S. Grutter. 2008. Ultraviolet sunscreens in reef fish mucus. Mar. Ecol. Prog. Ser. 353: 203–211.

Endler, J.A. 1990. On the measurement and classification of colour in studies of animal colour patterns. Biol. J. Linn. Soc. 41: 315–352.

Engström, K. 1963. Cone types and cone arrangements in teleost retinae. Acta Zool. 44: 179–243.

Fox, H.M. and G. Vevers. 1960. The Nature of Animal Colours. Sidgwick & Jackson, London.

Frédérich, B., S.C. Mills, D. Mathieu, E. Parmentier, C. Brié, R. Santos, V.P. Waqalevu and D. Lecchini. 2010. Colour differentiation in a coral reef fish throughout ontogeny: habitat background and flexibility. Aquat. Biol. 9: 271–277.

Fricke, H.W. 1973. Individual partner recognition in fish: field studies on *Amphiprion bicinctus*. Naturwissenschaften 60: 204–206.

Fujii, R., H. Kasukawa, K. Miyaji and N. Oshima. 1989. Mechanisms of skin coloration and its changes in the blue-green damselfish, *Chromis viridis*. Zool. Sci. 6(3): 477–486.

Gagliano, M. and M. Depczynski. 2013. Spot the difference: mimicry in a coral reef fish. PloS ONE 8(2): e55938.

Gerlach, G., J. Atema, M.J. Kingsford, K.P. Black and V. Miller-Sims. 2007. Smelling home can prevent dispersal of reef fish larvae. Proc. Natl. Acad. Sci. USA 104(3): 858–863.

Ghiradella, H., D. Aneshansley, T. Eisner, R.E. Silberglied and H.E. Hinton. 1972. Ultraviolet reflection of a male butterfly: interference color caused by thin-layer elaboration of wing scales. Science 178(4066): 1214–1217.

Goda, M. and R. Fujii. 1995. Blue chromatophores in two species of callionymid fish. Zool. Sci. 12(6): 811–813.

Goda, M., M. Ohata, H. Ikoma, Y. Fujiyoshi, M. Sugimoto and R. Fujii. 2011. Integumental reddish-violet coloration owing to novel dichromatic chromatophores in the teleost fish, *Pseudochromis diadema*. Pigment Cell Melanoma Res. 24(4): 614–617.

Hawryshyn, C.W., H.D. Moyer, W.T. Allison, T.J. Haimberger and W.N. McFarland. 2003. Multidimensional polarization sensitivity in damselfishes. J. Comp. Physiol. A 189(3): 213–220.

Hofmann, C.M., N.J. Marshall, K. Abdilleh, Z. Patel, U.E. Siebeck and K.L. Carleton. 2012. Opsin evolution in damselfish: convergence, reversal, and parallel evolution across tuning sites. J. Mol. Evol. 75(3-4): 79–91.

Horváth, G. and D. Varjú. 2004. Polarized light in animal vision: polarization patterns in nature. Springer, Berlin.

Jerlov, N.G. 1976. Marine Optics. Elsevier Scientific, Amsterdam, New York.

Job, S.D. and D.R. Bellwood. 2000. Light sensitivity in larval fishes: implications for vertical zonation in the pelagic zone. Limnol. Oceanogr. 45(2): 362–371.

Job, S.D. and J. Shand. 2001. Spectral sensitivity of larval and juvenile coral reef fishes: implications for feeding in a variable light environment. Mar. Ecol. Prog. Ser. 214: 267–277.

Jordan, T.M., J.C. Partridge and N.W. Roberts. 2014. Disordered animal multilayer reflectors and the localization of light. J. R. Soc. Interface 11: 20140948.

Kandel, F.L.M. 2012. Ultraviolet Sunscreen on the Coral Reef: From Coral to Fish. PhD, University of Hawai'i at Manoa.

Karplus, I., M. Goren and D. Algom. 1982. A preliminary experimental analysis of predator face recognition by *Chromis caeruleus* (Pisces, Pomacentridae). Z. Tierpsych. 58(1): 53–65.

Karplus, I., R. Katzenstein and M. Goren. 2006. Predator recognition and social facilitation of predator avoidance in coral reef fish *Dascyllus marginatus* juveniles. Mar. Ecol. Prog. Ser. 319: 215–223.

Kasukawa, H. and N. Oshima. 1987. Divisionistic generation of skin hue and the change of shade in the scalycheek damselfish, *Pomacentrus lepidogenys*. Pigm. Cell. Res. 1(3): 152–157.

Kasukawa, H., M. Sugimoto, N. Oshima and R. Fujii. 1985. Control of chromatophore movements in dermal chromatic units of blue damselfish—I. The melanophore. Comp. Biochem. Physiol. C 81(2): 253–257.

Kasukawa, H., N. Oshima and R. Fujii. 1986a. Control of chromatophore movements in dermal chromatic units of blue damselfish—II. The motile iridophore. Comp. Biochem. Physiol. C 83(1): 1–7.

Kasukawa, H., N. Oshima and R. Fujii. 1986b. Mechanism of light-reflection from fish skin possessing the motile iridophores. Zool. Sci. 3(6): 971–971.

Kasukawa, H., N. Oshima and R. Fujii. 1987. Mechanisms of light reflection in blue damselfish motile iridophore. Zool. Sci. 4: 243–257.

Katzir, G. 1981. Visual aspects of species recognition in the damselfish *Dascyllus aruanus* L. (Pisces, Pomacentridae). Anim. Behav. 29: 842–849.

Kelber, A. and D. Osorio. 2010. From spectral information to animal colour vision: experiments and concepts. Proc. R. Soc. B-Biol. Sci. 277(1688): 1617–1625.

Kelber, A., M. Vorobyev and D. Osorio. 2003. Animal colour vision—behavioural tests and physiological concepts. Biol. Rev. 78(1): 81–118.

Land, M.F. 1972. The physics and biology of animal reflectors. Prog. Biophys. Mol. Biol. 24: 75–106.

Lecchini, D., S. Planes and R. Galzin. 2005. Experimental assessment of sensory modalities of coral-reef fish larvae in the recognition of their settlement habitat. Behav. Ecol. Sociobiol. 58(1): 18–26.

Leclercq, E., J.F. Taylor and H. Migaud. 2010. Morphological skin colour changes in teleosts. Fish Fish. 11(2): 159–193.

Leis, J.M. 1991. Vertical-distribution of fish larvae in the Great-Barrier-Reef-lagoon, Australia. Mar. Biol. 109(1): 157–166.

Leis, J.M. 2004. Vertical distribution behaviour and its spatial variation in late-stage larvae of coral-reef fishes during the day. Mar. Freshw. Behav. Physiol. 37(2): 65–88.

Leis, J.M. and B.M. Carson-Ewart. 2000. The Larvae of Indo-Pacific Coastal Fishes—An Identification Guide to Marine Fish Larvae. Brill, Leiden.

Leis, J.M. and B.M. Carson-Ewart. 2003. Orientation of pelagic larvae of coral-reef fishes in the ocean. Mar. Ecol. Prog. Ser. 252: 239–253.

Leis, J.M., U. Siebeck and D.L. Dixson. 2011. How Nemo finds home: neuroecology of larva dispersal and population connectivity in marine, demersal fishes. Integr. Comp. Biol. 51: E79–E79.

Loew, E.R. and J.N. Lythgoe. 1978. The ecology of cone pigments in teleost fishes. Vision Res. 18(6): 715–722.

Longley, W.H. 1916. Observations upon tropical fishes and inferences from their adaptive coloration. Proc. Natl. Acad. Sci. USA 2: 733–737.

Longley, W.H. 1917. Studies upon the biological significance of animal coloration I—the colors and color changes of west indian reef-fishes. J. Exp. Zool. 23(3): 533–601.

Lorenz, K. 1962. The function of colour in coral reef fishes. Proc. R. Inst. GB 39: 282–296.

Losey, G.S. 2003. Crypsis and communication functions of UV-visible coloration in two coral reef damselfish, *Dascyllus aruanus* and *D. reticulatus*. Anim. Behav. 66: 299–307.

Losey, G.S., W.N. McFarland, E.R. Loew, J.P. Zamzow, P.A. Nelson and N.J. Marshall. 2003. Visual biology of Hawaiian coral reef fishes. I. Ocular transmission and visual pigments. Copeia 203(3): 433–454.

Lythgoe, J.N. 1979. The Ecology of Vision. Clarendon Press, Oxford.

Lythgoe, J.N. 1984. Visual pigments and environmental light. Vision Res. 24(11): 1539–1550.

Mahon, J.L. 1994. Advantage of flexible juvenile coloration in 2 species of labroides (pisces, labridae). Copeia (2): 520–524.

Marchiafava, P.L. 1985. Cell coupling in double cones of the fish retina. Proc. R. Soc. B-Biol. Sci. 226(1243): 211–215.

Marshall, N.J. 2000a. Communication and camouflage with the same 'bright' colours in reef fishes. Phil. Trans. R. Soc. Lond. B: 1243–1248.

Marshall, N.J. 2000b. The visual ecology of reef fish colours. pp. 83–120. *In*: Y. Espmark, T. Amundsen and G. Rosenqvist (eds.). Animal Signals. Adaptive Significance of Signalling and Signal Design in Animal Communication. Tapir Publishers, Trondheim.

Marshall, N.J., K. Jennings, W.N. McFarland, E.R. Loew and G.S. Losey. 2003. Visual biology of Hawaiian coral reef fishes. II. Colors of Hawaiian Coral Reef Fish. Copeia 203(3): 455–466.

Mason, D.S., F. Schafer, J.M. Shick and W.C. Dunlap. 1998. Ultraviolet radiation-absorbing mycosporine-like amino acids (MAAs) are acquired from their diet by medaka fish (*Oryzias latipes*) but not by SKH-1 hairless mice. Comp. Biochem. Physiol. A 120(4): 587–598.

McFarland, W.N. and F.W. Munz. 1975. Part III: The evolution of photopic visual pigments in fishes. Vision Res. 15: 1071–1080.

McFarland, W.N. and E.R. Loew. 1994. Ultraviolet visual pigments in marine fishes of the family pomacentridae. Vision Res. 34(11): 1393–1396.

Munday, P.L., P.J. Eyre and G.P. Jones. 2003. Ecological mechanisms for coexistence of colour polymorphisms in a coral-reef fish: an experimental evaluation. Oecologia 137: 519–526.

Munday, P.L., M.I. McCormick and G.E. Nilsson. 2012. Impact of global warming and rising CO_2 levels on coral reef fishes: what hope for the future? J. Exp. Biol. 215: 3865–73.

Muntz, W.R.A. 1990. Stimulus, environment and vision in fishes. pp. 491–511. *In*: R.H. Douglas and B.A. Djamgoz (eds.). The Visual System of Fishes. Chapman and Hall, London.

Mussi, M., T.J. Haimberger and C.W. Hawryshyn. 2005. Behavioural discrimination of polarized light in the damselfish *Chromis viridis* (family Pomacentridae). J. Exp. Biol. 208(16): 3037–3046.

Nelson, P.A., J.P. Zamzow and G.S. Losey. 2001. Ultraviolet blocking in the ocular humors of the teleost fish *Acanthocybium solandri* (Scombridae). Can. J. Zool. 79(9): 1714–1718.

Neumeyer, C. and K. Arnold. 1989. Tetrachromatic colour vision in the goldfish becomes trichromatic under white adaptation light of moderate intensity. Vision Res. 29(12): 1719–1727.

Nikolaou, N., A.S. Lowe, A.S. Walker, F. Abbas, P.R. Hunter, I.D. Thompson and M.P. Meyer. 2012. Parametric functional maps of visual inputs to the tectum. Neuron 76(2): 317–324.

Nilsson, G.E., D.L. Dixson, P. Domenici, M.I. McCormick, C. Sorenson, S.-A. Watson and P.L. Munday. 2012. Near-future carbon dioxide levels alter fish behaviour by interfering with neurotransmitter function. Nat. Clim. Change 2: 201–204.

Oliver, S.J. and P.S. Lobel. 2013. Direct mate choice for simultaneous acoustic and visual courtship displays in the damselfish, *Dascyllus albisella* (Pomacentridae). Environ. Biol. Fishes 96(4): 447–457.

Oshima, N., H. Kasukawa and R. Fujii. 1989. Control of chromatophore movements in the blue-green damselfish, *Chromis viridis*. Comp. Biochem. Physiol. C 93(2): 239–245.

Osorio, D. and M. Vorobyev. 2005. Photoreceptor spectral sensitivities in terrestrial animals: adaptations for luminance and colour vision. Proc. R. Soc. B-Biol. Sci. 272(1574): 1745–1752.

Pignatelli, V., C. Champ, J. Marshall and M. Vorobyev. 2010. Double cones are used for colour discrimination in the reef fish, *Rhinecanthus aculeatus*. Biol. Lett. doi: 10.1098/rsbl.2009.1010.

Randall, J.E., G.R. Allen and R.C. Steene. 1997. Fishes of the Great Barrier Reef and Coral Sea. Crawford House Publishing Pty Ltd., Bathurst, Australia.

Rilov, G., W.F. Figueira, S.J. Lyman and L.B. Crowder. 2007. Complex habitats may not always benefit prey: linking visual field with reef fish behavior and distribution. Mar. Ecol. Prog. Ser. 329: 225–238.

Sabbah, S., A. Lerner, C. Erlick and N. Shashar. 2005. Underwater polarization vision: a physical examination. pp. 123–177. *In*: S. Pandalai (ed.). Recent Research Developments in Experimental and Theoretical Biology. Transworld Research Network, Trivandrum.

Sabbah, S., R.L. Laria, S.M. Gray and C.W. Hawryshyn. 2010. Functional diversity in the color vision of cichlid fishes. BMC Biology 8: 133.

Seehausen, O., Y. Terai, I.S. Magalhaes, K.L. Carleton, H.D.J. Mrosso, R. Miyagi, I. van der Sluijs, M.V. Schneider, M.E. Maan, H. Tachida, H. Imai and N. Okada. 2008. Speciation through sensory drive in cichlid fish. Nature 455: 620–23.

Siebeck, U.E. 2002. UV vision and visual ecology of reef fish. PhD, University of Queensland.

Siebeck, U.E. 2004. Communication in coral reef fish: the role of ultraviolet colour patterns in damselfish territorial behaviour. Anim. Behav. 68: 273–282.

Siebeck, U.E. 2014. Communication in the ultraviolet: unravelling the secret language of fish. pp. 299–320. *In*: G. Witzany (ed.). Biocommunication of Animals. Springer, Heidelberg.

Siebeck, U.E. and N.J. Marshall. 2000. Transmission of ocular media in labrid fishes. Phil. Trans. R. Soc. Lond. B 355(1401): 1257–1261.

Siebeck, U.E. and N.J. Marshall. 2001. Ocular media transmission of coral reef fish—can coral reef fish see ultraviolet light? Vision Res. 41(2): 133–149.

Siebeck, U.E. and N.J. Marshall. 2007. Potential ultraviolet vision in pre-settlement larvae and settled reef fish—A comparison across 23 families. Vision Res. 47(17): 2337–2352.

Siebeck, U.E., S.P. Collin, M. Ghoddusi and N.J. Marshall. 2003. Occlusable corneas in toadfishes: light transmission, movement and ultrastruture of pigment during light- and dark-adaptation. J. Exp. Biol. 206(13): 2177–2190.

Siebeck, U.E., G. Losey and N.J. Marshall. 2006. UV communication in fish. pp. 423–456. *In*: B.G. Kapoor, F. Ladich, S.P. Collin and W.G. Raschi (eds.). Fish Communication. Science Publishers, Enfield, New Hampshire.

Siebeck, U.E., G.M. Wallis and L. Litherland. 2008. Colour vision in reef fish. J. Exp. Biol. 211: 354–360.

Siebeck, U.E., A.N. Parker, D. Sprenger, L.M. Mathger and G. Wallis. 2010. A species of reef fish that uses ultraviolet patterns for covert face recognition. Curr. Biol. 20(5): 407–410.

Siebeck, U.E., C. Braun, J. O'Connor and J.M. Leis. 2015. Do human activities influence survival and orientation abilities of larval fishes in the ocean? Integr. Zool. 10: 65–82.

Stevens, M. 2005. The role of eyespots as anti-predator mechanisms, principally demonstrated in the Lepidoptera. Biol. Rev. 80(4): 573–588.

Thresher, R.E. 1978. Territoriality and aggression in the threespot damselfish (Pisces; Pomacentridae): an experimental study of causation. Z. Tierpsych. 46: 401–434.

Thresher, R.E. 1979. The role of individual recognition in the territorial behaviour of the threespot damselfish, *Eupomacentrus planifrons*. Mar. Behav. Physiol. 6: 83–93.

Thresher, R.E. 1984. Reproduction in Reef Fishes. T.F.H. Publications, Neptune City, NJ.

Vorobyev, M. and D. Osorio. 1998. Receptor noise as a determinant of colour thresholds. Proc. R. Soc. B-Biol. Sci. 265(1394): 351–358.

Waller, S.J. 2005. Ontogenetic colour change and visual ecology of reef fish. PhD, The University of Queensland.

Waterman, T.H. 2006. Reviving a neglected celestial underwater polarization compass for aquatic animals. Biol. Rev. 81(1): 111–115.

Wucherer, M.F. and N.K. Michiels. 2012. A fluorescent chromatophore changes the level of fluorescence in a reef fish. PLoS ONE 7(6): e37913.

Wyszecki, G. and W.S. Stiles. 2000. Color Science: Concepts and Methods, Quantitative Data and Formulae, 2nd Edition. Wiley and Sons, New York.

Yokoyama, S. 2008. Evolution of dim-light and color vision pigments. Ann. Rev. Genom. Hum. Genet. 9: 259–282.

Zamzow, J.P. 2004. Effects of diet, ultraviolet exposure, and gender on the ultraviolet absorbance of fish mucus and ocular structures. Mar. Biol. 144(6): 1057–1064.

Zamzow, J.P. and G.S. Losey. 2002. Ultraviolet radiation absorbance by coral reef fish mucus: photo-protection and visual communication. Environ. Biol. Fishes 63(1): 41–47.

Zamzow, J.P., U.E. Siebeck, M.J. Eckes and A.S. Grutter. 2013. Ultraviolet-B wavelengths regulate changes in UV absorption of cleaner fish *Labroides dimidiatus* mucus. PLoS ONE 8(10): e78527.

Cerato-Mandibular Ligament: A Key Trait in Damselfishes?

Damien Olivier,* Bruno Frédérich[a] *and Eric Parmentier*[b]

Introduction

The evolutionary success of a natural population can sometimes be explained by a key character that enables the taxon to interact with its environment in a different and original way (Galis 2001, Losos 2010), and allows it to be more competitive in the use of resources (Baum and Larson 1991). Such a character is commonly named a "key innovation" (Miller 1949, Van Valen 1971, Levinton 1988, Baum and Larson 1991, Rosenzweigh and McCord 1991, Erwin 1992, Heard and Hauser 1995, Hunter 1998).

Usual examples of key innovations are the evolutions of the wings in birds, bats and pterosaurs. In each of these distantly related taxa, the new functional trait (wings) has allowed the taxon to use the environment in another way (flight) and has provided access to previously untapped niches. Other examples of key innovations are the adhesive toepads in geckos and anoles (Autumn et al. 2002, Losos 2010); the aerobic citrate utilization (Cit^+) in populations of *Escherichia coli* (Blount et al. 2008); the jaw-powered suction feeding in the fish-tetrapod transition (Heiss et al. 2013) or metamorphosis in insects (Carpenter 1953, Rainford et al. 2014). In teleosts, the most famous example is the modified pharyngeal jaws and associated muscles in cichlids that allow an efficient manipulation of food (particularly strong bite) (Liem 1973, Kaufman and Liem 1982, Galis and Drucker 1996, Wainwright 2006).

Laboratoire de Morphologie Fonctionnelle et Evolutive, AFFISH - Research Center, University of Liège, Quartier Agora, Allée du six Août 15, Bât. B6C, 4000 Liège (Sart Tilman), Belgium.
[a] Email: bruno.frederich@ulg.ac.be
[b] Email: E.Parmentier@ulg.ac.be
* Corresponding author: dolivier@ulg.ac.be

With 394 described species (see Chapter II), the Pomacentridae are an example of a highly successful radiation. Many characters at different organismal levels (physiological, biochemical, morphological, behavioural...) are likely involved in the success of this group. However, except a recent study that associated the radiation of clownfishes and their mutualism with sea anemones (Litsios et al. 2012), no hypothesis has been presented to explain such a success. The Pomacentridae possess at least two morpho-functional novelties. (1) Similar to cichlids, labrids and embiotocids, the pomacentrids have developed modified pharyngeal jaws that permit efficient food processing (Liem and Greenwood 1981, Stiassny and Jensens 1987, Galis and Snelderwaard 1997, Wainwright et al. 2012) which could be an important functional innovation. (2) Recent morpho-functional studies demonstrated a role for the cerato-mandibular ligament (c-md) during feeding and sound production (Parmentier et al. 2007, Olivier et al. 2014, 2015).

We focus in this chapter on the bucco-pharyngeal apparatus which is considered to have promoted the impressive evolutionary diversification and ecological success of teleosts (Wainwright and Bellwood 2002). We firstly summarize the data on the pharyngeal jaws and the c-md ligament of damselfishes. Secondly, we discuss the potential influence of the c-md on the diversification of damselfishes.

Morpho-functional Innovations in Pomacentridae

The pharyngeal jaws

Damselfishes have an articulation between the lower pharyngeal jaw and the pectoral girdle (Liem and Greenwood 1981, Stiassny and Jensens 1987, Galis and Snelderwaard 1997; Fig. 1). This articulation provides an original way to elevate the lower pharyngeal jaw because the contraction of the muscle *protractor pectoralis* (that joins the cleithrum to the skull) appears to lift the pectoral girdle, thus pushing up the lower pharyngeal jaw. The advantage of this mechanism could be a considerable increase in bite force (Galis and Snelderwaard 1997). Nonetheless, the functional advantage of having a powerful bite at the level of the pharyngeal jaw in Pomacentridae has never been tested. In addition, no species of Pomacentridae feed on hard-shelled prey (Chapter VII), so it is difficult to present hypotheses on the ecological advantage of a powerful pharyngeal bite in Pomacentridae.

Another hypothesis is that this novel biting mechanism of the pharyngeal jaw could also be involved in tasks other than feeding. During agonistic interactions, all clownfish species have evolved submissive postures (head shaking) that presumably serve to circumvent physical injury. This behaviour, consisting of a lateral quivering of the head, is associated with the production of submissive sounds composed of several pulses (Colleye and Parmentier 2012). Contrary to aggressive sounds (Chapter X), the mechanism allowing this kind of sound is not known but the pharyngeal jaws could be involved. This hypothesis is supported by the relationships between different characters such as sounds with multiple pulses, head shaking and pharyngeal anatomy. At this point, it is speculative but this assumption deserves to be tested because pharyngeal jaws are used during sound production in different species (Bertucci et al. 2014). To the best of our knowledge, submissive behaviours were not studied in other pomacentrids.

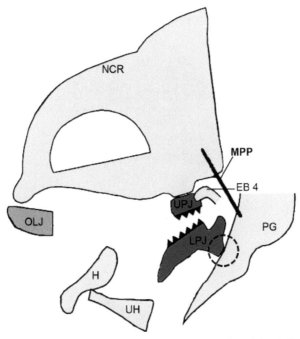

Fig. 1. The pharyngeal apparatus in pomacentrids. Schematic illustration of the skull of a pomacentrid (modified from Galis and Snelderwaard 1997). The articulation between the lower pharyngeal jaw and the pectoral girdle is indicated by the dashed circle. The thick black line represents the muscle protractor pectoralis (MPP). EB 4: epibranchial 4; H: hyoid bar; LPJ: lower pharyngeal jaw; NCR: neurocranium; OLJ: oral lower jaw; PG: pectoral girdle; UPJ: upper pharyngeal jaw; UH: urohyal.

The cerato-mandibular ligament

Another morphological characteristic of damselfish is the presence of a ligament, which joints the ceratohyal of the hyoid bar to the primordial process of the angular of the mandible (Fig. 2).

According to Stiassny (1981), this ligament, called the cerato-mandibular (c-md), is a synapomorphic trait within Pomacentridae. However, as phylogenetic hypotheses have greatly evolved since the eighties it was necessary to check this assertion. The current phyletic relationships of Pomacentridae are poorly supported but 14 groups (Fig. 3) are potentially damselfish sister taxa (see Wainwright et al. 2012). From the literature (Stiassny 1981), it is known that Cichlidae and Embiotocidae do not have a c-md ligament. We additionally checked the presence of this ligament in five putative closely-related groups: Plesiopidae, Pseudochromidae, Grammatidae, Opistognathidae and Blennioidei and found that only Pseudochromidae possess a c-md ligament (found in five species from four genera) (Fig. 3).

Contrary to Pomacentridae, the c-md ligament in Pseudochromidae inserts beneath the angulo-quadrate articulation. The c-md ligament has therefore an opposite function in the two families, i.e., it allows mouth closing in Pomacentridae and mouth opening in Pseudochromidae (Fig. 4). A c-md ligament is also present in other non-related

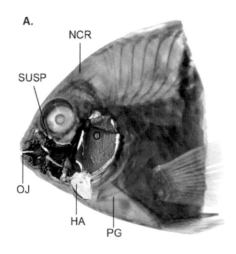

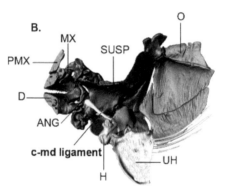

Fig. 2. Representation of the cerato-mandibular (c-md) ligament in Pomacentridae (modified from Olivier et al. 2014). A. Left lateral view of *Stegastes rectifraenum* skull and pectoral girdle. The left oral jaws, suspensorium, opercles and hyoid bar have been removed allowing a view of the right part of the hyoid apparatus in the buccal cavity. The opercles, the suspensorium, the oral jaws, the hyoid apparatus and the c-md ligament of the right side are 3-D reconstructed. B. Zoom on the 3-D reconstruction from a lateral view. ANG: angular; D: dentary; H: hyoid bar; HA: hyoid apparatus; MX: maxillary; NCR: neurocranium; OJ: oral jaw; O: Opercular series; PG: pectoral girdle; PMX: premaxillary; SUSP: suspensorium; UH: urohyal.

teleosts such as salmon (*Oncorhynchus mykiss*), cod (*Gadus morhua*) or catfish (*Chrysichthys nigrodigitatus, Clarias gariepinus*). However, it always inserts on the lower part of the angular, below the quadrate articulation and therefore allows mouth opening through the hyoid mechanism (Verraes 1977, Hunt Von Herbing et al. 1996, Diogo and Chardon 2000, Van Wassenbergh et al. 2005).

Although the c-md ligament is considered to be a synapomorphic trait of the Pomacentridae (Stiassny 1981), Frédérich et al. (2014) showed that it has been independently lost in several damselfish lineages (Table 1). All the species lacking the c-md ligament feed on elusive prey and present morphological adaptations to

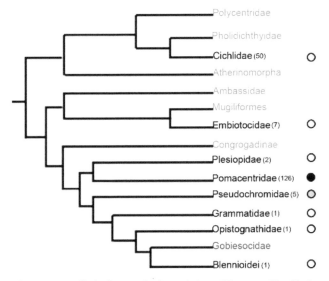

Fig. 3. Presence of a cerato-mandibular ligament in close-relatives of Pomacentridae. Phylogenetic relations come from Wainwright et al. (2012). Names in grey indicate that the situation is unknown. Empty circles indicate the absence of a cerato-mandibular ligament. Grey circle indicates a cerato-mandibular that inserts below the joint between the quadrate and the articulo-angular (opens the mouth). Black circle indicates a cerato-mandibular ligament that inserts above the joint between the quadrate and the articulo-angular (closes the mouth). In brackets either the number of species examined (for Plesiopidae, Pomacentridae, Pseudochromidae, Grammatidae, Opistognathidae and Blennioidei), or the number of genera examined (for Cichlidae and Embiotocidae) is indicated.

zooplanktivory (Coughlin and Strickler 1990, Liem 1993, Frédérich et al. 2014), i.e., a slender body with lengthened mandibles and a long ascending process of the premaxillary bone that improves the mouth protrusion distance.

Role of the Cerato-mandibular Ligament in Pomacentridae

The sound production in the clownfish Amphiprion clarkii

Parmentier et al. (2007) demonstrated that aggressive sounds emitted by the clownfish *Amphiprion clarkii* result from rapid mouth closing movements and that this fast jaw slam is due to the c-md ligament (Chapter X). The detailed mechanics of this system were determined using both manipulations of freshly euthanized fish and high-speed videos with X-ray and visible light sources (Parmentier et al. 2007, Olivier et al. 2015). The kinematic pattern during sound production in *Amphiprion clarkii* can be divided into three phases: initial, mouth-opening and mouth-closing (Fig. 5A). (1) During the initial phase, the mouth is closed, the neurocranium is held at rest and the hyoid apparatus is not depressed. At this moment, the cerato-mandibular ligament is loose and does not transmit any tension to the lower jaw (Fig. 5B) as revealed by manipulations of anesthetized individuals. (2) During the mouth-opening phase the lower jaw rotates and a relative movement between the neurocranium and the hyoid apparatus (RM) occurs because the first structure is elevated and the second is depressed

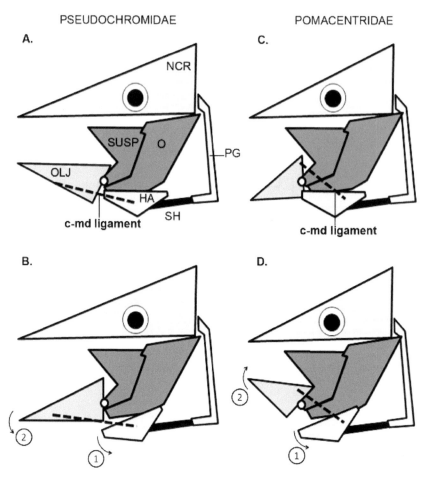

Fig. 4. Cerato-mandibular (c-md) ligaments in Pseudochromidae and Pomacentridae. (A) and (B) Biomechanical model of mouth opening *via* a cerato-mandibular ligament in Pseudochromidae. (C) and (D) Biomechanical model of mouth closing *via* a cerato-mandibular in Pomacentridae. HA: hyoid apparatus; NCR: neurocranium; OLJ: oral lower jaw; O: opercular series; PG: pectoral girdle; SH: *sternohyoideus* muscle; SUSP: suspensorium. The open circle indicates the articulation between the quadrate (bone of the suspensorium) and the lower jaw. In B and D, movement of the hyoid apparatus (1) induces either a mouth opening (2) in Pseudochromidae or a mouth closing (2) in Pomacentridae.

(α, Fig. 5A). As a result, the insertion points of the c-md ligament are moved away from one another (Fig. 5B). (3) At the transition between the mouth-opening and the mouth closing phases, the RM rapidly accelerates reaching maximum amplitude (MRM) at the end of the phase 3 (Fig. 5A, Table 2). This movement is accompanied by the slamming of the oral jaws with an angular speed reaching more than $3000°s^{-1}$ (Table 2). The confrontation of data between fish manipulations and videos allows us to interpret that the movement inducing a separation between the neurocranium and the hyoid apparatus strains the c-md ligament. This enables the ligament to transmit

Table 1. Absence-Presence of the cerato-mandibular ligament in Pomacentridae genera.

Genus	c-md ligament	no c-md ligament
Abudefduf	X	X
Acanthochromis	X	
Altrichtys	X	
Amblypomacentrus	X	
Ampblyglyphidodon	X	
Amphiprion	X	
Azurina		X
Cheiloprion	X	
Chromis	X	X
Chrysiptera	X	
Dascyllus	X	
Dischistodus	X	
Hemiglyphidodon	X	
Hypsypops	X	
Lepidozyginus	X	
Mecaenichthys	X	
Microspathodon	X	
Neoglyphidodon	X	
Neopomacentrus	X	
Nexilosus	X	
Parma	X	
Plectroglyphidodon	X	
Pomacentrus	X	
Pomachromis	X	
Premnas	X	
Pristotis		X
Similiparma	X	
Stegastes	X	
Teixeirichthys		X

force to the upper part of the lower jaw, which is forced to rotate around its articulation and to close (Fig. 5B). As the separation between the neurocranium and the hyoid apparatus is very abrupt, the c-md ligament is rapidly tightened which allows closing the mouth in a fast movement (Fig. 5A, Table 2). The lower jaw and the hyoid apparatus are linked by the c-md ligament; a fast tightening of the latter allows the acceleration of the hyoid apparatus and of the lower jaw to be well-synchronized (Table 2). It is well-known that neurocranial elevation and/or hyoid apparatus depression are involved

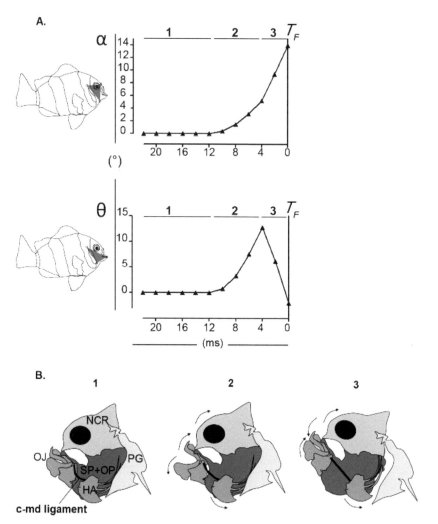

Fig. 5. Sound production mechanism in *Amphiprion clarkii* (modified from Olivier et al. 2015). (A) Representation of the kinematic pattern obtained *via* the analysis of the 500 fps movies. The full line indicates the time of the bite (final time, T_F); the illustrations 1, 2 and 3 represent respectively the initial, mouth opening and mouth closing phases. (B) Schematic illustration of the skull kinematics during the different phases. The illustrations 1, 2 and 3 represent respectively the initial, mouth opening and mouth closing phases. Arrows indicate the direction of the movement. The c-md ligament is in black. HA: hyoid apparatus; NCR: neurocranium; OJ: oral jaws; PG: pectoral girdle; SP + OP: suspensorium and opercular series.

in the lower jaw opening (Osse 1969, Ballintijn 1969, Schaefer and Lauder 1986). The kinematic pattern described here highlights that both structures can also induce mouth closing, at least during sound production. To our knowledge, this mechanism is unique and represents a novel way to close the mouth.

Table 2. Mean (±SD) of the kinematic variables measured for the different behaviours in three species of damselfishes. The kinematic patterns associated with the capture of planktonic prey, the seizure of fixed food item (biting-1 and biting-2), the seizure of filamentous algae and the sound production are indicated by the following abbreviations: PP, B-1, B-2, FA and SP. The kinematic patterns associated with the capture of planktonic prey and the seizure of fixed food items when the cerato-mandibular ligament has been cut are indicated by a*.

	PP	B-1	B-2	FA	SP	PP*	B*
Amphiprion clarkii							
1) MRM between neurocranium and hyoid apparatus (°)	8.6±1.9	11.3±2	17±3.4	-	14±1.5	7.8±0.73	10.7±0.9
2) Max. mandible depression (°)	13.5±1.8	20.4±4.3	12±1.7	-	12.2±5.2	14.6±3.9	22.4±4.6
3) Mean closing speed of lower jaw (°s⁻¹)	371±233	434±190	3013±562	-	3210±779	231±36	422±38
4) Max. acceleration of relative separation between neurocranium and hyoid apparatus (10^3 °s^{-2})	251±188	187±53	1904±390	-	1431±176	159±9	189±71
5) Time between (1) and the bite (ms)	30±24	26±13	0	-	0	41±22	28±7
6) Synchronization between hyoid apparatus and mandible accelerations (ms)	12±5	18±8	0.8±0.2	-	0.6±0.6	27±15	12±3
7) Lower jaw closing duration (ms)	38±24	46±16	4±0.8	-	4±0.3	62±26	43±7
Stegastes rectifraenum							
1) MRM between neurocranium and hyoid apparatus (°)	-	-	8.2±1.1	8.7±0.5	14.1±3.3	-	7.5±0.6
2) Max. mandible depression (°)	-	-	15.2±1.8	18±3.4	14.5±0.3	-	15.2±4.1
3) Mean closing speed of lower jaw (°s⁻¹)	-	-	1688±325	2771±376	3829±1202	-	396±65
4) Max. acceleration of relative separation between neurocranium and hyoid apparatus (10^3 °s^{-2})	-	-	580±42	1078±71	1602±531	-	85±18
5) Time between (1) and the bite (ms)	-	-	0	0	0	-	19±9
6) Synchronization between hyoid apparatus and mandible accelerations (ms)	-	-	1±1	0.4±0.4	0.3±0.7	-	17±9
7) Lower jaw closing duration (ms)	-	-	4±0.2	3±1	3±1	-	39±2

Table 2. contd....

Table 2. contd.

	PP	B-1	B-2	FA	SP	PP*	B*
Abudefduf troschelii							
1) MRM between neurocranium and hyoid apparatus) (°)	-	-	-	-	-	-	9.6±2.4
2) Max. mandible depression (°)	-	-	-	-	-	-	25.2±5.1
3) Mean closing speed of lower jaw ($°s^{-1}$)	-	-	-	-	-	-	542±206
4) Max. acceleration of relative separation between neurocranium and hyoid apparatus ($10^3 \, °s^{-2}$)	-	-	-	-	-	-	333±76
5) Time between (1) and the bite (ms)	-	-	-	-	-	-	23±8
6) Synchronization between hyoid apparatus and mandible accelerations (ms)	-	-	-	-	-	-	12±7
7) Lower jaw closing duration (ms)	-	-	-	-	-	-	34±8

The feeding mechanism in the clownfish Amphiprion clarkii

The kinematic patterns observed during feeding behaviours associated with the capture of planktonic prey, or the biting of fixed pieces of food were studied with a high-speed video camera and compared to the kinematic pattern used in sound production (Olivier et al. 2015). Clownfish use three kinematic patterns for these activities, one for planktonic prey and two for biting attached prey, named biting-1 and biting-2. Figure 6 shows the result of a PCA performed on several variables (Table 2, see Olivier et al. 2015 for a full description of all kinematic variables) used to describe the kinematic pattern of the different behaviours studied (*SP*: sound production; *PP*: planktonic prey; *B-1*: biting-1 and *B-2*: biting-2). Only one of the two patterns used in biting, *B-2*, is similar to the kinematic pattern used in sound production (Fig. 6). After the transection of the c-md ligament, fish are unable to perform either sound production or biting-2. Conversely, the kinematic patterns associated with the capture of planktonic prey and biting-1 are not altered (*PP** and *B** on Fig. 6). The role of the c-md ligament in the slamming of the jaws is also confirmed by the study of a species lacking the c-md ligament, *Abudefduf troschelii*, which shows a pattern similar to the biting-1 of the clownfish capturing fixed food items (Table 2).

In teleosts, the *adductor mandibulae* muscles are the most common way to close the mouth (Ferry-Graham and Lauder 2001). The *hyohyoideus* and *intermandibular* muscles can also participate in the buccal constriction (Osse 1969, Aerts 1991). In

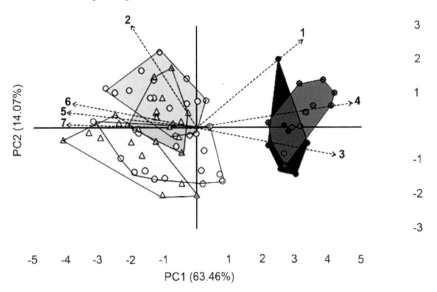

Fig. 6. Plot of principal components 1 and 2 for all behaviours studied in *Amphiprion clarkii* (from Olivier et al. 2015). The variables that load on each axis are indicated by the arrows, each arrow has a figure representing the variable (same figures as in Table 2). The different kinds of behaviours are color-coded; white: planktonic prey (*PP*); light grey: biting-1 (*B-1*); dark grey: biting-2 (*B-2*) and black: sound production (*SP*). The areas including all video sequences of one behaviour are also color coded from light grey for *PP* to very dark grey for *SP*. The data where the c-md ligament is cut are additionally separated by symbol type; white triangle: planktonic prey (*PP**); light grey triangle: biting fixed food item (*B**).

the cichlid *Haplochromis elegans*, the working-line of the *protractor hyoideus* can run above the angular/quadrate joint and thus adduct the lower jaw (Otten 1982). The c-md ligament represents a new way to close the mouth in teleosts and is, so far, unique to Pomacentridae. This ligament was shown to be a key trait for sound communication in clownfishes (Colleye and Parmentier 2012) and probably in all pomacentrids having this ligament. However, whether this trait is also crucial during feeding behaviour remains unknown. This morphological novelty, unquestionably, provides a diversification in the feeding mechanism allowing the development of a new kind of biting. However, the c-md ligament in *Amphiprion* seems to be more important for sound production than for feeding because it is mainly zooplanktivorous (Kuo and Shao 1991). In addition, approximately 80% of bites observed in the clownfish were associated with a biting-1 pattern.

Nonetheless, the c-md ligament is present in almost all the pomacentrids and the biting abilities of damselfishes could be important in species having different diets. A diversification in biting mechanisms allows taxa to explore new ecological zones and to use the environment in different ways. In pomacentrids, this functional novelty could be important for grazers that bite the substratum to grasp filamentous algae or benthic invertebrates. The next section is therefore devoted to the study of a farming species.

The cerato-mandibular ligament, a key trait for grazing filamentous algae?

As illustrated in Chapter VII, grazing damselfishes are often referred to as farmers because they defend small areas where they promote the growth of algae crops on which they feed. This behaviour consists of territory defence and the removal of unpalatable algae (Vine 1974, Montgomery 1980, Hata and Kato 2002, 2003, 2004, Ceccarelli 2007). In the grazer *Stegastes rectifraenum*, a recent study revealed that the same kinematic pattern is used during biting filamentous algae/fixed food items and during sound production (Olivier et al. 2014; Fig. 7). All these behaviours are characterized by a slamming of the oral jaws as in the sound production (and biting-2) in clownfish.

The c-md ligament allows *S. rectifraenum* to slam its lower jaw with an average speed that can reach more than 3000° s^{-1}, similar to *A. clarkii* (Table 2). After the transection of the c-md ligaments, fish are muted and are no longer able to bite on filamentous algae. They are however still able to bite fixed food items but use a different pattern (Fig. 7).

Coral reefs harbor at least two ecological groups of fish that rely on algae as a main food source. First, roving grazers that feed in single or multi-species schools such as Acanthuridae, Scaridae and Siganidae and second, grazers that are highly site-attached such as Pomacentridae and Blennidae (Ceccarelli et al. 2005). Grazing damselfishes defend small areas from conspecifics and all other fish (Vine 1974, Montgomery 1980, Hata and Kato 2002, 2003, 2004, Ceccarelli 2007). Herbivorous fishes can also be categorized on the basis of their mechanics of feeding. Non-selective grazing fishes such as scarids engulf large amounts of substrate per bite (Clements and Bellwood 1988, Russ and John 1988, Bellwood and Choat 1990, Bruggemann et al. 1994). Some acanthurids practice heavy grazing on a larger surface like a lawnmower (pers. obs., Vine 1974, Hixon 1997) and also ingest large amounts of sediment along

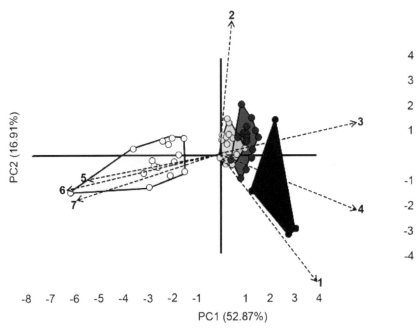

Fig. 7. Plot of principal components 1 and 2 for all behaviours studied in *Stegastes rectifraenum* (modified from Olivier et al. 2014). The variables that load on each axis are indicated by the arrows, each arrow has a figure representing the variable (same figures than in Table 2). The different kinds of behaviours are color-coded; light grey: bite on fixed food item (*B-2*); dark grey: bite on filamentous algae (*FA*); black: sound production (*SP*) and white: bite on fixed food item after c-md ligament transection (*B**). The areas including all video sequences of one behaviour are also color coded from light grey for *B** to very dark grey for *SP.*

with the algal filaments (Hiatt and Strasburg 1960, Russ and John 1988). On the other hand, damselfishes are highly selective grazers, picking small filamentous algae of sometimes only a few mm long (Montgomery 1980, Hata and Kato 2002). Bites of damselfishes are like surgical strikes that enable the fish to catch filamentous algae, while preventing or decreasing damage to the substratum supporting the crops. Accurate bites are likely also useful for weeding out the undesirable algae without scraping the palatable filamentous algae. We hypothesize that fast mouth closure is required to perform these accurate strikes efficiently. Indeed, the targeted filamentous algae are only a few mm long and a slow mouth-closing movement would decrease the success of seizure because it would increase the probability of the algae gliding along the lips. Other grazing species (e.g., blennids, acanthurids and scarids) engulf/scrape a large part of the substrate to ingest their food. This strategy is incompatible with farming because it inevitably causes damage to the substrate supporting the crops. This assumption is supported by the destruction of the farm by other herbivorous species once the resident pomacentrid is excluded (Lobel 1980, Hata and Kato 2003). The very fast mouth closing mechanism of damselfish possibly enables them to be highly selective and to perform accurate strikes on tiny filamentous algae, and probably allowed the evolution of such farming behaviour.

The cerato-mandibular ligament: a structure to mitigate the force-speed trade-off?

Grazing damselfishes have to cope with two different mechanical demands: (1) the quick closing of the mouth on filamentous algae and (2) the development of enough force to extract the algae or to take undesirable objects out of the territory. The *adductor mandibulae* muscles, which are the main mechanism for closing the mouth in teleosts (Ferry-Graham and Lauder 2001), have evolved differently in response to varying selection pressures. Manipulators that crush, scrape, excavate or tear their food have a more powerful force-generating capacity of the *adductor mandibulae*. Conversely, the *adductor mandibulae* configuration of suction feeders allows for a better speed transmission to promote the capture of elusive prey (Wainwright and Bellwood 2002, Sonnefeld et al. 2014).

Complex systems enable the mitigation of trade-offs (Holzman et al. 2011). For example, the force that a suction-feeding fish exerts on its prey can be enhanced by an expanded supraoccipital crest without a reduction in gape size which would decrease the spatial reach of the flow field (Holzman et al. 2011). Moreover, the evolutionary bundle duplications of muscle clearly provide opportunity for increases in functional complexity (Friel and Wainwright 1998). In the majority of teleost fishes, the *adductor mandibulae* are divided into four main bundles: A_1, A_2, A_3 and A_ω (Wu and Shen 2004). In Labridae, the geometry of the A_3 muscles is better suited for fast closing, whereas the A_2 is designed for more forceful closing (Westneat 2003). In Pomacentridae, in addition to these bundle duplications (Gluckmann et al. 1999), the c-md ligament provides a new way to close the mouth, increasing the complexity of the system even more. The functional redundancy in mouth closing mechanisms in damselfish (*adductor mandibulae* and c-md ligament) allows them to circumvent the constraints induced by the velocity-force trade-off because the *adductor mandibulae* could evolve towards a configuration better suited for force transmission without decreasing the ability of the fish to slam its jaws rapidly.

Sound Production and Feeding Mechanism in Damselfishes: How to Kill Two Birds with One Stone?

Exaptation refers to a functional character previously shaped by natural selection for a peculiar function and that has been then coopted for a new use (Gould and Vrba 1982). This term has not become widely used in the biological sciences (Larson et al. 2013), mainly because only few concrete examples have been demonstrated (Ostrom 1979, Patek et al. 2006, Cullen et al. 2013). In damselfish, comparisons of kinematic data highlight that the same kinematic pattern is used in sound production and in the biting of fixed food items. This brings a new element to the exaptation hypothesis suggested by Parmentier et al. (2007), i.e., the c-md ligament provides first, a mouth closing mechanism for feeding; then this specific mechanism appears to have been coopted for sound production. It is, however, difficult to answer the question of which was first: biting or sound production? Indeed no damselfish species is known to be non-vocal. Feeding behaviour can seem more primitive in comparison to sound production; sound has been recorded in 109 phylogenetically unrelated fish families

(Slabbekoorn et al. 2010, Fine and Parmentier 2015) whereas all fish need to eat. In the pomacentrid *Dascyllus flavicaudus*, males were shown to produce pulsed sounds during different behaviours (Parmentier et al. 2010). All these sounds were made on the basis of the same mechanism, which most probably involves the c-md ligament. Amongst these behaviours, one is related to fighting and is mainly made of only one pulse, corresponding to a single slam. The origin of the sound could be found in biting because fighting sounds usually occur during the display of aggressive behaviour with biting (Parmentier et al. 2010). Moreover, one-pulsed sounds are also recorded in *Stegastes rectifraenum* during grazing filamentous algae and do not appear to be related to communication purposes; they are clearly by-products of the feeding mechanism. These sounds are, however, similar to those recorded during agonistic behaviour (Olivier et al. 2014). In conclusion, it is more likely that sound production is an exaptation of a feeding mechanism than the contrary. Calls probably first consisted of single pulses before evolving into sounds made of multiple pulses that correspond to other elaborated behaviours.

The Cerato-mandibular Ligament Allows Easy Shift among a Small Number of Trophic Niches during Evolution

It is interesting to note that the high species diversity of damselfishes is not accompanied by high morphological and trophic diversities. The same trophic and morphological groups appear several times throughout the evolution of the family, which is described as a reticulated radiation or iterative ecological radiation (Cooper and Westneat 2009, Frédérich et al. 2013; Chapter IX).

The c-md ligament may be involved in this radiation. On one hand, the c-md ligament confers advantages in grazing filamentous algae and could facilitate the colonization of this trophic niche. However, this trait does not prevent an evolution towards zooplanktivory; many damselfishes with the c-md ligament feed mainly on small and elusive prey. The c-md ligament could facilitate a shift to algivory without impeding zooplanktivory. The loss of the c-md ligament impedes that versatility; consequently, all the pomacentrid lineages lacking it are zooplanktivorous (Frédérich et al. 2014). On the other hand, the c-md ligament probably acts as a constraint on the morphological diversification of damselfishes. Indeed, species lacking the c-md ligament have evolved towards a morphospace unoccupied by species with the c-md ligament (Frédérich et al. 2014). These species have an elongated body form and lengthened buccal jaws, which are recognized as adaptations to a pelagic lifestyle and a specialization to zooplanktivory (Webb 1984, Coughlin and Strickler 1990, Liem 1993). Although the c-md ligament could facilitate easy shifts among some trophic niches, it also acts as a constraint to the colonization of new adaptive zones and could be the motor of the iterative ecological radiation among a small number of trophic niches.

Conclusion

The great majority of damselfishes possess a c-md ligament, which provides them with a second way to close the mouth. This new mouth closing mechanism was first described to explain how the clownfish, *Amphiprion clarkii*, produces sounds. It turns out that this new mechanism is also indispensable in catching filamentous algae in grazing damselfishes. The c-md ligament allows them to perform accurate and very fast strikes on small filamentous algae. This motion could be essential for the farming activity of many damselfishes, an unusual behaviour in teleosts. Key innovations have been defined as attributes of organisms that potentially allow the taxon to occupy new adaptive zones and it seems that the c-md ligament fulfils this criterion. Moreover, this ligament allows damselfish to produce sounds and thus contributes to the diversity of sound communication in this family. This morphological trait plays a key role in two fundamental behaviours for survival: to feed and to communicate, reinforcing our hypothesis that the c-md ligament can be considered as a key character in the success of the Pomacentridae. Other biochemical, physiological and morphological traits could also have promoted the diversification of Pomacentridae, for example it would be interesting to perform future studies on the role of the innovation in the pharyngeal jaw apparatus in damselfishes.

Acknowledgements

We thank E. Balart for facilitating our field work in Mexico on *Stegastes rectifraenum*, Pablo Monsalvo and Gabriel Robles (CIBNOR) for helping to maintain aquarium and fish conditions throughout the study on *Stegastes rectifraenum* and Enrique Calvillo and Milton Spanopoulos-Zarco for helping to capture specimens. The authors would like to thank the two reviewers for their constructive comments of a previous version of the manuscript. This research was supported by FRS-FNRS grants (no. 2.4.535.10) and project EPO.02 of CIBNOR. DO is Research Fellow at the FRIA. BF is a Postdoctoral Researcher at the F.R.S.-FNRS (Belgium).

References

Aerts, P. 1991. Hyoid morphology and movements relative to abducting forces during feeding in *Astatotilapia elegans* (Telesotei: Cichlidae). J. Morphol. 208: 323–345.

Autumn, K., M. Sitti, Y.A. Liang, A.M. Peattie, W.R. Hansen, S. Sponberg, T.W. Kenny, R. Fearing, J.N. Israelachvili and R.J. Full. 2002. Evidence for van der Waals adhesion in gecko setae. Proc. Natl. Acad. Sci. USA 99: 12252–12256.

Ballintijn, C.M. 1969. Functional anatomy and movement co-ordination of the respiratory pump of the carp (*Cyprinus carpio* L.). J. Exp. Biol. 50: 547–567.

Baum, D.A. and A. Larson. 1991. Adaptation reviewed: a phylogenetic methodology for studying character macroevolution. Syst. Zool. 40: 1–18.

Bellwood, D.R. and J.H. Choat. 1990. A functional analysis of grazing in parrotfishes (family Scaridae): the ecological implications. Environ. Biol. Fish. 28: 189–214.

Bertucci, F., L. Ruppé, S. Van Wassenbergh, P. Compère and E. Parmentier. 2014. New insights into the role of the pharyngeal jaw apparatus in the sound-producing mechanism of *Haemulon flavolineatum* (Haemulidae). J. Exp. Biol. 217: 3862–3869.

Blount, Z.D., C.Z. Borland and R.E. Lenski. 2008. Historical contingency and the evolution of a key innovation in an experimental population of *Escherichia coli*. Proc. Natl. Acad. Sci. USA 105: 7899–7906.

Bruggemann, J.H., J. Begeman, E.M. Bosma, P. Verburg and A.M. Breeman. 1994. Foraging by the toplight parrotfish *Sparisoma viride*. II. Intake and assimilation of food, protein and energy. Mar. Ecol. Prog. Ser. 106: 57–71.

Carpenter, F.M. 1953. The geological history and evolution of insects. Am. Sci. 41: 256–270.

Ceccarelli, D.M. 2007. Modification of benthic communities by territorial damselfish: a multi-species comparison. Coral Reefs 26: 853–866.

Ceccarelli, D.M., G.P. Jones and L.J. Mc Cook. 2005. Foragers versus farmers: contrasting effects of two behavioural groups of herbivores on coral reefs. Oecologia 145: 445–453.

Clements, K.D. and D.R. Bellwood. 1988. A comparison of the feeding mechanisms of two herbivorous labroid fishes, the temperate *Odax pullus* and the tropical *Scarus rubroviolaceus*. Aust. J. Mar. Freshwater Res. 39: 87–107.

Colleye, O. and E. Parmentier. 2012. Overview on the diversity of sounds produced by clownfish (Pomacentridae): importance of acoustic signals in their peculiar way of life. PLoS ONE 7: e49179.

Cooper, W.J. and M.W. Westneat. 2009. Form and function of damselfish skulls: rapid and repeated evolution into a limited number of trophic niches. BMC Evol. Biol. 9: 1–17.

Coughlin, D. and R. Strickler. 1990. Zooplankton capture by a coral reef fish: an adaptive response to evasive prey. Environ. Biol. Fish. 29: 35–42.

Cullen, J.A., T. Maie, H.L. Schoenfuss and R.W. Blob. 2013. Evolutionary novelty versus exaptation: oral kinematics in feeding versus climbing in the waterfall-climbing hawaiian goby *Sicyopterus stimpsoni*. PLoS ONE 8: e53274.

Diogo, R. and M. Chardon. 2000. Anatomie et fonctions des structures céphaliques associées à la prise de nourriture chez le genre *Chrysichthys* (Téléostéi: Siluriformes). Belg. J. Zool. 1: 21–37.

Erwin, D.H. 2002. A preliminary classification of evolutionary radiations. Historical Biology: An International Journal of Paleobiology 6: 133–147.

Ferry-Graham, L.A. and G.V. Lauder. 2001. Aquatic prey capture in ray-finned fishes: a century of progress and new directions. J. Morphol. 248: 99–119.

Fine, M.L. and E. Parmentier. 2015. Mechanisms of fish sound production. pp. 77–126. *In*: F. Ladich (ed.). Sound Communication in Fishes. Springer-Verlag, Berlin Heidelberg, Germany.

Frédérich, B., L. Sorenson, F. Santini, J.S. Graham and M.E. Alfaro. 2013. Iterative ecological radiation and convergence during the evolutionary history of damselfishes (Pomacentridae). Am. Nat. 181: 94–113.

Frédérich, B., D. Olivier, G. Litsios, M.E. Alfaro and E. Parmentier. 2014. Trait decoupling promotes evolutionary diversification of the trophic and acoustic system of damselfishes. Proc. R. Soc. Biol. Sci. Ser. B 281: 20141047.

Friel, J.P. and P.C. Wainwright. 1998. Evolution of motor patterns in Tetraodontiform fishes: does muscle duplication lead to functional diversification? Brain. Behav. Evol. 52: 159–170.

Galis, F. 2001. Key innovations and radiations. pp. 581–605. *In*: P. Wagner (ed.). The Character Concept in Evolutionary Biology Academic Press, San Diego, CA.

Galis, F. and E.G. Drucker. 1996. Pharyngeal biting mechanics in centrarchid and cichlid fishes: insights into a key evolutionary innovation. J. Evol. Biol. 9: 641–670.

Galis, F. and P. Snelderwaard. 1997. A novel biting mechanism in damselfishes (Pomacentridae): The pushing up of the lower pharyngeal jaw by the pectoral girdle. Neth. J. Zool. 47: 405–410.

Gluckmann, I., J.C. Bussers, M. Poulicek and P. Vandewalle. 1999. Preliminary study of the morphology of the head in Pomacentridae: *adductor mandibulae* organization in *Dascyllus aruanus* (Teleostei: Perciformes). Proceedings of the 5th Indo-Pacific Fish Conference: 89–97.

Gould, S.J. and E.S. Vrba. 1982. Exaptation-a missing term in the science of form. Paleobiology 8: 4–15.

Hata, H. and M. Kato. 2002. Weeding by the herbivorous damselfish *Stegastes nigricans* in nearly monocultural algae farms. Mar. Ecol. Prog. Ser. 237: 227–231.

Hata, H. and M. Kato. 2003. Demise of monocultural algal farms by exclusion of territorial damselfish. Mar. Ecol. Prog. Ser. 263: 159–167.

Hata, H. and M. Kato. 2004. Monoculture and mixed-species algal farms on a coral reef are maintained through intensive and extensive management by damselfishes. J. Exp. Mar. Biol. Ecol. 313: 285–296.

Heard, S.B. and D.L. Hauser. 1995. Key evolutionary innovations and their ecological mechanisms. Hist. Biol. 10: 151–173.

Heiss, E., N. Natchev, M. Gumpenberger, A. Weissenbacher and S. Van Wassenbergh. 2013. Biomechanics and hydrodynamics of prey capture in the Chinese giant salamander reveal a high-performance jaw-powered suction feeding mechanism. J. R. Soc. Interface 10: 20121028.

Hiatt, R.W. and D.W. Strasburg. 1960. Ecological relationships of the fish fauna on coral reefs of the Marshall Islands. Ecol. Monogr. 30: 65–127.

Hixon, M.A. 1997. Effects of reef fishes on corals and algae. pp. 230–248. *In*: C. Birkeland (ed.). Life and Death of Coral Reefs. Chapman and Hall, New York.

Holzman, R., D.C. Collar, R.S. Mehta and P.C. Wainwright. 2011. Functional complexity can mitigate performance trade-offs. Am. Nat. 177: E69–E83.

Hunt Von Herbing, I., T. Miyake, B.K. Hall and R.G. Boutilier. 1996. Ontogeny of feeding and respiration in larval atlantic cod *Gadus morhua* (Teleostei, Gadiformes): II. Function. J. Morphol. 227: 37–50.

Hunter, J.P. 1998. Key innovations and the ecology of macroevolution. Trends Ecol. Evol. 13: 31–36.

Kaufman, L.S. and K.F. Liem. 1982. Fishes of the suborder Labroidei (Pisces: Perciformes): phylogeny, ecology and evolutionary significance. Breviora. 472: 1–19.

Kuo, S.R. and S.P. Shao. 1991. Feeding habits of damselfishes (Pomacentridae) from Southern part of Taiwan. J. Fish. Soc. Taiwan 18: 165–176.

Larson, G., P.A. Stephens, J.J. Tehrani and R.H. Layton. 2013. Exapting exaptation. Trends Ecol. Evol. 28: 497–498.

Levinton, J.S. 1988. Genetics, Paleontology, and Macroevolution. Cambridge University Press, Cambridge.

Liem, K.F. 1973. Evolutionary strategies and morphological innovations: cichlid pharyngeal jaws. Syst. Zool. 22: 425–441.

Liem, K.F. 1993. Ecomorphology of the teleostean skull. pp. 422–452. *In*: B.H.J. Hanken (ed.). The Skull: Functional and Evolutionary Mechanisms. The University of Chicago Press, Chicago.

Liem, K.F. and P.H. Greenwood. 1981. A functional approach to the phylogeny of the pharyngognath teleosts. Am. Zool. 21: 83–101.

Litsios, G., C.A. Sims, R.O. Wüest, P.B. Pearman, N.E. Zimmermann and N. Salamin. 2012. Mutualism with sea anemones triggered the adaptive radiation of clownfishes. BMC Evol. Biol. 12: 212.

Lobel, P.S. 1980. Herbivory by damselfishes and their role in coral reef community ecology. Bull. Mar. Sci. 30: 273–289.

Losos, J.B. 2010. Adaptive radiation, ecological opportunity, and evolutionary determinism. Am. Nat. 175: 623–639.

Miller, A.H. 1949. Some ecologic and morphologic considerations in the evolution of higher taxonomic categories. pp. 84–88. *In*: E. Mayr and E. Schüz (eds.). Ornithologie als Biologische Wissenschaft. Carl Winter, Heidelberg, Germany.

Montgomery, W.L. 1980. Comparative feeding ecology of two herbivorous damselfishes (Pomacentridae: Teleostei) from the gulf of california, Mexico. J. Exp. Mar. Biol. Ecol. 47: 9–24.

Olivier, D., B. Frédérich, M. Spanopoulos-Zarco, E.F. Balart and E. Parmentier. 2014. The cerato-mandibular ligament: a key functional trait for grazing in damselfishes (Pomacentridae). Front. Zool. 11: 63.

Olivier, D., B. Frédérich, A. Herrel and E. Parmentier. 2015. A morphological novelty for feeding and sound production in the yellowtail clownfish. J. Exp. Zool. 323A: 227–238.

Osse, J.W.M. 1969. Functional morphology of the head of the perch (*Perca fluviatilis* L.): an electro-myographic study. Neth. J. Zool. 19: 289–392.

Ostrom, J.H. 1979. Bird flight: how did it begin? Am. Sci. 67: 46–56.

Otten, E. 1982. The development of a mouth-opening mechanism in a generalized *Haplochromis* species: *H. elegans* Trewavas, 1933 (Pisces: Cichlidae). Neth. J. Zool. 32: 31–48.

Parmentier, E., O. Colleye, M.L. Fine, B. Frederich, P. Vandewalle and A. Herrel. 2007. Sound production in the clownfish *Amphiprion clarkii*. Science 316: 1006–1006.

Parmentier, E., L. Kéver, M. Casadevall and D. Lecchini. 2010. Diversity and complexity in the acoustic behaviour of *Dascyllus flavicaudus* (Pomacentridae). Mar. Biol. 157: 2317–2327.

Patek, S.N., J.E. Baio, B.L. Fisher and A.V. Suarez. 2006. Multifunctionality and mechanical origins: ballistic jaw protrusion in trap-jaw ants. Proc. Natl. Acad. Sci. USA 103: 12787–12792.

Rainford, J.L., M. Hofreiter, D.B. Nicholson and P.J. Mayhew. 2014. Phylogenetic distribution of extant richness suggests metamorphosis is a key innovation driving diversification in insects. PLoS ONE 9: e109085.

Rosenzweigh, M.L. and R.D. McCord. 1991. Incumbent replacement: evidence for long-term evolutionary progress. Paleobiology 17: 202–213.

Russ, G.R. and S.T. John. 1988. Diets, growth rates and secondary production of herbivorous coral reef fishes. Proceeding of the 6th International Coral reef Symposium, Australia 2: 37–43.

Schaefer, S.A. and G.V. Lauder. 1986. Historical transformation of functional design: evolutionary morphology of feeding mechanisms in loricarioid catfishes. Syst. Zool. 35: 489–508.

Slabbekoorn, H., N. Bouton, I. van Opzeeland, A. Coers, C. ten Cate and A.N. Popper. 2010. A noisy spring: the impact of globally rising underwater sound levels on fish. Trends Ecol. Evol. 25: 419–427.

Sonnefeld, M.J., R.G. Turingan and T.J. Sloan. 2014. Functional morphological drivers of feeding mode in marine teleost fishes. Adv. Zool. Bot. 2: 6–14.

Stiassny, M.L.J. 1981. The phyletic status of the family Cichlidae (pisces, perciformes): a comparative anatomical investigation. Neth. J. Zool. 31: 275–314.

Stiassny, M.L.J. and J.S. Jensens. 1987. Labroid intrarelationships revisited: Morphological complexity, key innovations, and the study of comparative diversity. Bull. Mus. Comp. Zool. 151: 269–319.

Van Valen, L.M. 1971. Adaptive zones and the orders of mammals. Evolution 25: 420–428.

Van Wassenbergh, S., A. Herrel, D. Adriaens and P. Aerts. 2005. A test of mouth-opening and hyoid-depression mechanisms during prey capture in a catfish using high-speed cineradiography. J. Exp. Biol. 208: 4627–4639.

Verraes, W. 1977. Postembryonic ontogeny and functional anatomy of the ligamentum mandibulo-hyoidem and the ligamentum interoperculo-mandibulare, with notes on the opercular bones and some other cranial elements in *Salmo gairdneri* Richardson, 1836 (Teleostei: Salmonidae). J. Morphol. 151: 111–119.

Vine, P.J. 1974. Effects of algal grazing and aggressive behaviour of the fishes *Pomacentrus lividus* and *Acanthurus sohal* on coral-reef ecology. Mar. Biol. 24: 131–136.

Wainwright, P.C. 2006. Functional morphology of the pharyngeal jaw apparatus. pp. 77–101. *In*: R. Shadwick and G.V. Lauder (eds.). Biomechanics of Fishes. Elsevier, Chicago, USA.

Wainwright, P.C. and D.R. Bellwood. 2002. Ecomorphology of feeding in coral reef fishes. pp. 33–55. *In*: P.F. Sale (ed.). Coral Reef Fishes. Dynamics and Diversity in a Complex Ecosystem. Academic Press, San Diego, CA.

Wainwright, P.C., W.L. Smith, S.A. Price, K.L. Tang, J.S. Sparks, L.A. Ferry, K.L. Kuhn, R.I. Eytan and T.J. Near. 2012. The evolution of pharyngognathy: a phylogenetic and functional appraisal of the pharyngeal jaw key innovation in labroid fishes and beyond. Syst. Biol. 61: 1001–1027.

Webb, P.W. 1984. Body form, locomotion and foraging in aquatic vertebrates. Am. Zool. 24: 107–120.

Westneat, M.W. 2003. A biomechanical model for analysis of muscle force, power output and lower jaw motion in fishes. J. Theor. Biol. 223: 269–281.

Wu, K. and S. Shen. 2004. Review of the teleostean *adductor mandibulae* and its significance to the systematic positions of the Polymixiiformes, Lampridiformes, and Triacanthoidei. Zool. Stud. 43: 712–736.

General Index

Species Index

T - #0380 - 071024 - C340 - 234/156/15 - PB - 9780367782887 - Gloss Lamination